PRODUCTION AND UTILIZATION OF PROTEIN IN OILSEED CROPS

WORLD CROPS:
PRODUCTION, UTILIZATION, AND DESCRIPTION

volume 5

Other volumes in this series:

1. Stanton WR, Flach M, eds: SAGO The equatorial swamp as a natural resource. 1980. ISBN 90-247-2470-8.
2. Pollmer WG, Phipps RH, eds: Improvement of quality traits of maize for grain and silage use. 1980. ISBN 90-247-2289-6.
2. Bond DA, ed: Vicia faba: Feeding value, processing and viruses. 1980. ISBN 90-247-2362-0.
4. Thompson R, ed: Vicia faba: Physiology and breeding. 1981. ISBN 90-247-2496-1.

Series ISBN: 90-247-2263-2

PRODUCTION AND UTILIZATION OF PROTEIN IN OILSEED CROPS

Proceedings of a Seminar in the EEC Programme of Coordination of Research on the Improvement of the Production of Plant Proteins organised by the Institut für Pflanzenbau und Pflanzenzüchting at Braunschweig, Federal Republic of Germany, 8-10 July 1980

Sponsored by the Commission of the European Communities, Directorate-General for Agriculture, Coordination of Agricultural Research

edited by

E.S. BUNTING

1981

MARTINUS NIJHOFF PUBLISHERS
THE HAGUE / BOSTON / LONDON
for the
COMMISSION OF THE EUROPEAN COMMUNITIES

Distributors

for the United States and Canada
Kluwer Boston, Inc.
190 Old Derby Street
Hingham, MA 02043
USA

for all other countries
Kluwer Academic Publishers Group
Distribution Center
P.O. Box 322
3300 AH Dordrecht
The Netherlands

ISBN 90-247-2532-1 (this volume)
ISBN 90-247-2263-2 (series)

Publication arranged by
Commission of the European Communities,
Directorate-General Information Market and Innovation, Luxembourg

EUR 6894 EN

© ECSC, EEC, EAEC, Brussels-Luxembourg, 1981
All rights reserved. No part of this publication may be reproduced, stored in a retrieval system or transmitted in any form or by any means, mechanical, photocopying, recording, or otherwise, without the prior written permission.

Proceedings prepared by:
Janssen Services, 33a High Street, Chiselehurst, Kent BR7 5AE, UK

LEGAL NOTICE

Neither the Commission of the European Communities nor any person acting on behalf of the Commission is responsible for the use which might be made of the following information.

PRINTED IN THE NETHERLANDS

CONTENTS

Preface	IX
Opening Remarks	XI

SESSION I: GENETIC AND BREEDING ASPECTS OF RAPESEED, SUNFLOWER AND SOYABEAN I

POTENTIALS AND RESTRICTIONS OF BREEDING FOR PROTEIN IMPROVEMENT IN RAPESEED G. Röbbelen	3
DISCUSSION	12
THE CROP PHYSIOLOGY OF RAPESEED N.J. Mendham	17
DISCUSSION	36
PROBLEMS IN SIMULTANEOUS BREEDING FOR OIL AND PROTEIN CONTENT IN SUNFLOWER E. Alba and I. Greco	43
ASPECTS AND PROGRESS OF OIL CROP BREEDING IN SPAIN J. Fernandez-Martinez and J. Dominguez-Gimenez ..	50
DISCUSSION	56
THE USE OF TISSUE CULTURE IN BREEDING OILSEED CROPS G. Mix	58
ASPECTS OF PLANT REGENERATION FROM SINGLE CELLS AND PROTOPLASTS OF *Brassica napus* E. Thomas	66
SHORT COMMUNICATION F. Schenk	77
SHORT COMMUNICATION G. Röbbelen	78
APPLICATION OF PROTOPLAST FUSION IN THE IMPROVEMENT OF RAPESEED YIELD G. Pelletier	79
DISCUSSION	80

SESSION II: GENETIC AND BREEDING ASPECTS OF RAPESEED, SUNFLOWER AND SOYABEAN II

BREEDING FOR LOW CONTENT OF GLUCOSINOLATES IN RAPESEED G. Röbbelen	91

NEW METHODS OF QUANTITATIVE ANALYSIS OF GLUCOSINOLATES
H. Sørensen 107

DISCUSSION 124

CONTENT AND PATTERN OF GLUCOSINOLATES IN RESYNTHESISED RAPESEED
Astrid Gland 127

DISCUSSION 134

BREEDING ASPECTS OF SUNFLOWER IN MIDDLE EUROPE
W. Schuster and I. Kübler 136

DISCUSSION 152

EXPERIENCE IN SOYABEAN BREEDING IN MIDDLE EUROPE
W. Schuster and J. Böhm 158

DISCUSSION 172

SESSION III: AGRICULTURAL ASPECTS OF RAPESEED, SUNFLOWER AND SOYABEAN

REGULATION OF POD AND SEED NUMBERS IN OILSEED RAPE
D.G. Morgan 179

DISCUSSION 187

PLANT-WATER RELATIONS OF RAPESEED
A. Bramm 190

DISCUSSION 200

SHORT COMMUNICATION
G.J. de Jong 203

MAIN FACTORS LIMITING SUNFLOWER YIELD IN DRY AREAS
R. Blanchet, J-R. Marty, A. Merrien and J. Puech .. 205

DISCUSSION 224

YIELD STABILITY IN SOYABEAN IN NORTH-EASTERN ITALY
L. Toniolo and G. Mosca 227

DISCUSSION 238

SESSION IV: ASPECTS OF ANIMAL NUTRITION

NUTRITIVE VALUE OF PROTEIN FEEDINGSTUFFS FROM OILSEED CROPS
E. Schulz and U. Petersen 243

USE OF OIL MEALS FOR DETERMINING PROTEIN REQUIREMENTS OR AS SUPPLEMENTS FOR RAT AND PIG DIETS
A.A. Rérat 263

DISCUSSION 287

SHORT COMMUNICATION - NUTRITIONAL VALUE OF RAPESEED
MEAL
C. Calet 291

NUTRITIONAL PROBLEMS RELATED TO DOUBLE LOW RAPESEED
IN ANIMAL NUTRITION
B.O. Eggum 293

DISCUSSION 306

RAPESEED MEAL IN POULTRY RATIONS
H. Vogt 311

DISCUSSION 344

RAPESEED MEAL AS A PROTEIN SUPPLEMENT FOR DAIRY COWS -
RESULTS FROM FEEDING EXPERIMENTS IN SWEDEN
L. Lindell 347

DISCUSSION 362

SESSION V:

    CLOSING SESSION 367

    FINAL DISCUSSION 372

List of Participants 378

PREFACE

This publication contains the proceedings of a seminar on 'Production and Utilization of Protein in Oilsead Crops', held at Braunschweig from 8 to 10 July, 1980. The meeting was held under the auspices of the Commission of the European Communities, as part of the EEC Common Research Programme on Plant Protein Improvement.

Methods for the intensive production of meat and milk have been adopted to an increasing extent in EEC countries over the past two decades, their success is based on animal diets of high quality, balanced for energy and protein contents. The substantial improvements in cereal yields in EEC over this period has kept pace with the increasing demand for dietary energy in concentrated animal foodstuffs, but provision of the necessary protein supplementation has required ever-increasing imports of soybean products. Grain legumes and oilseed meals are the two main sources of concentrated protein for the animal feeds industry, and there is an urgent need for increased EEC production of both. Seminars on grain legumes have been held at regular intervals since 1976; the meeting at Braunschweig, now reported, was the first to consider in detail the potential contribution from oilseed crops suitable for EEC conditions.

Local arrangements for the seminar were made by Professor Dr. M. Dambroth, Dr. C. Sommer, and their colleagues at the Institut für Pflanzenbau und Pflanzenzüchtung, Braunschweig - Volkenröde. Participants will hope that their gratitude for the extremely efficient organisation is, to some extent at least, reflected in the contributions made to the important subject under consideration.

OPENING REMARKS

M. Dambroth *(FRG)*

It is a great honour for me to welcome you to this conference on the 'Production and Utilisation of Protein in Oilseed Crops'. I am very pleased that the response to our invitations was so good and I hope that your journey here was pleasant.

I am speaking to you as Vice-President of the FAL and also as Director of the Institute of Crop Science and Plant Breeding. As Vice-President of the FAL, I would like to convey the greetings of the synod and the executive committee and also to give you a brief survey of the FAL itself. The FAL appreciates the honour of your visit and thanks you for coming. We hope that the conference will be successful and that you have an enjoyable stay in Braunschweig. We also hope that as a result of your visit the various international studies of the FAL will be extended and that a decisive step will be taken towards greater scientific co-operation within the EEC. The research projects currently being undertaken are interesting for all member states and this should again encourage further co-operation. Therefore, we must thank the CEC for their research support, the financing of research programmes and seminars. I would like to ask Dr. Gillot to convey our thanks to the various organisational bodies of the CEC and I would like to thank him personally for participating in this conference, allowing us to have an opportunity to discuss various aspects of administration.

I will now provide some information about the FAL. It was established in 1947 in this area which from 1935 belonged to German Aviation Research. The FAL is one of 13 research centres of the Ministry of Food, Agriculture and Forestry in Bonn. At the moment the FAL consists of 14 Institutes, 12 of which are in this area. Not every research centre of the FRG is of such a high standard and large size as the FAL. Some of them are more like Institutes. You can perhaps imagine the size of the FAL if I tell you that we receive 25% of the total budget for research provided by the Ministry of Food and Agriculture -

at present about DM 70 million. Its task is to advise the
Government which means that data have to be collated for legislation, for example in the field of environmental protection,
animal behaviour etc. The 14 Institutes of the FAL can be divided into four groups. The first group, made up of four
Institutes, is working in the field of soil and land research.
The second group, with three Institutes, works in the field of
animal production. The third group, with four Institutes, works
in the field of agricultural techniques. Finally, there is the
economic group consisting of three Institutes. From the structure of the FAL you can see that it is possible to deal with all
aspects of research into agricultural production.

Let me now introduce the Institute of Crop Science and
Plant Breeding. This institute was established in 1947 under
the name of the Institute of Crop Science and Seed Production.
Since then there have been various alterations and due to the
changing requirements in the field of plant production the
research aims of the Institute have recently been reformulated.
In addition, extra technical facilities have resulted in new
investigations in the field of breeding research. For these
reasons we changed the name of the Institute six months ago to
the Institute of Crop Science and Plant Breeding.

The Institute's activities concentrate mainly on the problems of breeding research and plant ecology. In the field of
breeding research the Institute aims to provide the necessary
scientific assistance for the work of the practical plant breeders. Our main interests are:

1) To guarantee the reliability of plant genetic potential
   of wild and primitive forms of crop plants and
   economic plants.

2) To quantify selection methods on the basis of physiology of yield and metabolism.

3) To find new methods for breeding rapeseed using a
   combination of protoplast cell tissue and microspore
   culture.

4) To search for new industrial plants.

In relation to plant ecology, the Institute's aim is to quantify the position of plant production with regard to the tension that exists between ecology and economics.

Further information concerning the various fields of research and the objectives of our research work is provided in your folders.

I would here like to point out the connections that exist between the Institute's work and the research support activities of the CEC. At present the Institute is working in the following areas:
- plant resistance and the better use of gene banks
- plant protein programme
- irrigation of field crops
- alternative foods.

The plant protein programme is a good example of the connections that exist between the Institute's work and the research activities of the CEC. It is an interesting field since the plants in question supply both protein and oil and I have heard that the next programme will mainly concentrate on this type of plant. Our complete support for this programme is assured. In the future we must ensure the protein supply in the EEC, but it will be increasingly important for the stability of the rural economy to develop alternatives. One alternative is the production of industrial basic resource material from biomass. I am not only thinking of ethanol but also the production of oil, fibres, tanning materials, pharmaceutical materials etc., especially since industry is so interested in such products as these. Unfortunately, potentially suitable plants have not been subject to any breeding programme and therefore there is strong pressure for such supporting programmes to receive essential assistance.

Oil plants are ideally suited for an international research programme since they can be utilised both as food and industrial basic material. For two years the Institute has concentrated its activities on the utilisation of biomass as a resource for industrial production and we are prepared to participate in any such activities. In our opinion, breeding research should not aggravate existing problems in agriculture but, on the contrary, help to minimise these problems.

Every industrial firm is always searching for new products. However, the number of agricultural products is decreasing - about 50 species have disappeared from production in this century alone, many of them oilseeds. More research work should be dedicated to these species so that their yield potential will be increased to meet current needs. International knowledge is a necessary basis for this and with CEC assistance progress can be achieved. In our opinion, no more research work is needed on the influence of nitrogen fertilisation on the quality of wheat, or on the colour of potato chips. Planning of research concepts must be focussed on future problems and such a working group as this one on plant protein can help in this regard.

I hope that we will have the opportunity to show you some aspects of our work in this field, especially in the field of tracing industrial resources based on biomass production, and to discuss this complex matter during the conference.

Before finishing, I would like to express my thanks to Dr. Sommer for the organisation of this conference and I am sure that his organisational ability will ensure that you will enjoy your stay here in Braunschweig and that the conference will be a profitable one. Once again I would like to thank you for coming and I look forward to the forthcoming papers and discussions.

SESSION I

GENETIC AND BREEDING ASPECTS OF RAPESEED,
SUNFLOWER AND SOYABEAN I

Chairman : M. Dambroth

POTENTIALS AND RESTRICTIONS OF BREEDING FOR PROTEIN IMPROVEMENT IN RAPESEED

G. Röbbelen
Institute of Agronomy and Plant Breeding,
Georg-August University, Göttingen,
Federal Republic of Germany

Oil and protein are the main components of rapeseed. So far, breeding programmes for protein improvement have been mainly devoted to cereals. This is well justified, since cereals are dominant in world agriculture, and provide two-thirds of the calculated food protein production (Table 1). Only one-third is contributed by grain legumes, tuber crops, oilseeds, and others; and even soya belongs to this minor group.

On the other hand, cereal seeds are by no means specialised to store protein (Table 2); their main component is starch. Grain legumes are the real specialists in protein storage and their seeds have the highest protein percentages. Oilseeds such as rape follow closely in percentage protein content, and if we consider the sum of oil and protein, then the food value of oilseeds may equal or even exceed that of certain grain legumes.

The highest protein yields/ha are not recovered from seeds but from green matter, e.g. clover/grass (Table 2): but the concentration of this protein per unit of feedstuff is low and for efficient utilisation special digestive devices are needed, provided naturally by the rumen of cattle and sheep or technologically by procedures of protein concentration which are highly energy demanding.

Even under the latter conditions, protein production via green matter offers specific advantages, because the translocation and deposition of plant protein into seed is not only an energy requiring, but also a time consuming process. Wherever time is insufficient for seed ripening, as in northern latitudes or for late-sown second crops in warmer areas, harvest in the vegetative

TABLE 1
WORLD PRODUCTION OF PLANT FOOD PROTEIN IN MILLION TONS, 1968 (AFTER JALIL AND TAHIR 1973)

	Cereals	Grain legumes	Root and tuber crops	Oilseeds	Others	Total
Europe	15.59	0.63	2.40	0.28	1.10	20.00
North-America	21.55	11.72	0.28	0.74	0.40	34.69
Latin-America	6.78	1.56	0.53	0.72	1.70	11.29
Near-East	4.09	0.35	0.05	0.51	0.31	5.31
Far-East	19.35	4.13	0.67	0.69	1.80	26.64
Africa	4.82	1.45	0.74	0.19	0.32	7.52
Oceania	2.06	0.02	0.02	0.01	0.06	2.17
UdSSR	15.47	1.84	1.74	1.58	0.25	20.88
China, mainland	15.78	6.15	2.23	0.58	0.61	25.35
World, total	105.49	27.85	8.66	5.30	6.55	153.85

TABLE 2

CONTENT AND PRODUCTION OF PROTEIN IN SOME CROPS (FEDERAL REPUBLIC OF GERMANY)

	Crude protein %	Grain yield dt/ha	Protein yield kg/ha
Grain legumes	30	25	750
(Soybean, US)	(40)	(22)	(880)
Winter rapeseed	25	24	600
Cereals	12	40	480
Potatoes	2	450	900
Clover/grass	2	800	1600

P 3/80

stage is the best choice. Cruciferous species are well suited for rapid green matter production, and are often sown by farmers after cereal harvest. It should be remembered that eight times more cruciferous seed is sown in Germany for such green matter production than for grain harvest of rapeseed. The second advantage of green matter over seed is the high quality of its protein (Table 3). As is well known, the nutritional value of a protein is determined by the quality of its 20 amino acid constituents. Most of the various proteins essential for cell metabolism in higher organisms are of similar composition in plants and in animals. This obviously reflects evolutionary relations. Animals, however, are unable to synthesise 8 of the 20 amino acids required for production of these vital proteins, and must obtain them from food of plant origin. Such so-called essential amino acids are contained in the required amounts in enzyme proteins and in the total proteins of green leaves or potatoes.

TABLE 3

ESSENTIAL AMINO ACIDS (mg/16 g N) IN PROTEINS OF VARIOUS PLANT MATERIALS AS COMPARED TO FAO/WHO MINIMUM REQUIREMENTS FOR HUMANS

			Seed			
Amino acids	Fract. I chloroplast	grass	corn	horse bean	rape	FAO/WHO
ile	3.4	3.6	3.8	4.3	3.7	3.6
leu	6.5	6.1	10.6	8.3	6.3	4.1
lys	8.2	8.0	2.7	6.6	5.8	3.6
met	1.4	1.1	2.4	0.8	1.8	1.9
phe	3.4	3.4	4.5	4.4	3.5	2.4
thr	4.9	3.9	4.0	3.3	3.8	2.4
trp	2.1	-	-	1.0	-	1.2
val	6.2	5.5	5.0	3.9	4.8	3.5

But the main source of plant protein is seeds, which contain protein for storage and not for immediate use in cell functions. The synthesis of seed protein is less strictly protected from mutational changes by selection, as its main purpose of nitrogen accumulation is best achieved by the production of large amounts of the amides asparagine and glutamine. For these reasons, cereal grains lack lysine, and legume seeds lack methionine, although other minor deficiencies are also apparent (Table 3). Oilseeds, however, in general yield high quality protein in their meal after oil extraction, soyabean being the pioneer in the nutritional plant protein industry.

Differences are also evident in the content of seed protein between those species which have starch and those that have oil as their other main storage compound. In starchy seeds there is a high negative correlation between content of protein and carbohydrate, and protein rich seeds tend to lack starch deposition. In oilseeds, however, species with the larger seeds tend to be richer in oil. In rapeseed, Stefansson suggested that it would be better to select for the sum of

'oil % + protein %' rather than to select separately for each
component and he was indeed rather successful in effectively
increasing the protein and oil contents of his material
(cf. Röbbelen and Rakow, 1979).

In order to understand this basic difference, we should
analyse in more detail the cellular metabolism and localisation
of the storage products in question. What we know from barley
(Breidert and Schön, 1979) is that the main accumulation of
substrate in the endosperm occurs between 14 and 30 days after
fertilisation. Starch, of course, is synthesised in the
plastids; but the primary amination of amino acids is also
exclusively performed in the chloroplast matrix, although
later the storage protein is composed at the rough ER membranes
of the cytoplasm. By electron micrographs Bergfeld et al.,
(1980) have shown details of this process in *Sinapis*. Protein-
like material appears within Golgi vesicles and within vacuoles
formed by transfer of substrate into the inner space of the
ER membranes. The normal pathways of membrane flow in the
development of central vacuoles in post-meristematic cells,
finally leads to the striking protein accumulation in the
vacuole.

In contrast, oil bodies appear in the cytoplasm without
any visible connection to the ER in *Crambe* (Smith, 1974). Only
after reaching a certain diameter (Bergfeld et al., 1978) do
the naked droplets in *Sinapis* become coated by a thin layer of
proteins which are trapped and interconnected in a stable
lamellar structure at the lipid/water interface. This coat
is most probably a *de-novo* product of the ER. It prevents
coalescence of the lipid oleosome contents during seed
desiccation and thus provides a large surface for the attack
of the incorporated lipases after the initiation of germination
(cf. Röbbelen and Thies, 1980 a, b).

It is difficult to suggest plausible hypotheses in order
to explain the observed differences in protein ratios between
starchy seed and oilseed. But the above data indicate that it

is the same plastid organelle which is involved in the primary
syntheses towards starch and protein, while fatty acid and
triglyceride molecules originate immediately in the cytoplasm,
which is an independent genetic system. A second possible
direction of reasoning is the fact that during germination the
remobilisation of starch for growth-requirements is a rather
direct process, while the enzymatic turnover to convert fat into
cellular compounds for early plant development is much more
complex and needs more protein capacity or metabolic energy.

No precise answer is possible at present to the breeder
who asks at which point of his work physiological restrictions
may occur. Within one species, of course, oil and protein
contents are negatively correlated (Figure 1) and it may be
advisable in future to develop separately high-oil and high-
protein cultivars.

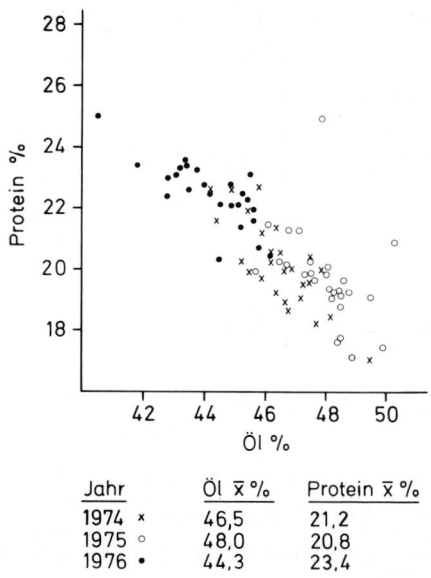

Jahr	Öl x̄ %	Protein x̄ %
1974 ×	46,5	21,2
1975 ○	48,0	20,8
1976 •	44,3	23,4

Fig. 1. Contents of oil and protein (in percent of dry matter) in
performance tests over three years (after Röbbelen and Rakow, 1979)

An effective and efficient selection for protein content, however, requires adapted analytical methods for protein determination. Oil determination was revolutionarily improved ten years ago by the introduction of low resolution nuclear magnetic resonance spectrometry (NMR). This new physical method not only reduced the expense, material and time requirements of the earlier chemical extraction techniques, but even more advantageously, it enabled the measurement to be made without any injury to the seeds. Such non-destructive methods are not yet available for protein analysis, although a great many automatic instruments have been developed recently, such as the N_2 elementary analysis after Dumas, infra-red reflection spectroscopy, or proton activation, to support the more traditional methods based on chemical methods like Kjeldahl, or staining with ninhydrine or Biuret (Röbbelen et al., 1976).

Rapeseed is usually most abundantly fertilised by mineral nitrogen. Some experiments indicate that 40% of the total nitrogen taken up by a winter rapeseed plant is already lost before the beginning of flowering. This is less so in forage types of rapeseed, which however are inferior in seed yield, possibly because leaves compete with inflorescences for essential nutrients. If this is true, breeding should make a greater distinction between grain cultivars and forage cultivars, and develop two very distinct ideotypes for these different uses. Little is yet known as to what extent cultivars or species (cf. *B. napus* and *B. juncea*) react differently with respect to their fertilisation requirements (low-input varieties) or whether such agronomic measures influence protein yield and quality as they do in cereals after late N fertilisation

Nutritional acceptability of proteins in a food, or feed, is highly dependent also on the conditions of seed processing in the oilmill. Dehulling before crushing the seed greatly increases the protein content in the final meal. The procedure of oil extraction so far has been exclusively directed to maximize technical efficiency and oil yields, but we should know much more of the effects of extraction and heating on the quality of protein in the meal. Again, seed from different cultivars or times of harvest may behave differently, a higher amount of soluble carbohydrates may increase the danger of Maillard reactions with proteins.

The availability of the proteins for such reactions will certainly differ with their given physical structure (i.e. crystalline, gel or sol form). Almost nothing is known for rapeseed in this field, which seems to be rather important for a better understanding of protein quality in rapeseed meals. This is especially true if more sophisticated techniques of chemical protein concentration or isolation are considered, and they are routinely used in the soya industry already.

Finally, it must always be kept in mind that rapeseed meal or protein as a feed, or food, will never be the only component of a diet. From the fine experiments of Kofrányi (1970) who accurately analysed diets with two defined components only, we know that extensive complementations are possible. Thus it is dubious whether breeding for a single feed component is the best way to arrive at the ideal composition of the total diet, especially as such diets must be quite different, for different animal species and also for different production aims (i.e. growth, milk, meat or fat).

Usually, selection in protein breeding is done between living plants and living seeds, but more feedback information is needed from the processor and nutritionist on the determining parameters of protein quality in the extracted oilseed meals.

REFERENCES

Bergfeld, R., Hong, Y.N., Kuhnl, T. and Schopfer, P., 1978. Formation of oleosomes (storage lipid bodies) during embryogenisis and their breakdown during seedling development in cotyledons of *Sinapis alba* L. Planta 143. 297-307.

Bergfeld, R., Kühnl, T. and Schopfer, P., 1980. Formation of protein storage bodies during embryogenesis in cotyledons of *Sinapis alba* L. Planta 148, 146-156.

Breidert, D. and Schön, W.J., 1979. Die Bildung und Speicherung von Kohlenhydraten und Proteinen in reifenden Gerstenkaryopsen. I. und II. Angew. Botanik 53, 65-81, 145-159.

Kofrányi, E., 1970. Physiologie des Eiweisshaushaltes. In: Forschung 71. Ber. aus Wiss.u.Technik. Umschau-Verlag, Frankfurt/M. pp 41-55.

Röbbelen, G. and Rakow, G., 1979. Eiweissaat Raps: Züchterische Erfolge und Möglichkeiten. Fette - Seifen - Anstrichmittel 81, 197-200.

Röbbelen, G., Schön, W.J. und Thies, W., 1976. Ziele und Wege landwirtschaftlicher Pflanzenzüchtung zur Verbesserung des ernährungsphysiologischen Wertes pflanzlicher Nahrungs- und Futtermittel. Berichte über Landwirtschaft 54, 9-37.

Röbbelen, G. and Theis, W., 1980a. Biosynthesis of seed oil and breeding for improved oil quality in rapeseed. In: Brassica crops and Wild Allies - Biology and Breeding. Edit. S. Tsunoda, K. Hinata and C. Gómez-Campo. Japan. Scient. Soc. Press, Tokyo. pp 253-283.

Röbbelen G. and Thies, W., 1980b. Variation in rapeseed glucosinolates and breeding for improved meal quality. In: Brassica Crops. (See above). pp 285-299.

Smith, C.G., 1974. The ultrastructural development of spherosomes and oil bodies in the developing embryo of *Crambe abyssinica*. Planta 119, 125-142.

DISCUSSION

L.G.M. van Soest

I would like to ask a question about the genetics of protein. You said that we need to know more about the genetical background of protein. As yield is polygenically determined, I think you can also say that protein will be. Do you think that the use of some related wild species, in rapeseed for instance, could increase the yield of protein? I will just mention the example of potatoes: the use of *Solanum demissum* increased the yield of the potato. Therefore, some variability in the gene pool of rapeseed could also increase the yield of protein.

G. Röbbelen *(FRG)*

I think that the situation with potatoes is slightly different. The percentage of protein in potato is 2 or 3 at the maximum. Our best chance of getting a high protein yield per hectare is to use the most productive species we have. Indeed, the yield level in the various species called 'rapeseed' in industry is rather different. For example, *Brassica juncea*, which is dominant in India, is at a lower level of production than our *Brassica napus* is at present. To answer the question of whether a related species would give us a better chance is almost impossible. As a plant breeding adviser, I would be hesitant to advise anyone in this direction. However, I would not hesitate to indicate the difficulty which we will find using the most efficient analytical technique in the oilseeds which we have - that is NMR. The introduction of this will speed up greatly the increase in oil percentage because it is so very easy. In our group ten years ago, before the introduction of this machine, we averaged 1 000 analyses per year; now there are 50 000. Of course this increases the chance of improving oil. As soon as you improve oil, one way or another protein is reduced. Therefore, we must apply the utmost pressure to get efficient analytical methods for protein.

We are still talking about oilseeds in the Common Market, knowing that their protein component is of the utmost importance; perhaps of dominant importance since projections of market tendencies show that, in the 1980s, we are expecting a good world supply of fat, when the oil palm production in Malaysia, etc., gets underway, but not of protein. In the long run we should consider many of the oilseeds as being protein producers. In what class would soyabeans be counted?

B.O. Eggum (Denmark)

You mentioned that dehulling was a possible way of improving the nutritional value. Do you believe this to be a solution? What is the potential for the plant breeder to reduce the crude fibre level?

G. Röbbelen

Many techniques are being developed for dehulling.

We have a project where dehulling is done by a form of milling technique: there are two turning wheels with a certain amount of surface roughness. They 'squash' the seeds, breaking them up like the plant breeders used to do with the 'half-seed' technique. This works quite well as a specific way of conditioning: aspiration then separates the embryos from the husks. Canadian techniques used to blow the rapeseed against a wall which caused dehulling, followed by separation by aspiration. There may be other, different, methods.

For sophisticated techniques of protein preparation, as with protein concentrates or isolates, such dehulling is rational and even economical. With regard to the plant breeder, we all know that in *Brassica campestris*, which is also rapeseed, yellow seed can also be used for reducing the crude fibre content; this has already been accomplished in the Canadian variety, Candle. Other varieties are in the process of development. There is intensive work going on in plant breeding programmes to introduce this characteristic into rapeseed by inter-specific crossing.

This appears to be difficult in the amphidiploid, *Brassica napus*, because of the epistatic effects of black colour from *Brassica oleracea*. However, by resynthesising *Brassica napus* rapeseed from very light coloured *oleracea* and *campestris* types, we were able to make some progress in getting a brownish type. So this can be done. The same is true of *Brassica juncea*: there is yellow *juncea* available and there is yellow *carinata* available also. Therefore, we can go on in this direction. However, there is an analytical question, and I am sure that if we had better methods for determining crude fibre we could use the variation available within the black-coated rapeseeds; they may have some of the advantages that we find with the yellow seed. There is little difficulty in the oil processing to remove the phenols. We have answered some of the questions raised in crude fibre determination; perhaps I can go into it in more detail later.

E. Thomas (UK)

Professor Röbbelen mentioned that one of the limiting problems in rapeseed development was due to the analytical methods of protein determination. Recently, I had the good fortune to go to a conference in Paris where some extremely new results were presented by Professor Sussex from his student, Martha Louise Crouch, at Yale University. She has been studying the extent to which embryo development in *Brassica napus* is controlled by interactions with non-embryogenic tissues, by comparing storage protein accumulation in embryos isolated from the seed with the normal pattern of accumulation. First she isolated the two major storage proteins - 12S and 1.7S - and purified them from the seed extracts. Then she raised antibodies in rabbits against purified 12S protein, and used rocket immuno electrophoresis to detect and to quantify - very exactly - the 12S protein in crude extracts. It showed that the limit of detection was 4 nanogrammes/mg of tissue for 12S protein, and 20 nanogrammes/mg of tissue for the 1.7S protein. I think this opens up the way to molecular biological techniques in rapeseed. I thought you might be interested in these techniques.

D.G. Morgan *(UK)*

Could you tell me if there are any obvious differences in the nitrogen metabolism of the plants with the high protein seeds? Where does the nitrogen come from that goes into the seeds later on? Is it redistribution within the plant? Or is it continued uptake?

G. Röbbelen

I do not know specifically for the nitrogen; I only know about the dry matter. The dry matter accumulation is similar to what we know from cereals: one third comes from the uppermost leaves. One third comes from the stems and branches of the influorescence, and the final third comes from the pod wall. Very little comes from the seed itself; we were interested in this because our C14 feeding was done to show that with linoleic acid there was a major contribution from the seed. For that, we traced the assimilates from the plant into the seed. I do not know how transformation occurs, or details of the branching into fat on the one hand and into protein synthesis on the other. I would be happy if someone else could contribute in this direction.

U. Petersen *(FRG)*

You presented some results from the research of Dr. Kofranyi and stated that in practical nutrition a proper combination of different protein sources is used. Would you go so far as to say that the quality of a single protein is of minor interest?

Protein, as it is analysed conventionally, is a very heterogeneous substance with many non-protein compounds in it. Would it not be better to select for the whole group of amino acids, or for special amino acids? There has been increasing evidence, in recent times, to suggest that monogastric animals have a small but significant potential to use non-protein nitrogen, for instance, for their non-specific needs.

G. Röbbelen

There are others here today who would be more suited to answer such a question. I hope that we can go into these details on Thursday.

I have a comment, as a plant breeder, for the nutritionists. The different plants available for a food or feed do not all need to have the optimum combination which we want in the diet. It may well be easier to compromise in one component of the diet if we have the complementary component in another. Of course the feed manufacturers get a certain degree of freedom in usage if they have a near-optimum nutritional quality in one fraction. It is price which is the dominant factor in determining the ratios which they are going to use. However, from a nutritional standpoint, or from the standpoint of breeding efforts, I would question whether we should take the FAO values or the chloroplast fraction 1 protein and say that this is the ideal amino acid composition and therefore breed towards it.

Lastly, it is not only the amino acid composition but, in many aspects, it is the structure of the proteins. Although I am almost a layman in this connection, I tried to emphasise the point that the soluble carbohydrates, for example, in an oilseed may well interact during processing with the quality which I, as a plant breeder, produce. The processing may destroy all that I produce before it gets into the feeding trough of the pig or cow. These are the interactions; and the measures we must take to understand them are really very complex.

M. Dambroth *(FRG)*

Thank you very much.

THE CROP PHYSIOLOGY OF RAPESEED

N.J. Mendham
School of Agriculture, University of
Nottingham, U.K.*

Improvement of protein and oil quality in rapeseed will do much to allow increased acceptance and utilisation of the crop by feed and food manufacturers. However, yields must be raised to ensure profitability for farmers, and to reduce dependence on subsidies to oilseed crushers within the EEC. Current seed yields are not high compared to those being achieved with winter wheat, even allowing for the greater energy and protein content of the oilseed.

A series of experiments at the University of Nottingham, in the English Midlands, carried out between 1969 and 1977 has allowed us to examine in detail the way rapeseed grows and develops, and how these two processes interact to produce yield. We can now begin to identify yield-limiting stages in the crop's life cycle, upon which we can concentrate breeding and agronomic research.

We sowed Victor, a variety widely grown when the experiments began, on a range of dates in the autumn from August to October or even November (Figure 1). While Victor is no longer grown commercially, it is representative of the general type of European winter rape, the main changes since being in product quality. In four of the seasons (Figure 1a) yield declined when sowing was delayed after about mid-September, in accord with research results and general experience in Europe and Britain. In the other three years, however (Figure 1b), yields either changed little or even increased with later sowing. It is obviously of great importance to understand why some crops, sown far later than normal, actually outyielded the best of the early sowings.

* Now at University of Tasmania, Hobart, Tasmania, Australia

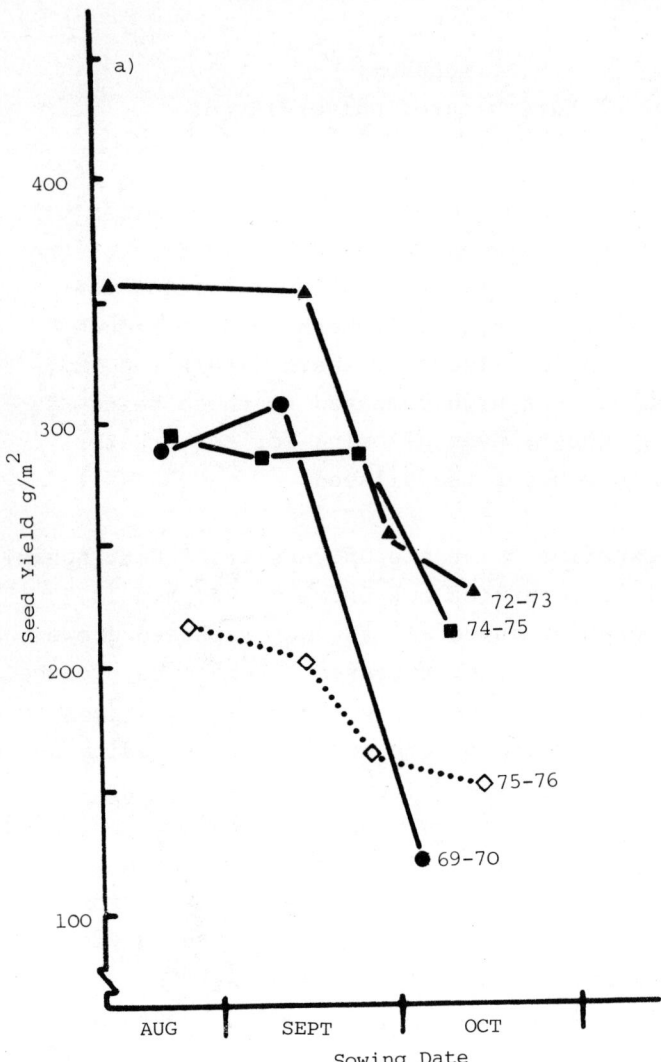

Fig. 1. The effect of delayed sowing on yield of Victor, 1969-70 to 1976-77.
a) Years when yield declined with later sowing.

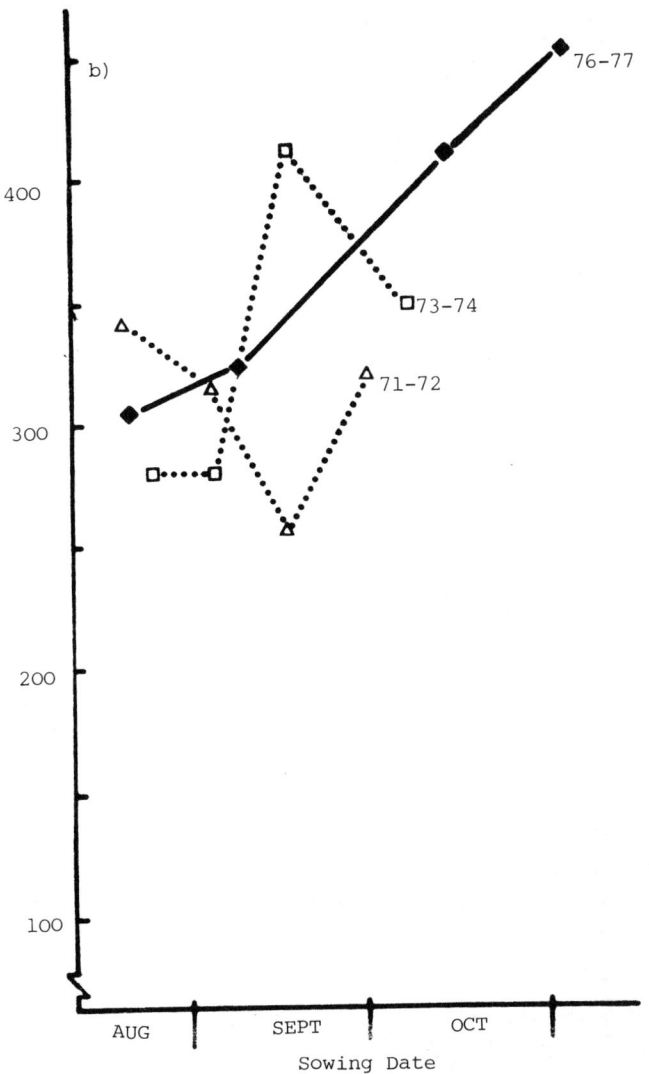

Fig. 1. The effect of delayed sowing on yield of Victor, 1969-70 to 1976-77.
   b) Years with a different response.

Apart from the wide range in yields of late sowings (equivalent to 1.2 - 4.5 t seed/ha), the other outstanding feature of Figure 1 is the consistency of the yields from early sowings; all were around 3 t/ha except in the very dry season of 1975/76 when seed growing was curtailed. Why, in favourable years when late sowings yield well, do early sowings not perform even better? To explain why yields of late sowings can vary so much, when early sowings vary so little, we will first examine the growth of these crops before flowering, and then follow pod and seed development.

In our experiments, pests and diseases were either of negligible importance or controlled effectively, and the standard fertiliser treatments applied should have removed nutrition as a limiting factor. Crop response should, therefore, have been related largely to weather, solar radiation, temperature and water supply.

Growth before flowering

The great variation in yields of late sowings had its origin in the amount of growth made by these crops before they flowered in spring. Figure 2 compares the leaf area and growth of crops in two seasons: 1969/70 when the late (October) sowing yielded poorly, and 1971/72 when the equivalent sowing yielded well. Results from early sowings in each of these seasons are also included for comparison. The late sowings in both 1969 and 1971 made little growth before winter, and then remained almost dormant until warmer weather in spring. In 1972, warmer weather began in mid-March, and the crop sown late in 1971 then began expanding leaves and growing rapidly. In 1970, however, spring did not begin effectively until mid-April. Both crops flowered at similar times in mid-May (indicated in Figure 2), so the net effect was that the crop sown late in 1969 had only about half the time available in the following spring for leaf expansion and growth before flowering as the crop sown late in 1971. While the 1971/72 crop rapidly reached a large leaf area capable of intercepting most of the incoming solar

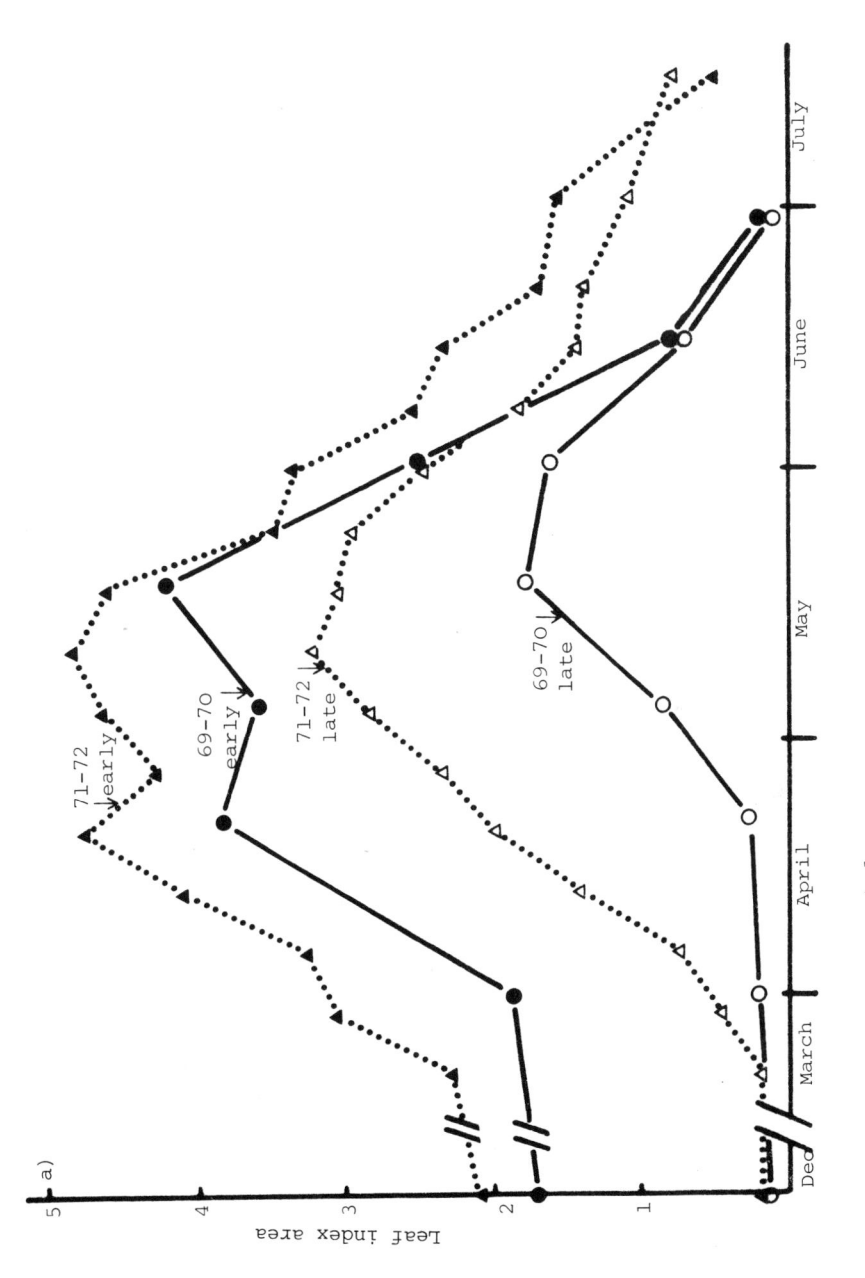

Fig. 2a. Changes in leaf area index.
(●22/8, ○ 2.10) and 1971-72 (▲ 20/8, △ 1/10). ↓ = 50% of plants in flower.

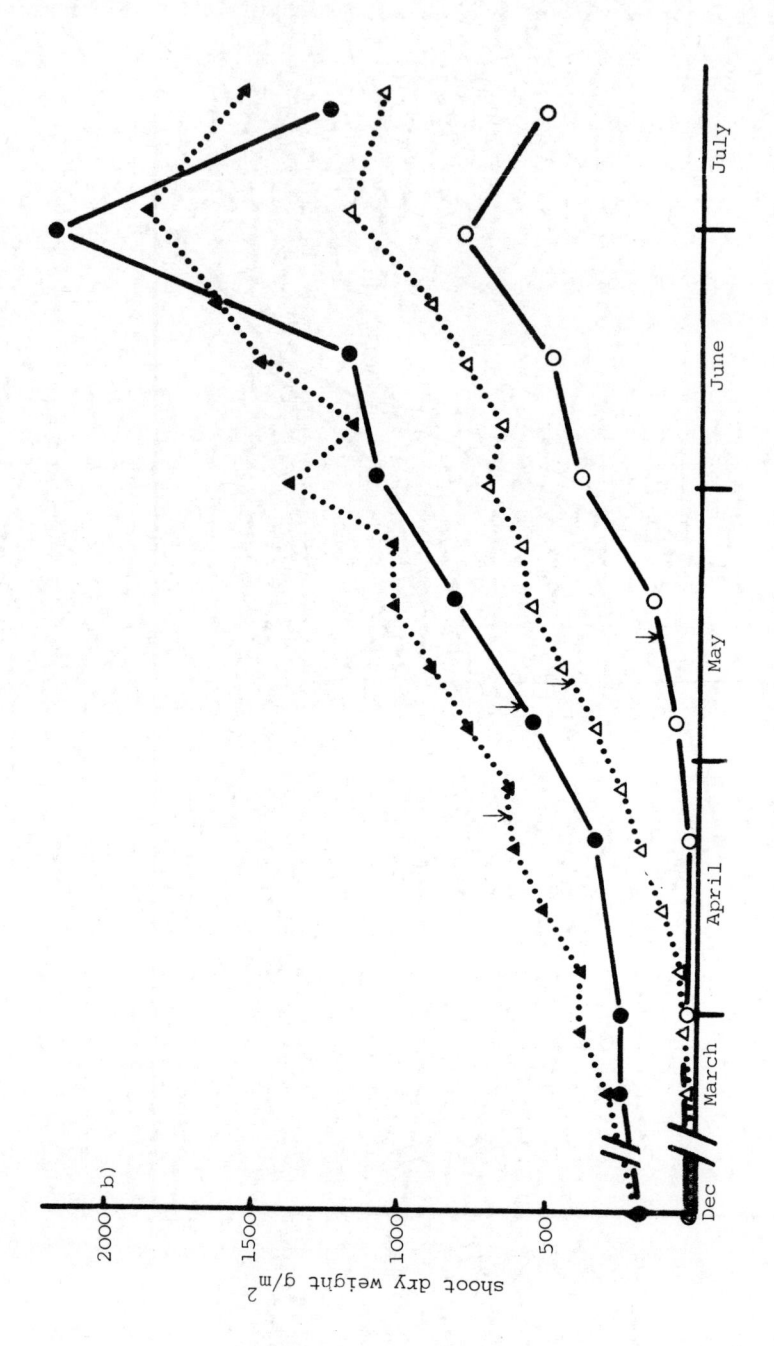

Fig. 2b. Changes in shoot dry weight in crops sown in 1969-70 (● 22/8, ○ 2/10) and 1971-72 (▲ 20/8, △ 1/10). ↓ = 50% of plants in flower.

radiation, and therefore made good growth, the 1969/70 crop was unable to utilise a large proportion of the available solar energy, which fell on bare ground.

In both crops, leaves stopped expanding at 'full flower' when most plants were in flower, and about a week after 50% of plants had at least one flower open, as is shown in Figure 2. The total area of leaf then declined rapidly, being shaded by flowers and then by pods. Rapeseed is unusual among arable crops in that at flowering it produces a dense mass of non-photosynthetic yellow flowers at the top of the crown. Measurements at Nottingham (J.L. Monteith, personal communication) have shown that crops during flowering can reflect 40 - 45% of total solar radiation, as compared to 20 - 25% reflected by green crops before flowering. Further radiation is absorbed by the flowers, with the result that the leaves underneath become heavily shaded and virtually non-functional.

Growth made by flowering time is therefore likely to be of greater importance in determining yield potential in rapeseed, than in most other crops, where leaves persist and make a positive contribution to crop and seed growth after flowering. While several studies have shown that green rape pods produce assimilate to support growth of their seeds, the potential yield has already been largely set by seed abortion shortly after flowering, discussed in a later section.

The early (August) sowings in the two years, illustrated in Figure 2b, showed a rather different response to seasonal conditions than the late sowings. In both years, the early sowings made good growth in autumn, and then remained more or less dormant over the winter. Both crops were already large and with an effective leaf area when warmer weather began in spring, and the delay of a month in 1970 had little effect on growth of the crop in that year by comparison with 1972. Yields were similar, and not restricted in either year by the amount of growth made in spring.

The length of time for growth in spring was thus important for late sowings, but not for early sowings. Table 1 shows that, for the late sowings in different years, the length of time from the beginning of spring growth to the start of flowering varied from about 40 - 90 days. The main difference was in the date when growth began in spring, as most crops reached full flower between 17 - 23 May. In the 1976/77 crop, however, flowering was delayed because cold weather in winter had delayed inflorescence initiation. Combined with an early spring this gave a particularly long period for growth.

TABLE 1

TIME AVAILABLE IN LATE-SOWN CROPS FOR GROWTH IN SPRING BEFORE FLOWERING. 'FULL FLOWER' = 7 DAYS AFTER 50% OF PLANTS IN FLOWER. 'BEGINNING OF SPRING' = DATE WHEN MEAN TEMPERATURES ROSE CONSISTENTLY ABOVE 5°C.

Season	Date of sowings	(1) 'Beginning of Spring'	(2) 'Full flower'	Number of days, 1-2
1969-70	2/10	13/4	21/5	38
1971-72	1/10	14/3	17/5	64
1972-73	13/10	19/3	21/5	63
1973-74	8/10	15/3	23/5	69
1974-75	9/10	11/4	22/5	41
1975-76	13/10	25/4	20/5	56
1976-77	13/10	3/3	5/6	94
1976-77	3/11	3/3	6/6	95

As well as the length of time for growth in spring, other important factors are the amount of incoming solar radiation and the leaf area available to intercept it. Estimates of 'intercepted radiation' combine all these factors, and relate closely (Figure to the amount of growth made in spring by later sown crops. The poorest growth was made in years when spring began late (1970 and 1975), and also in 1976 when increasing water stress slowed leaf expansion. The best growth was in 1977, as discussed above.

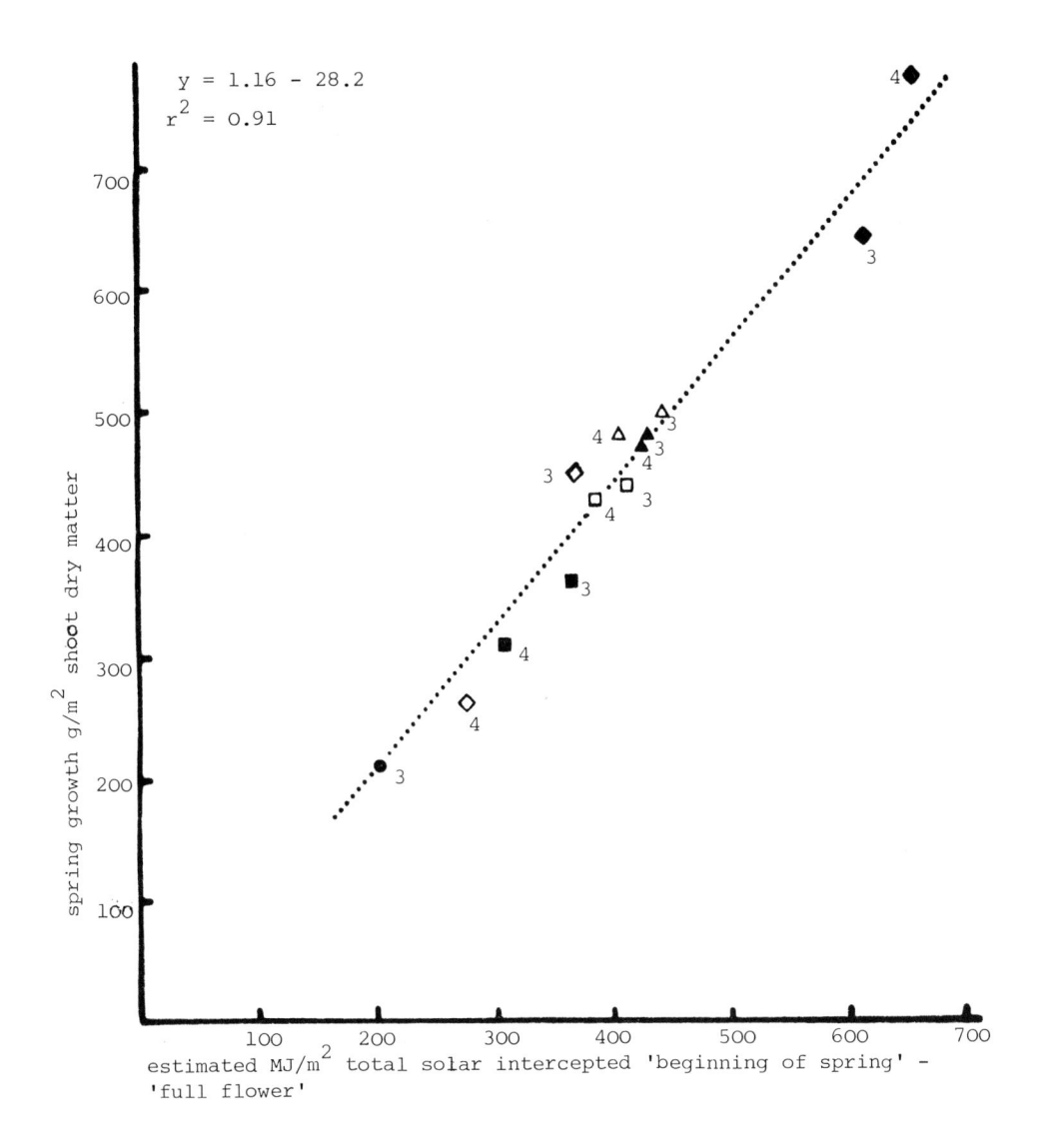

Fig. 3. The relationship between estimates of intercepted radiation and crop growth between the 'beginning of spring' and 'full flower', for late (3rd and 4th) sowings in ● 69-70, △ 71-72, ▲ 72-73, ◻ 73-74, ◼ 74-75, ◇ 75-76, ◆ 76-77.

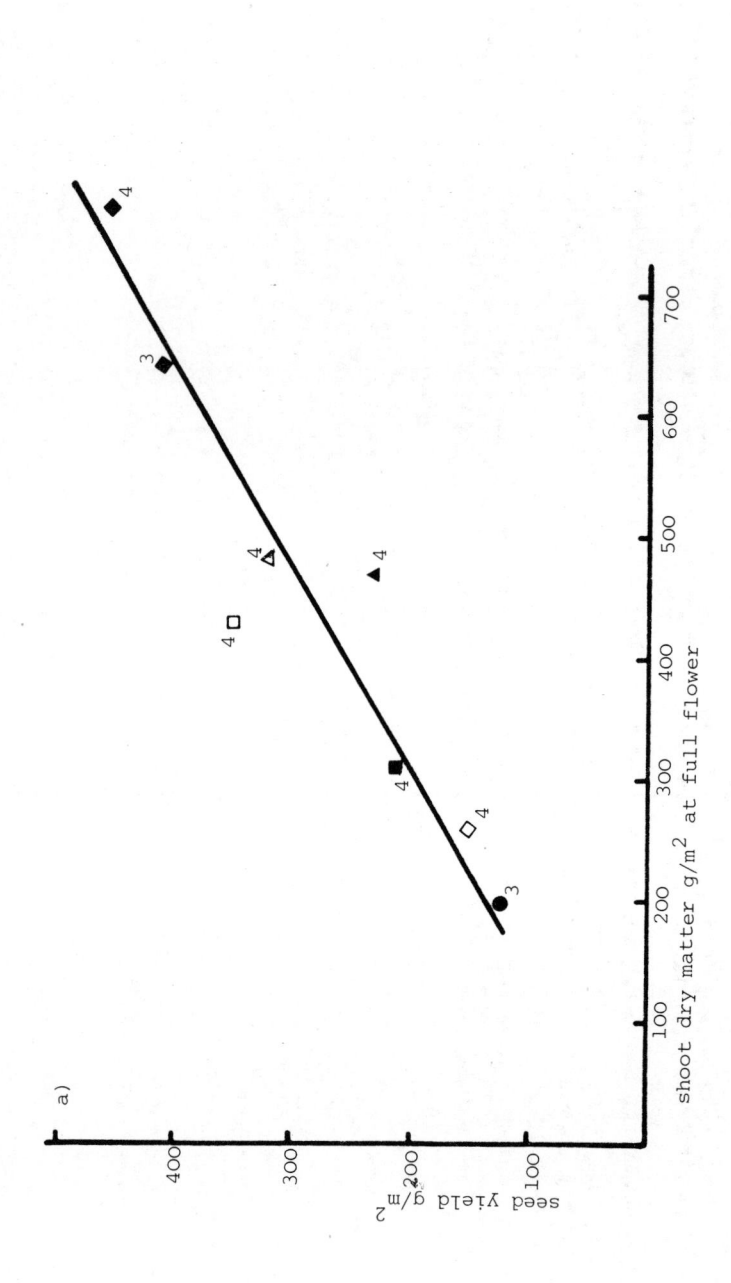

Fig. 4a. The relationship between crop size at flowering and seed yield for late sowings (Oct.-Nov.)

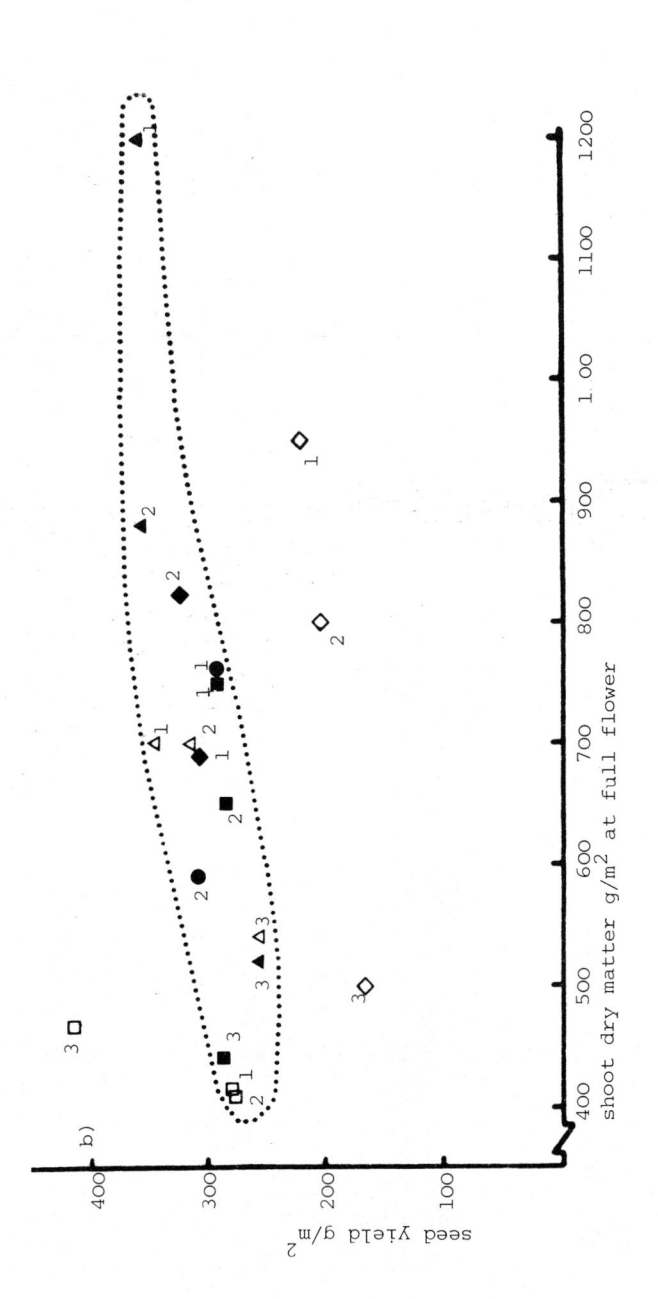

Fig. 4b. The relationship between crop size at flowering and seed yield for early sowings, also including late September (some 3rd) sowings. 1-4 = sowings made in ● 69-70, △ 71-72, ▲ 72-73, ☐ 73-74, ■ 74-75, ◇ 75-76, ◆ 76-77.

The amount of growth that late (October) sowings made before flowering, closely determined their yield potential (Figure 4a), the only substantial changes to potential being in 1972 - 73 when seed numbers per pod were low, and in 1973 - 74 when they were high (see later section). There was some evidence (Figure 4b) that seed yield from crops sown late in September (third sowings mainly) was also related to crop dry weight at flowering, particularly in season 1973/74, but the early sowings (before mid-September) showed little increase in yield over the full range of crop sizes at flowering. In 1975/76, crops were below the general relationship due to curtailed seed growth, but seed numbers were comparable with other years. What, then, was restricting the yield of the other early sowings?

YIELD COMPONENTS

An examination of the components of yield gives us some clues as to why the yields of late sowings can closely reflect the potential set up by the amount of growth made before flowering, whereas the early sowings vary so little in spite of widely different crop sizes and weather conditions. Figure 5 shows the effect of sowing date on the three yield components, and seed yield itself, for four of the seasons under consideration. While delayed sowing affected yield very differently, the components showed remarkably consistent trends in all seasons, as listed below:

1)   <u>Pod number always declined with later sowing</u>. This was related to the reduced primary leaf and branch number of late sown plants, as well as to their ability to support pods. While early sowings carried up to 12 000 pods (with seeds)/m^2 to final harvest, late sowings had as few as 3 - 4 000, and this included 1976/77 when yields were highest. Early sowings would appear to carry too many pods.

2)   <u>Seed number per pod increased with later sowings</u>. Early sowings always had very low seed numbers/pod, but the increase with delayed sowing varied from very small in 1969/70 to very

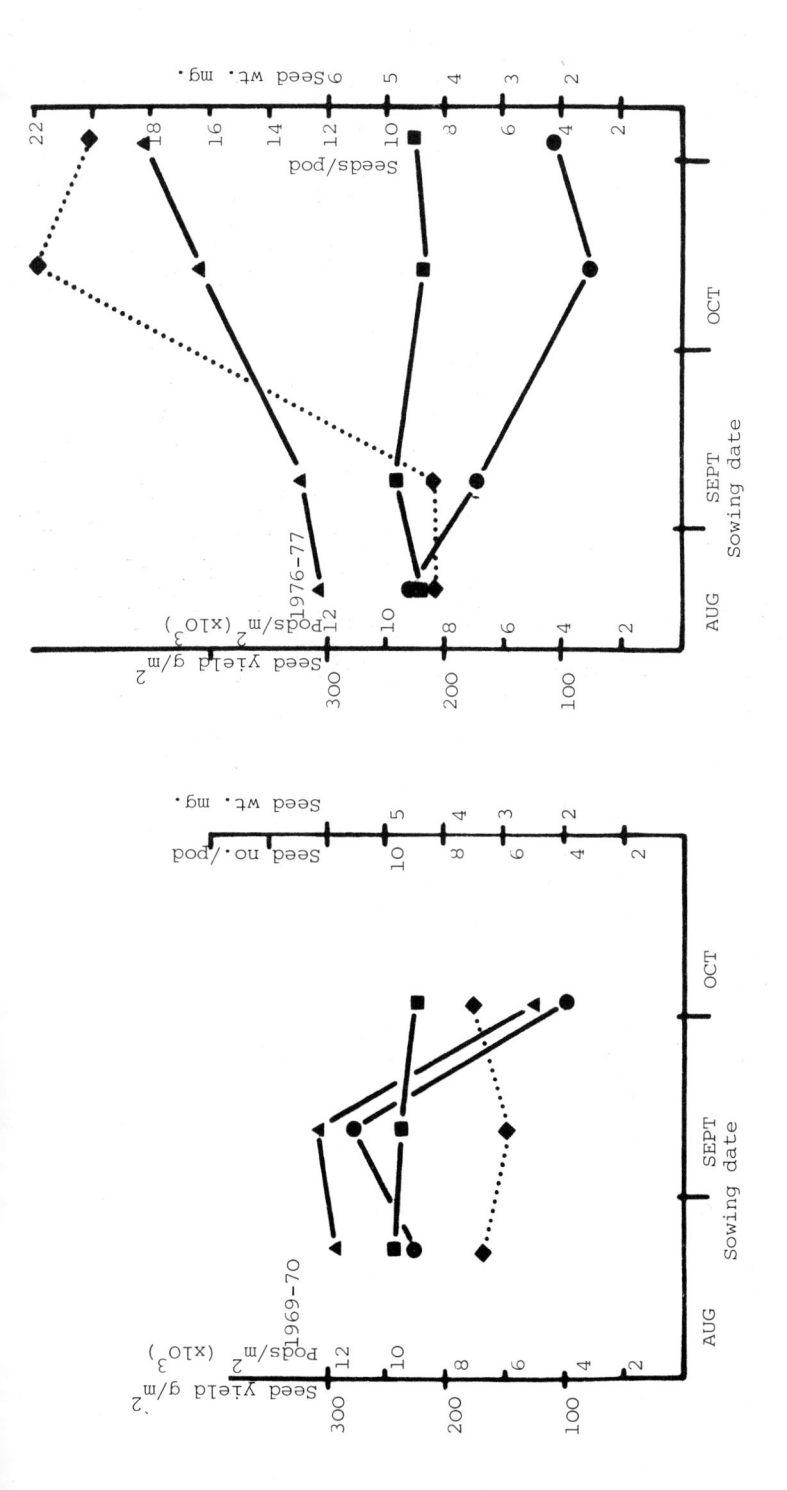

Fig. 5a. The effect of sowing date in two seasons: ▲ seed yield, ● pod number, ◆ seed number/pod and ■ mean seed weight.

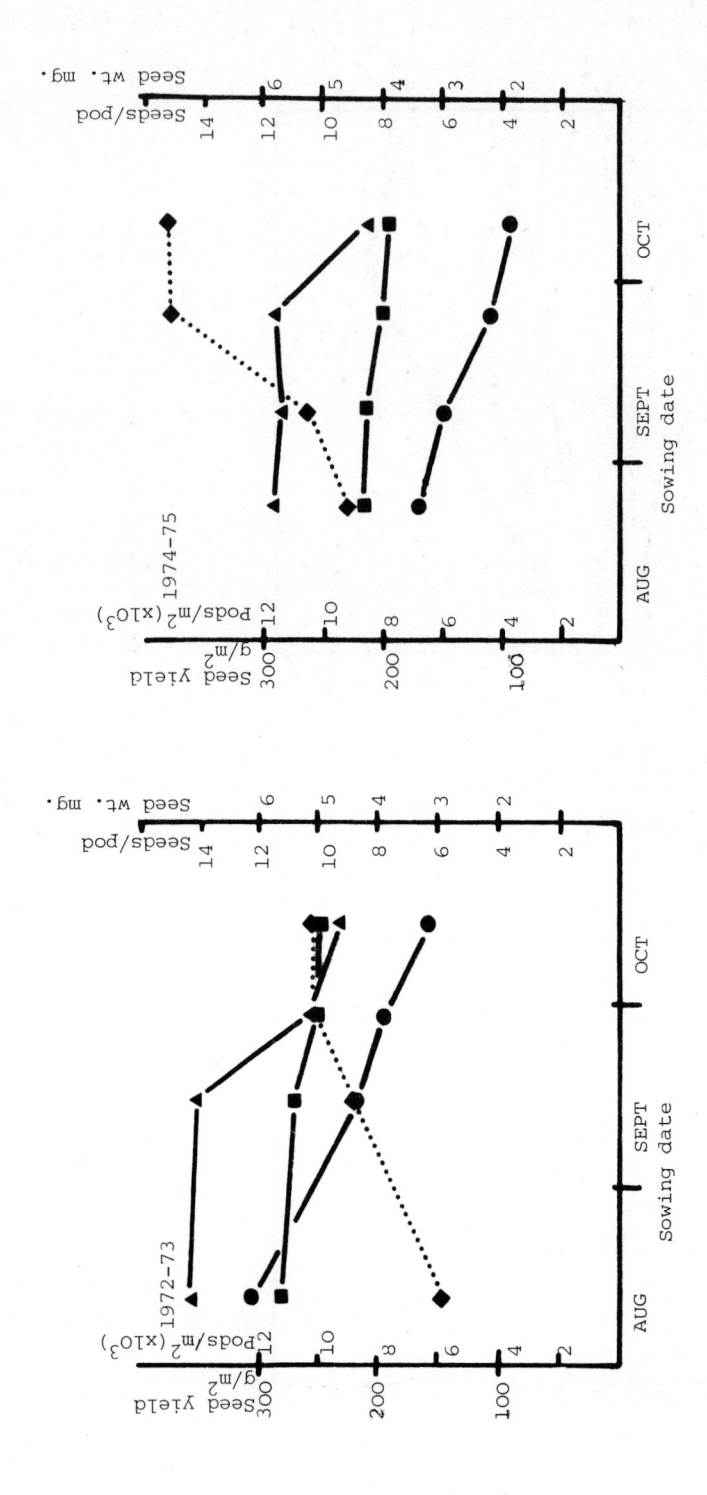

Fig. 5b. The effect of sowing date in two seasons: ▲ seed yield, ● pod number, ◆ seed number/pod and ■ mean seed weight.

large in 1976/77 - the years of greatest contrast in yield of late sowings. Clearly, the key to the success of later sowings was the degree to which the decline in pod number was offset by an increase in seed number. Where one balanced the other, as in the first three sowings in 1974/75, yield remained unchanged. Late sown crops which grew well, particularly in 1976/77, compensated for few pods by a large increase in seed number/pod. The late sowing in 1969/70, however, with the same number of pods but only a third of the plant dry-matter yield at flowering as the 1976/77 crop, only retained a third as many seeds/pod. The early sowings in all years invariably retained few seeds per pod, presumably associated with their apparently excessive pod production, and this must be the cause of their consistent, but never very high, yield.

3)   Mean seed weight was hardly affected by delayed sowing, at least by comparison with the other two components, although there was a small decline in some years.

The most important difference between early and late sowings, therefore, was in their ability to respond in seed number/pod. When is this component determined, and what factors restrict seed numbers in early sowings? Figure 6 shows an example of changes in seed number from flowering onwards in an early- and a late-sown crop. Samples of pods from upper, middle and lower sections of mainstems and primary branches were dissected to give an idea of the effect of position on the plant or in the pod canopy. The total number of ovules or sites for seed development was about 30, and represents the maximum 'potential' number. Those still developing in an apparently healthy state were classed as 'surviving'.

Few seeds were lost at pollination, but after about a week, when pod hulls started to grow rapidly, seeds began to abort. Losses were heavier in the early sowing, and in the lower parts of the canopy (in general, the first pods to set seed), where less than 5 seeds survived. In the later sowing, losses were less marked, with little difference between sections of the canopy, although the average surviving

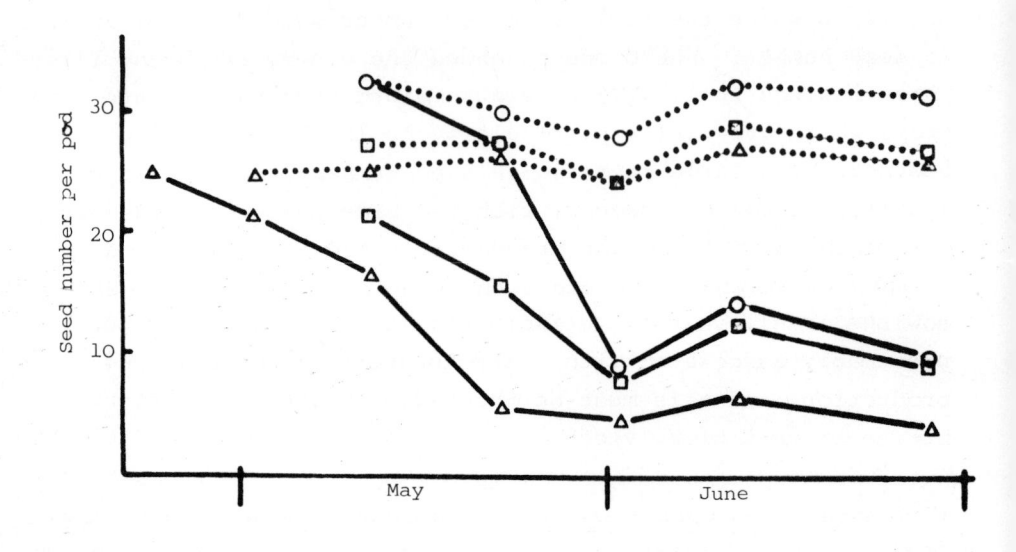

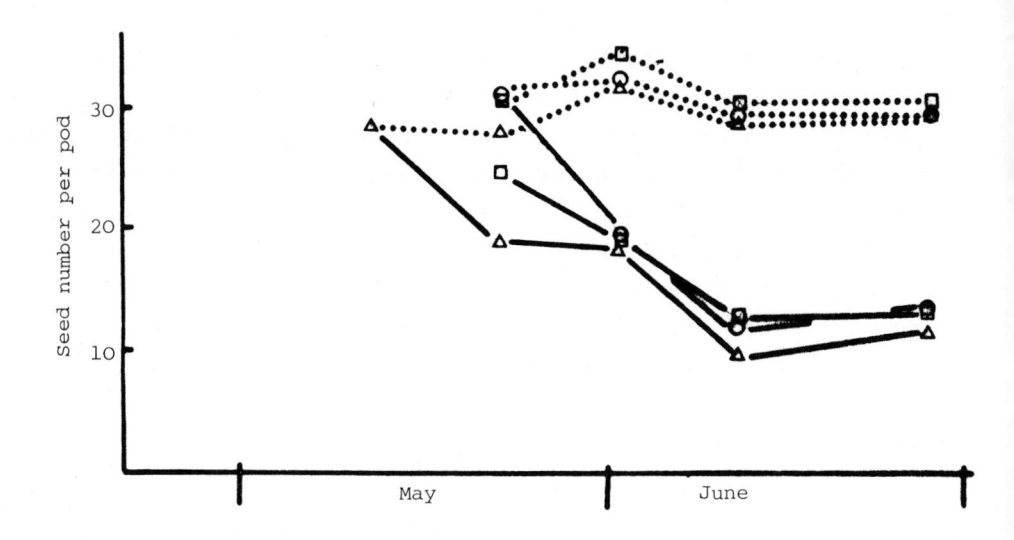

Fig. 6. Potential .... and surviving ———— seed numbers in
a) an early sowing and b) a late sowing; on o upper,
□ middle and Δ lower sections of the plants.

number was still only 12. In the later sowings in 1976/77, seed numbers (20 - 22) were much nearer to the potential.

As seen in Figure 6, an important feature of the abortion phase is that it is largely finished by early- or mid-June, and thus the critical period lasts for about three weeks from full flower. During the first part of this period, radiation is being reflected or absorbed by flowers, and in the second part the dense canopy of pods is developing rapidly. There is intense competition for assimilate between pods, with the result that in the lower, more shaded parts of the canopy, pods receive low levels of radiation and are under considerable stress. Both the shading and competition effects are much more acute in early sowings, with very large numbers of flowers and pods, and as a result there is considerable seed abortion. Many pods also absciss, or abort all their seeds, and thus contribute to the loss in yield potential.

The effects of low radiation levels on seed and pod abortion can be artificially intensified by shading the whole crop. When we did this, allowing only about a quarter of the natural radiation to reach the crop over the critical three week period, those pods which did carry seeds to final harvest contained only 3 - 5 on average, and many others lost all seeds. After the heavy losses the remaining seeds were able to compensate to a limited extent by growing to a larger size, but only up to about 20% greater than normal. Seed yields were reduced to between a third and a half of the control plots.

Seed growth is the last stage in the formation of yield, and took place in our crops mainly after the seed abortion phase had finished (early or mid-June). Compensatory growth by the remaining seeds after heavy seed abortion was observed in some crops, for example in 1972/73, when seed numbers were low but mean weight quite high (Figure 5). Seed weight is, however, a conservative character and increases are never likely to be enough to compensate fully for the kind of pod and seed losses observed.

CONCLUSIONS AND IMPLICATIONS FOR BREEDING

Particular physiological problems arise in rapeseed, because it produces flowers and then pods in a mass at the top of the crop. The two important and inter-related effects of this in our crops can be explained as follows:

1) Further leaf expansion, and continued photosynthesis of existing leaves, was largely prevented, as flowers either reflected or absorbed a large proportion of the solar radiation. Crops which had not developed an adequate leaf area and crop size by flowering time were thus restricted in their potential yield. Such crops resulted when leaf expansion had been slowed either by low temperatures early in spring (March or April), or by water shortage in May.

2) Heavy seed abortion took place, particularly in the earlier pods to set seed in lower parts of the canopy, as they were shaded by later flowers and pods, and as the large number of pods competed with each other for assimilate. The critical period for seed abortion was during the three weeks after full flower, when leaves were largely non-functional, and average radiation levels on pods were low due to shading. Losses were most severe on early sowings which usually produced very large numbers of flowers and then pods. This inefficient crop canopy appeared to be the main reason why yields of early sowings, although consistent, were never as high as the best of the late sowings.

The best way to minimise these two effects would appear to be to control flower and pod production, by either breeding new cultivars or developing new management techniques. In normal varieties and under the usual range of management practices, a more vigorous crop produces more leaf area, more branches and more pods. This link needs to be broken, so that a well grown crop can produce fewer pods. Each would then be better supplied with assimilate both from its own production and from leaves, and would be able to retain all or nearly all, of the potential number (25 - 30) of seeds per pod.

A crop of this type was produced by chance with late sowings in 1976/77, but late sowing cannot be recommended as a normal practice because low yields, or even crop failure, are likely. Breeders may be able to make a major improvement in yield if they can select types with vigorous growth but limited pod production. Jet Neuf may be a step in the right direction, as selection there was reported to be for 'high seed weight per pod' (i.e. a combination of high seed number and large seeds). Selection was also quite probably against pod number, which may be negatively correlated with the other two components.

The experiments discussed here are dealt with in greater detail in papers to be submitted to the Journal of Agricultural Science, Cambridge, and are the joint work of the author, Mr. P.A. Shipway and Dr. R.K. Scott.

DISCUSSION

D.G. Morgan   (UK)

There are a number of questions which spring to mind because there is a lot of overlap with much of what I want to talk about tomorrow. However, there are certain questions I would like to put to Dr. Mendham at this stage and others which I will reserve for later.

Could you explain why you think seed numbers went up so dramatically when there was a slight reduction in pod number in 1976/1977?

N.J. Mendham   (UK)

The reason why seed numbers went up so dramatically that year was because the crop produced several hundred grams of DM/m^2 at flowering, whereas some of the other crops produced only 300 g - 400 g. In 1976/1977 we had a crop with only 3 000 $pods/m^2$ but a great amount of dry matter, a lot of leaf area, and a lot of basic crop structure to support seed numbers. Therefore, instead of the abortion pattern that I showed in the other crops, where number of seeds/pod drops from 25 - 30 down to 10, here it has only gone down to 22. Probably the most important thing to come out of that particular crop is that you do not need a lot of pods - 3 000 $pods/m^2$ should be enough. With 25 seeds each they would produce yields of 4 - 5 t/ha.

D.G. Morgan

When the flower canopy is being developed, how long does the shading last? Is that period of time critical, and is what comes afterwards not so important in determining yield?

N.J. Mendham

Yes, that is the critical time. From the time that the crop is in full flower there is a dense mass of yellow flowers reflecting up to half the incoming radiation for about two or

three weeks - this is the important stage. At the end of this
time seed numbers have been largely determined, though the
seeds themselves are still very small, and it is a job to dissect
them out and weigh them. It is from about mid-June onwards
that the seeds grow rapidly. Therefore, we can separate seed
development into two phases. First, seed number determination
- complete by about three weeks after full flower. Then, for
another six or seven weeks, seed growth. At the moment we are
trying to look at these phases separately. In the phase of
seed number determination we are considering its relationship,
within the canopy, to the amount of light coming in; the effect
of shading of flowers and pods giving this abortion of seeds.
From then on, once flowering has finished, the number of seeds
is determined and the seeds grow. Other factors effect it
then.

D.G. Morgan

One thing worries me about your sampling into layers.
The lower axilliaries are the later developing ones, and so
tend to be lower yielding than the top ones anyway. When you
sample, how can you make sure that you are taking that into
account and that you are getting a uniform division of your
crop? And how do you take into account the origin of the
branches and the origin of the pods?

N.J. Mendham

It is not just a shading effect. It is also very much
a competition effect. The main stem develops first of course;
the first branch is always a little smaller than the main stem
because the main stem is so dominant. The other branches are
larger later on. To divide the crop in this way is looking at
both factors: the position in the canopy and the position on
the plant. In some of our later work we have stratified the
canopy: taking it off in layers. We have used tube solarimeters
in each layer to measure the amount of light reaching it. We
have harvested the layer and we have looked at pod and seed
numbers in it. We are still in the process of putting all that
together, but that is the basic approach.

G. Röbbelen (FRG)

I found these data fascinating. In Germany there is not the same degree of freedom in sowing date as in England. For example, late sowing always gives a smaller chance of high yields because of low survival. Overwintering requires a critical plant mass; growth in your late-sown crops would surely be drastically reduced.

Secondly, during the years of your experiments, were you able to determine these things up to yield without any interference from secondary influences? For example, have diseases never come into the picture?

N.J. Mendham

We like to think that we are good farmers and know how to grow the crop.

G. Röbbelen

I think with canker etc., you would possibly have problems.

We know that pollination definitely determines seed numbers. I am fascinated when you say that abortion of fertilised seed is the dominating factor. We know this from *Vicia faba*, for example, and there you can see it. However, with rapeseed, I wonder whether or not primary fertilisation should also come into the picture.

Finally, one of your Figures showed that earlier sowing gave more stable yields. Perhaps it was not the topmost, but it was always quite good. Stable yields for the practical farmer have great advantages over very high and then low yields.

N.J. Mendham

I will take the last point first because it is the most important. What I have put forward here are some ideas on the physiology of the crop which would help us to improve the yields from early-sowings, which are obviously far better from the point of view of getting the crop established, surviving the winter and competing against weeds, pigeons and such.

There are factors in these late-sowings which may help breeders to think about ways in which the pod production of the plant could be controlled to make it produce fewer pods in a better arrangement in the pod canopy, with more light getting to each pod, and so enable those 25 - 30 seeds to be fertilised and set.

On the point of population density: in some years we have also had winter kill; there were as few as 8 plants/m^2 on some late-sowings. With only 8 plants/m^2 one cannot expect a very high yield, but we still got about 2.5 t/ha when no other factors intervened, such as frost or pigeon damage. Even with only 8 plants/m^2 there was still the possibility for a substantial crop because each of the plants could develop to quite a large extent, producing a large leaf area which, when it flowered, was not under the same dense yellow mass - it was more like the plants you can see growing by the side of the road where light can penetrate to the leaves and to the lower pods. Therefore, surprisingly, on these low density crops we can get quite a respectable yield.

G. Röbbelen

There was also the matter of pollination.

N.J. Mendham

As far as we could see from our dissections, with the sort of conditions under which we worked, pollination was not really a problem. About ten days later, just as the pod husks start to grow very rapidly, the seeds are competing totally with the pods for assimilate. From then until about two weeks later, they abort progressively.

D.G. Morgan

Professor Röbbelen is quite right; there is a pollination factor. You can spot it if you follow the change in seed number per pod. We have done this within the inflorescence; you get the ovular number and, within the first few days, there are always some which are not fertilised - then it stabilises. However, with some of those at the top end, or with some of the lower ones if they are in stress, you can still get a subsequent abortion. Therefore, I think there are two factors operating.

N.J. Mendham

I neglected pollination completely because we have never seen any evidence of it affecting more than about the first five ovules - down to 25. If we could get 25 there would be no problem; it is 25 down to 5 or 6 that is the worry. Most of our early-sown crops have between 5 - 10 seeds per pod which is a gross waste of resources.

T.M. Thomas (Ireland)

In the year when you got a higher yield from the late sowings, did you get a lower plant population in the earlier sowings?

N.J. Mendham

We have not had yield increases with a lower population. It just emphasises the compensatory ability of the crop. I do not think that lowering the population would help much, at least not with the levels we were working with. But many farmers are ending up with 100 - 150 plants/m^2 which I think is too many; I think that about 60 is enough. If there are more than that and they all come into flower at once, the leaves are completely shaded out. The flowering must be controlled, or the crop adjusted for a higher population with good seed numbers.

We do not have all the answers to this. The experiments were done to try and understand the physiology of the crop. We would not like to say that we counted everything; there were only a couple of us working on these experiments every year hoping to get some results.

C. Paul *(FRG)*

Have you had any thoughts about the differences in the root growth between the early and the late sowings? What influence would the root growth have had in the interpretation of your results?

N.J. Mendham

Because most of the growth of the late-sown crops is in the spring, it depends on what the spring is like as to how much root growth there is. In years like 1976, when there was a dry spring, root growth was very restricted so the late-grown crops were the ones that suffered from water stress. The early-sown crops had their roots right down and did not suffer from water stress until seed fill. But with the late-sown crops, with a small root system which was unable to reach the water, the leaves did not expand, did not intercept enough light and did not grow, producing a poor yield.

C. Paul

Have you tried to regress seed yield on the available water, immediately following pollination? You were trying to point out that, in your opinion, incoming radiation is a limiting factor. However, available water during that time period, even in England, could be a critical factor.

N.J. Mendham

It should be a critical factor, but in none of the springs did we get sufficient deficit to affect yield. Some of the years were quite good for seed numbers: 1974 was a dry spring but we had 15 - 16 seeds per pod. This spring we had almost no rain in April and May and it will be interesting to see what sort of seed numbers we get. So far, apart from this restriction of radiation into the canopy we have seen nothing obviously affecting yield in the experiments we have done. These were field experiments and we were not looking specifically for water stress, although in two years we used the neutron probe to give us data - but we did not find any link between water stress and seed numbers. After all, it is still early in the season - we are talking about May; it is not as if it was happening in June or July with a crop that flowers in the summer.

M. Dambroth (FRG)

Perhaps we could continue this discussion tomorrow after the paper of Dr. Bramm. He will be talking about the relationship of the water regime, at different stages of rapeseed growth.

N.J. Mendham

The purpose of my being in England at the moment is to write up this work for publication. It was carried out for so long, and became so complex that it got a bit beyond us, but I hope the detailed papers will appear some time this year.

M. Dambroth

Thank you very much.

PROBLEMS IN SIMULTANEOUS BREEDING FOR OIL AND PROTEIN CONTENT IN SUNFLOWER

E. Alba and I. Greco

Istituto di Miglioramento Genetico
Università degli Studi, Via Amendola, 165/A
70126 Bari, Italy.

ABSTRACT

The possibilities for utilising the oil and protein content of sunflower seed are analysed.

The main factors determining protein quality, and the prospects for obtaining two qualitatively different types of oil, are also discussed.

A survey is then made of the main correlations existing among the qualitative and quantitative components.

Finally, a breeding programme is proposed to consider the quantitative and qualitative characteristics of seed oil and protein and to find new morpho-physiological parameters to be utilised when breeding for a new seed ideotype.

INTRODUCTION

In sunflower crop development, so far, consideration has been given only to the increase in seed production and relative oil content, whereas other nutritional possibilities, including sunflower as a source of protein, have been neglected. The sunflower seed is of economic importance in human nutrition for its oil content and in animal nutrition for the protein content in the defatted flour. Moreover, the latter could also constitute a valid protein source for man.

Many studies confirm the satisfactory chemical and biological qualities of sunflower protein. For example, it has been shown that sunflower protein is soluble in low or high concentration of sodium and calcium chloride, salts that are common constituents in the aqueous phase of many food systems (Sosulski 1979). Nevertheless, sunflower protein has some factors, such as its low lysine content and high chlorogenic acid percentage, which limit its biological value.

The total protein content of the seed and the relative contributions of the various amino-acids are not, however, constant in the Helianthus species, but vary with genotypic differences within, and between, populations (Alba et al., 1976; Baudet et al., 1971; Morozov, 1972). For different commercial cultivars, the total protein content of the defatted flour varies from 50 - 60%, the lysine content from 2.5 - 3.4%, and the chlorogenic acid content from 1.5 - 4.9%.

Chlorogenic acid (CA) causes problems in the preparation of protein concentrates and isolates; indeed, in the concentration of defatted flour and in alkaline environment, the CA produces, with the proteins, some green and brown coloured compounds which are very stable with lysine in peptide fraction. This decreases the digestibility of products based on defatted sunflower flour. Recently, methods have been developed for the almost total removal of CA (Sodini et al., 1977; Lanzani et al., 1978), and for its possible agricultural and chemical use, (Lener, 1979). Ivanov (1975) has also reported remarkable success in obtaining lines almost devoid of CA.

However, once the genetical control and technological utilisation of CA has been established, the problem remains of improving the lysine content of the protein.

The possibility of a qualitative improvement in fatty acids to respond to specific nutritional needs is being evaluated in breeding programmes. Two different types of oil could be obtained, one with a high linoleic acid content, to be used for specific diets and for industrial purposes, the other with a better oleic/linoleic rate for wider and more general nutritional purposes.

At present, however, even though appreciable results have been reported on some of the components determining qualitative and quantitative improvements in oil and protein (Alba, et al., 1979), further studies should be carried out before utilising these data, since results concerning the relation between oil, lysine and protein contents show some disagreements. According to some researchers, breeding of lines having a high lysine content do not necessarily bring about a decrease in oil content (Girault et al., 1970; Alba et al, 1978).

In effect, oil and protein contents are correlated and selection aiming at high oil content results in a decrease in protein. A negative correlation between lysine and protein content, due to a modification in the contents of albumins and globulins, has also been detected (Girault et al., 1970). In Marinescu's opinion (1978) there is no negative correlation between protein quantity and quality, while Dorrell (1976) states that there is a positive correlation between oil and CA. It must be noted, however, that seed yield is influenced by the environment (Alba et al., 1978) and by the association of different morpho-physiological characters (Alba et al., 1979).

It will be related to plant efficiency, to particular agroecological situations (Gatto et al., 1980), to certain requisites the plant must have concerning: stem height; duration of developmental stages (reproductive and vegetative); resistance to diseases; leaf number, their angle of insertion and surface area so as to facilitate the photosynthetic process; and leaf duration and photosynthetic efficiency during the most intensive period of seed dry matter production. Recent studies on the ripening of sunflower seeds (Pomenta et al., 1971; Dorrell, 1976; Canella et al., 1978) have shown how quantitative and qualitative characteristics of seed oil and protein vary during seed development. These variations show that the biosynthetic processes of the lipidic fraction are completed around the 35th day after flowering. During this period, while the total protein content is constant in the different ripening phases, the amino-acids composition undergoes variations. In particular, there is a decrease in albumins in favour of globulins.

A detailed plant physiological and morphological study is necessary to establish the genes or genetic complex which control structural characteristics, and specific biochemical and physiological processes of the plant responsible for the production of certain types of seed.

Relating to this, a programme with the aim of obtaining a simultaneous quantitative and qualitative improvement of sunflower seed production is required. The goal should be to define a new seed ideotype.

RESEARCH STUDIES

In the framework of studies on sunflower, the Institute of Plant Breeding of Bari is carrying out research aiming to produce synthetic cultivars and hybrids having both high oil and protein content, and protein and fatty acids of high quality. This research is carried out in collaboration with other institutions (Venturi, 1980; Alba et al., 1978).

The plan of this research includes the increase of seed yield/ha and the following:

1) Assessments of oil content, total protein and relative amino-acid fractions in populations, varieties and inbred lines of *Helianthus annuus* L.

2) Development of new lines, cultivars and hybrids of sunflower in order to determine and investigate the mechanisms of control of seed yield, and content and quality of oil and protein (especially where negative correlations exist).

3) Study of protein content and relative amino-acid fractions in the germ-plasm of the *Helianthus* genus.

At this stage of the programme answers cannot be given to such questions as a) which parameter influences the amino-acids composition and which the composition of fatty acids, b) which characters, when transferred to a different genetic context, retain their effectiveness, c) to what extent incorporation of quality characters affects productivity.

These problems might be resolved by finding new genetic-morphophysiological parameters to utilise in the breeding of biotypes capable of producing a seed of improved quality.

When these parameters have been obtained, they should help us to understand the individual contribution of a character and the genetic control mechanisms involved in biosynthetic processes.

REFERENCES

Alba, E., and Greco, I., 1976. Primi risultati di una ricerca in linee ed ibridi di girasole per incrementare la frazione proteica contenuta nei semi. Ann. Fac. Agraria. Univ. Bari. Vol.XXVIII: 201-209.

Alba, E., and Vannella, S., 1978. Studio della stabilità dei principali caratteri in alcuni genotipi di girasole. Proc. of International Sunflower Conf., PISA, 14-15 December; 64-75.

Alba, E., Greco, I., Fiorentino, S., and Lener, M., 1978. Analisi della variabilità in due popolazioni composite di girasole per i caratteri: contenuto in olio, in proteine totali e in lisina. Proc. of International Sunflower Conf., PISA, 14-15 December: 96-106.

Alba, E., and Greco, I., 1979. An analysis of association of the factors influencing seed yield in Sunflower. Sunflower Newsletter, N.2 April: 13-15.

Alba, E., Benvenuti, A., Tuberosa, R., and Vannozzi, G.P., 1979. A path coefficient analysis of some yield components in Sunflower. Helia (in press).

Baudet, J., Leclercq, P., and Mosse, J., 1971. Sur la richesse en lysine des graines de Tournesol en function de leur teneur in protéines. C.R. Acad. Sci., PARIS, 273: 1112-1115.

Canella, M., and Lener, M., 1978. Effetto della maturazione sulla composizione chimica dell'olio e della farina di girasole. Proc. of International Sunflower Conf., PISA, 14-15 December: 342-361.

Dorrell, D.G., 1976. Chlorogenic acid content of Sunflower seed flour as affected by seeding and harvest date. Can. J. Plant Sci., 56: 901.

Gatto, L., Greco, I., and Alba, E., 1980. The effect of seasonal irrigation water regimes in Sunflower characters in Southern Italy. IX Conferencia Internacional del Girasol. 8-13 June, Torremolinos (in press).

Girault, A., Baudet, J., and Mosse, J., 1970. Etude des proteines de la graine de Tournesol en vue de l'amelioration de leur teneur en lysine. Plant Protein Nucl. Tech. Symp. IAEA, VIENNA: 275-284.

Ivanov, P., 1975. Variation of the protein, lysine and chlorogenic acid content in some Sunflower selfed lines. Plant. Sci. XII, n.10, SOFIA: 23-27.

Lanzani, A., Cardillo, M., and Petrini, M.C., 1978. Le farine da semi di girasole nella prospettiva di impieghi alimentari. Proc. of International Sunflower Conf., PISA, 14-15 December: 44-63.

Lener, M., 1979. A personal communication.

Marinescu, R., 1978. Determination of protein composition in Sunflower. Probleme de Genetica teoretica si aplicata. 10(5): 527-531.

Morozov, V.K., 1980. Research on increasing protein content in sunflower seeds. Selektsiya i Semenovodstro USSR. No. 1, 18.

Pomenta, J.V., and Burus, E.E., 1971. Factors affecting chlorogenic, quinic and caffeic acid levels in Sunflower kernels. J. Food Sci., 36: 490.

Sodini, G., and Canella, M., 1977. Acidic butanol of colour-forming phenols from Sunflower meal. J. Agric. Food Chm., 25: 822.

Sosulski, F., 1979. Food uses of Sunflower proteins. J. Am. Oil Chemists' Soc. 56(3): 438-442.

Venturi, G., 1980. Il progetto del Ministero dell'agricoltura per una ricerca sulle oleaginose. Informatore Agrario, Anno XXXVI: 10351.

ASPECTS AND PROGRESS OF OIL CROP BREEDING IN SPAIN

J. Fernandez-Martinez and J. Dominguez-Gimenez
CRIDA 10 INIA, Finca de la Alameda
Apartado de Correos del Obispo 240, Cordoba,
Spain.

SUNFLOWER

Sunflower is now a major oilseed crop in Spain, having expanded its area of cultivation considerably in the last ten years. It is quite important in southern Spain, where high temperatures and low soil moisture often coincide with the period of blooming and maturation. Because of these adverse conditions, the sunflower seed yield in Spain is low (700 - 800 kg/ha) compared with that in other countries where sunflower is widely grown.

The following breeding objectives are being considered in our programme:

a) Seed yield: As mentioned above, the main limiting factors for yield are high temperatures and low level of moisture in the soil during the flowering-ripening period. Two strategies can be considered in breeding for higher yields under these conditions, 1) Selection of material tolerant or resistant to drought. 2) Development of varieties or hybrids that can avoid the adverse conditions. This may be achieved by selecting early types and/or planting earlier than is traditional, which involves selection for tolerance to the low temperatures likely to occur during early stages of growth.

More emphasis has been given to the second strategy and the main objectives of our breeding programme have been earliness, and tolerance to low temperatures.

When this programme was started in 1974, all the early material that we had available was low yielding and had a low oil content. A population was made using germplasm from different origins with different growth cycles. Male-sterile cytoplasm, normal cytoplasm and fertility-restorer material were included. After several generations of recombination, we carried out several cycles of recurrent selection for the characters already mentioned.

Experimental early hybrids made with lines derived from this programme were tested in 1979 using commercial medium-late hybrids and open-pollinated varieties as controls.

Results of this test are given in Table 1.

TABLE 1

YIELD, OIL CONTENT AND OTHER CHARACTERISTICS OF EXPERIMENTAL EARLY HYBRIDS AND MEDIUM-LATE HYBRIDS GROWN UNDER DRYLAND CONDITIONS AT SEVILLA.

Cultivar	No.	Yield (Qu/ha) x	Range	% Oil x	Range	Days to 50% flowering x	Range
Exp. Hybrids	55	21.2	14 - 29	50	45 - 54	73	67 - 78
Com. Hybrids	9	20.8	18 - 22	47	42 - 50	89	79 - 95

Although, under optimal environmental conditions, later maturing hybrids show greater yield potential, the data of Table 1 indicates that some early material obtained in our programme may perform better under dryland conditions, probably because it avoids stress factors during critical periods of growth.

The data given in Table 1 also show that the oil content in selected experimental hybrids is reasonably high. This is the result of strong selection for this character in the population developed initially. The evolution of oil content during three cycles of selection is given in Table 2.

TABLE 2

EVOLUTION OF THE MEAN OIL CONTENT OF CYTOPLASMIC MALE-STERILE (C.M.S.), NORMAL (N) AND FERTILITY-RESTORER (R) POPULATIONS IN THREE CYCLES OF SELECTION

Cycles	C.M.S. population $\bar{x}$	σ	N population $\bar{x}$	σ	R population $\bar{x}$	σ
I	41.9	3.7	42.3	3.9	43.3	4.0
II	43.9	1.7	43.2	2.1	44.3	2.2
III	51.2	4.3	47.9	4.0	48.5	5.7

It is important to point out that the strong positive correlation between oil content and length of growing cycle reported by several authors does not occur in the selected early material, suggesting a possible breaking of the linkage between the two characters through the process of recombination and selection.

Since 1975 we have carried out a programme to develop high oleic acid material in sunflower. Traditionally Spain consumes high oleic oil (olive oil) which has advantages for deep-frying purposes.

At the same time we have searched for high linoleic types. These two fatty acids are negatively correlated and make up about 90% of the total fatty acid content of sunflower oil.

The range of variability found by us in wild *H. annuus* and *H. exilis* species was 4 - 85% for oleic acid and 8 - 88% for linoleic acid.

High sources of oleic acid (>70%) were crossed to cultivated material (20 - 35% oleic acid). The F_2 generation did not segregate in clearly defined classes, although the range of variation for oleic acid was from 25% to 85%. Oleic acid content is greatly affected by temperature conditions, low temperature decreasing the content considerably, but in the sources used it has always been higher than in the cultivated typ

The high linoleic source seems to be more stable under different ranges of temperature and, although it comes from *H. exilis*, it appears to be a good source of linoleic acid for breeding purposes.

SAFFLOWER

As in sunflower, and for similar reasons, seed yield of safflower in Spain is lower than in other countries.

A breeding programme was initiated in 1978 with three main objectives: yield, oil content and oil quality. Adapted local varieties have been crossed to germplasm from different sources with a low hull percentage, which gives higher oil and protein contents, with a high oleic acid content (> 70%), and with improved yield components.

The 'single seed descent' method of selection has been used in some crosses, with close plant spacing, speeding up the breeding programme. Four generations per year have been obtained under growth chamber conditions. F_4 lines derived from this method (some of which have a high oleic acid content) are in initial tests this year.

We are also evaluating the effect of the spineless character and of appressed type of branching on yield and on oil content. Efforts in the past to develop superior spineless cultivars have not been successful. However, there is need for more precise data on the effect of the spineless character on yield components and oil content. Appressed types of branching, having branches closely against the central stem, may permit the use of more plants/ha, and possibly increase yields.

RAPESEED AND BRASSICA spp.

This crop is in the introductory phase in Spain. In most areas, best results have been obtained by planting spring cultivars in winter.

Breeding research in progress, has been mostly concentrated on developing cultivars more adapted to our conditions. Better yielding low-erucic cultivars have been crossed with double-low material to evaluate double-low segregates, with the intention of developing better double-low cultivars for Spanish conditions.

Two sources of cytoplasmic-male-sterility are being evaluated in our conditions. These sources are not completely male-sterile under field conditions, although one of them seems to be very promising.

A hundred entries of introductions of *B. juncea*, *B. carinata*, *B. campestris* and *B. napus* have been tested. Many of them were too early, too tall, or presented other undesirable characteristics. Thirty eight were selected and tested for yield in replicated plots; the results are given in Table 3.

TABLE 3

YIELD, HEIGHT AND DAYS TO FLOWERING IN 38 ENTRIES OF FOUR BRASSICA SPECIES.

Spp	No.	Yield (Qu/ha) x	Range	Height (cm) x	Range	Days to flowering x	Range
B. juncea	8	17.5	10 - 30	127	75 - 198	103	90 - 132
B. carinata	15	31.4	24 - 41	128	111 - 166	126	115 - 132
B. campestris	6	18.0	13 - 24	119	100 - 140	97	85 - 118
B. napus	9	14.8	10 - 22	120	100 - 132	112	95 - 131

The controls included cultivars yielding well in Spain in previous years of test: Cresor, Gulliver, Midas, Erglu, etc.

Late introductions of *B. juncea* and (especially) *B. carinata*, had very good yields as compared with the *B. napus* cultivars used as controls. Provisional results of tests in the present year seem to confirm these results, especially for *B. carinata*.

Some late introductions of *B. campestris* also yielded well. All this material showed high erucic acid and high glucosinolate contents.

Interspecific crosses are being made between *B. juncea* and *B. carinata* and double-low *B. napus* cultivars, to select for lower levels of erucic acid that would come from the genomes *campestris* and *oleracea* respectively, in the hope that meanwhile, and elsewhere, genes for low-erucic acid content in the *nigra* genome might be discovered for possible incorporation into *B. juncea* and *B. carinata* species.

We are also crossing late, high-yielding, *B. campestris* introductions with the double-low, yellow seeded, cultivar Candle to select high-yielding, double-low, yellow seeded *B. campestris* cultivars for our conditions.

DISCUSSION

N.J. Mendham (UK)

I have a comment. The conditions in Spain are very similar to the conditions in mainland Australia where there has been a lot of breeding work for the types of spring rapeseed which are sown in the autumn and grow through the winter, ripening early in the spring. It could be beneficial to examine some of the new Australian lines under Spanish conditions. Perhaps Dr. Buzza - a breeder in Victoria - could be approached to assist with this.

G. Röbbelen (FRG)

It is rather surprising to see safflowers, sunflowers and rapeseed all grown in one area. There is no doubt that ecological conditions for these three crops are very different. These developments in Spain are technical to some extent. In the southern part of Spain, oil crop production has been known for some time, but this is not so in northern Spain.

The introduction of rapeseed into Spain, a few years ago, started in the area where oilseed production was well known already, both from sunflower and the olive oil industry. Several expert committees, including the Germans and Canadians, were of the opinion that rapeseed, finally, would be produced in the northern parts as opposed to the south.

I would never start a rapeseed growing campaign in an area where I could grow sunflowers. I would also hesitate to advise someone to grow safflowers in a sunflower area.

What are the gene pools that you are using for safflower breeding? Is it mainly from California, and do you have the high oleic acid mutant.

J.M.F. Martinez (Spain)

We are using a safflower variety which has been developed in Spain. We are crossing this material with cultivars from

California and also from the Middle East, mainly because they are disease resistant. We have received a complete collection from the United States, mainly of the types I have been talking about. And, of course, we have received this mutant of high oleic. I did not tell you that the high oleic mutant is very stable under different temperature conditions. Some cultivars lack certain characteristics but, with material from all over the world, we increase our gene pool every year.

J.P. Dunne

Have the tubers of *Helianthus tuberosus* been examined for protein content?

J.M.F. Martinez

I cannot tell you. We have not done so. Fatty acids have been examined in relation to disease resistance, but I cannot say for protein.

M. Dambroth *(FRG)*

I think we can end the discussion here. Thank you very much.

THE USE OF TISSUE CULTURE IN BREEDING OILSEED CROPS

G. Mix

Federal Research Centre for Agriculture
Braunschweig-Völkenrode (FAL), Bundesallee 50,
Federal Republic of Germany

Plant breeding is an essential and continuing enterprise. Plant breeding usually involves recurrent cycles of hybridisation and selection. Most often this consumes time and resources, despite the development of means to short-circuit the time interval between one cycle and the next. Also the spectrum of genetic variability available to the plant breeder is usually restricted. The ability to manipulate chromosomes, the use of polyploidy, the induction of mutations and the theory of quantitative inheritance have each contributed to the breeder's armoury

For some years aseptic cultures of plant cells, tissues and organs, or simply plant tissue culture, was merely the experimental tool for a few biologists to produce disease-free and disease-resistant plants, multiply various genotypes and store or maintain stocks of valuable plants or to grow organs, which in nature will not have been able to develop.

More recently, several technical developments in plant tissue culture have led to the hope that *in vitro* techniques may make a significant contribution to plant breeding. *In vitro* methods can help plant breeders in two ways; firstly, some techniques can serve as aids in attaining the traditional objectives of breeding; secondly, novel approaches that are currentl under development might result in genotypes that are not attainable by conventional methods, thus increasing genetic variability.

The intention of this short review is to describe, with some examples, the present situation in tissue culture in rape and soybean.

Here it has been demonstrated that nearly every part of the plant is applicable for any kind of tissue and/or cell culture. But up to now not all these techniques are involved in improving rape and soybean plants.

Brassica napus is a plant which offers concrete breeding aims. The level of glucosinolates and of erucic acid needs to be decreased if utilisation of rapeseed is to be achieved, and for this the transfer of characters from spring types into winter types is necessary.

The discovery by Guha and Maheshwari in 1964 that large numbers of haploid plants could be obtained from individual immature pollen grains makes it possible to produce homozygous plants.

Thomas and Wenzel started to plate rape anthers, and gave in 1975 the first report that anthers of *Brassica napus* were able to regenerate calluses which can give rise to plantlets. The embryogenesis from microspores occured on a modified Linsmaier and Skoog Medium with growth substances tested at various concentrations and combinations. On the medium with BAP and NAA and Gibberelic acid a typical embryoid proliferated further and gave rise to numerous plants after transferring to a medium without growth substances. The chromosome counts made sure that these were amphihaploid (n = 19) plants. Trying an isolated microspore culture, it was just possible to obtain multicellular structures, as Kameya et al. did in 1970. Thompson (1974) concluded that naturally occuring haploids are quite common in many winter and spring rape varieties, and he was the first to produce a new variety from a haploid rape. In trials from 1971 to 1973 'Maris Haplona' always gave a higher yield of oil than 'Ora', the cultivar from which the haploid came (Table 1).

Hansson reported in 1978 that a temperature - shock treatment could increase the frequency of embryoids. But in the same year Keller and Armstrong succeeded in greatly increasing the frequency of regenerated plants from plated anthers by treating the culture with different extreme temperatures. Embryo yield was increased when anthers were cultured for 14 or 21 days at 30°C, followed by 25°C; a 40° treatment for one day also had a good effect. So it was possible to obtain more than 500 plants from anther derived embryos. The frequency of spontaneous doublings can be high, but this depends considerably on the culture conditions and the strains used, because all genotypes do not react in the same way. Anther-derived doubled haploid lines from *Brassica napus* are now available and under field tests in Saskatoon.

TABLE 1

OIL RATES AND OIL PERCENTAGES OF *BRASSICA NAPUS* VARIETY 'ORO' AND THE HOMODIPLOID VARIETY 'MARIS HAPLONA', DEVELOPED FROM 'ORO' (THOMPSON, 1974)

Variety	1971	Oil rate 1972	1973	Oil percentage
Oro	100*	100	100	41.5
Maris Haplona	165	112	123	42.8

* 'Oro' badly lodged

Now it seems that it is possible to include the haploid technique in a practical breeding programme, because the system also offers the possibility of increased efficiency in early selection through chemical analysis of cotyledons from microspore derived embryos. Keller and his colleagues found that the level of erucic acid production in the resulting embryos is comparable to that found in mature seed embryos.

Another very important point in Brassica breeding is the content of glucosinolates, some split products of these are responsible for a toxic effect of Brassica. Hoffmann (1977) made an interesting observation on microspore-derived plants; finding that spontaneous diploids which have doubled up during early stages in culture have a high glucosinolate level, while plants with a low content remained haploid. This should offer a powerful selection system for economically important plants with low glucosinolate content.

Ahmad et al. (1977) published a report that haploid soybeans are a rare occurence in twin seedlings. He tested about 36 000 seeds of three different cultivars. I could not find a report about anther culture in soybean but perhaps in this audience someone could refer me to some appropriate paper or research.

Plant regeneration is an important aspect of many tissue culture studies. It depends on the ability of undifferentiated cells *in vitro* to undergo regeneration, but not all kinds of calluses respond to regeneration. Kartha et al. (1974) reported on the first morphogenesis of Brassica in culture where plantlet formation from diploid stem explants occurred. Stringam showed that stem explants of spontaneous haploids could give rise to plants in culture, and in 1979 he reported that isolated callusses of leaf discs from spontaneous haploid plants of rapeseed produced plantlets with a high frequency of spontaneous chromosome doublings.

Legumes seem more difficult. Oswald et al. in 1977 did not succeed in regenerating soybean plantlets from callusses of cotyledon explants. Nor did Beaversdorf et al. in the same year; although they cultured hypocotyls and ovaries of 56 soybean cultivars on semi-solid and liquid media, none developed a plantlet. The report of Kinbell and Bingham in 1973 described that somatic soybean cells from hypocotyl sections can differentiate and form adventitious buds, which at least develop into plants. In 1976 a report of a Chinese group working on soybean tissue culture described the production of plantlets from cotyledon, leaf, stem and hypocotyl-callusses in order to enlarge the extent of soybean material of new cultivars with high photosynthetic rates, good quality and high yield.

Theoretically it is possible to establish a cell-culture or suspension-culture from all kinds of callusses or isolated cells directly from the plant, but this method has to be improved so that it may become suitable for all important crops. Many reports have been made on this point.

Rapidly growing forms of soybean lines have been selected during growth on a maltose medium by Limberg et al. (1979). Hemphill et al. (1977) cultured different chlorophyllous callus phenotypes on various carbohydrate sources to determine the influence of the growth rate, and after two years in culture these phenotypes showed a certain degree of differentiation. Saus reported an increase of soybean callus growth when he had tried out different cytokinine concentrations. Oswald et al. (1977) came to the same conclusion, that cytokinine promotes the growth of *in vitro* cultures of soybean. For measuring the growth rate Kubek et al. (1978) developed an optical density method, which makes it possible to determine the rapid and accurate growth of soybean cell suspension culture and perhaps all other varieties as well.

Furthermore, these techniques will allow us to produce new strains of plants with considerable resistance to herbicides, fungicides, insecticides and pathogen toxins. There are several reports of isolated cells from tissue culture with increased tolerance or resistance.

The fungal parasite *Plasmodiophora brassica* (club root disease) remains in the host cells during its life cycle. So Sacristan and Hoffmann infected stem embryo culture obtained from regenerated plantlets of protoplast culture from haploid androgenetic rape - to select embryos resistant to the parasite, and to isolate, in connection with mutagenic treatment, homozygous mutant brassica plants. Flack and Collin isolated a herbicide ('Asulam') tolerant rape-strain. Davis et al. (1977) succeeded in regenerating tolerant soybean plants for the herbicide chlorproham and Zenk in 1974 reported an atrazine-resistant strain selected from a diploid soybean cell suspension culture. The routine field selection for herbicide resistance is often found to be inefficient for many reasons, selection in tissue culture could perhaps reduce the problems and make it possible to screen a large number in less space and time.

The work of plant breeding has always been slow and often tedious. The new techniques which are now proposed will permit a speeding up of the selection processes; however, they also need a great deal of patience.

REFERENCES

Ahmad, Q.W., Britten, E.J. and Byth, D.E., 1977. Haploid soybeans, a rare occurrence in twin seedlings. Journal of Heredity, 12: 68 ff.

Chinese Soybean Tissue Culture Group, 1976. Successful induction of plantlets from callus culture of soybean hypocotyl. Acta Botanica Sinica, pp 258-262.

Davis, D.G., Hoerauf, R.A., Dusbabek, K.E. and Dongall, D.K., 1977. Isopropyl m-chlorocarbanilate and its hydroxylated metabolites: their effects on cell suspension and cell division in soybean and carrot. Physiol. Plantarum, 40: 15-20.

Flack, J. and Colling, H.A., 1978. Selection of resistance to asulam in oil seed rape. Abstr. 4th Intern. Congr. Plant Tissue and Cell Culture, Calgary, No. 1744.

Guha, S. and Maheshwari, S.C., 1964. *In vitro* production of embryos from anther of datura. Nature, 204: 497 ff.

Hansson, B., 1978. Temperature shock - a method of increasing the frequency of embryoid formation in anther culture of rape (*Brassica napus*). Sveriges Utsädesförenings Tidshrift, 88 (3): 141-148.

Hemphill, J.K. and Venketeswaran, S., 1977. Growth studies of three chlorophyllous callus phenotypes of Glycine max. American Journal of Botany, 64: 658-663.

Hoffmann, F., 1978. Mutation and selection of haploid cell culture systems of rape and rye. In: A.W. Alfermann, E. Reinhard (Eds.), Production of natural compounds by cell culture methods, pp 319-326.

Kameya, T. and Hinata, K., 1970. Induction of haploid plant from pollen grains of Brassica. Jap. J. Breed., 20: 82-87.

Kartha, K.K., Garnborg, O.L. and Coustable, F., 1974. *In vitro* plant formation from stem explants of rape (*Brassica napus* cv. Zephyr.). Physiol. Plant, 31: 217-220.

Keller, W.A. and Armstrong, K.C., 1978. High frequency production of microspore-derived plants from *Brassica napus* anther cultures. Z. f. Pflanzenzüchtung, 80: 100- 108.

Kimbell, S.L. and Bingham, E.T., 1973. Adventitious bud development of soybean. Hypocotyl sections in culture. Crop Science, 13: 758-760.

Kubek, O.J. and Shuler, M.L., 1978. A rapid quantitative method to measure growth of plant cell suspension cultures. Canadian Journal of Botany, 56: 2340-2343.

Limberg, M., Cress, D. and Lark, K.G., 1979. Variants of soybean cells which can grow in suspension with maltose as a carbon-energy source. Plant Physiology, 63: 718-721.

Oswald, T.H., Smitz, A.E. and Phillips, D.V., 1977. Callus and plantlet regeneration from cell cultures of ladino clover and soybean. Physiol. Plantarum, 39: 129-134.

Sacristan, M.D. and Hoffmann, F., 1979. Direct infection of embryogenic tissue cultures of haploid *Brassica napus* with resting spores of *Plasmodiophora brassicae*. Theoretical and Applied Genetics, 54: 129-132.

Saus, F.L. and Blaydes, O.F., 1978. Effect of chloramphericol and acidione on soybean callus growth. Proceedings of the West Virginia Academy of Science, 50: S. 12.

Stringam, G.R., 1979. Regeneration in leaf-callus cultures of haploid rapeseed (*Brassica napus* L.). Z. f. Pflanzenphysiologie, 92: 459-462.

Thomas, E. and Wenzel, G., 1975. Embryogenesis from microspores of *Brassica napus*. Z. Pflanzenzüchtung, 74: 77-81.

Zenk, M.H., 1974. Haploids in physiological and biochemical research. In: Haploids in higher plants: advances and potential. Univ. of Gnelph, 339-354.

ASPECTS OF PLANT REGENERATION FROM SINGLE CELLS AND PROTOPLASTS OF *Brassica napus*

Emrys Thomas

Biochemistry Department, Rothamstead Experimental Station
Harpenden, Herts., UK

ABSTRACT

Over the past decade various advances in cell and tissue culture technology have opened up new approaches for the study of plant cell genetics, physiology and biochemistry. Potentially, the techniques also provide new tools for generating, selecting and propagating new, economically important plant varieties (Thomas and Wernicke, 1978). Unfortunately, progress in this area is still hampered by numerous technical obstacles such as the failure of a large number of crop plants to respond to current tissue culture techniques and a lack of knowledge of the biochemistry of agronomically important characters which may be used for selection procedures at the cellular level (Thomas et al., 1979_a). The Agricultural Research Council of Britain has recently established a project aimed at plant regeneration from protoplasts and single cells of several crop plants including Brassica *napus (rape). Our ability to achieve this reproducibly could pave the way for the genetic manipulation of crops using current techniques of biochemistry and molecular biology. In this lecture the current status of research in microspore, protoplast and cell cultures of* Brassica *napus will be presented. Some of the ways in which these cultures could contribute to our knowledge and possibly provide breeders with a new approach for generating variability within this crop are briefly described.*

Ideally, for maximum use to be made of current technology, the crop plant involved should be responsive to the tissue culture manipulations depicted in Figure 1 which comes from the work of Melchers (1974).

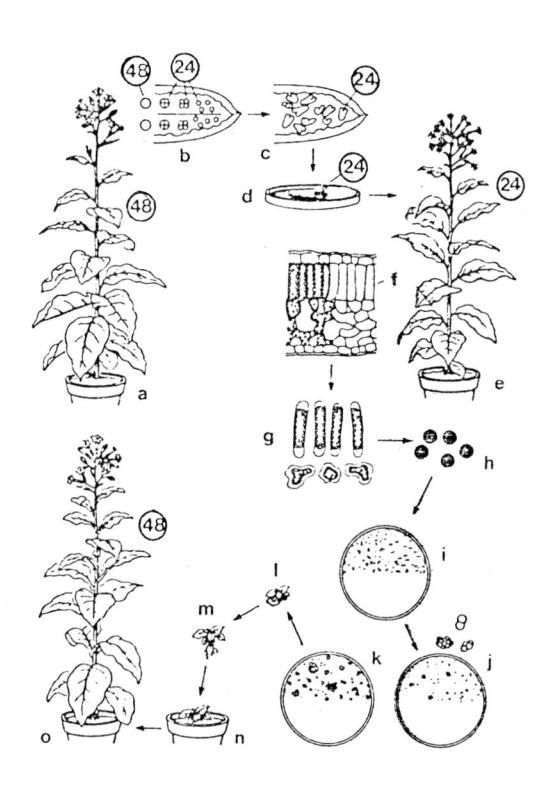

Fig. 1. Schematic summary of a combined microbial technique with conventional crossings for plant breeding. In circles are the number of chromosomes e.g. 48 = amphidiploid *Nicotiana tabacum*; 24 = haploid *N. tabacum*: (a) Normal *N. tabacum* plant with petiolated leaves. (b) Meiosis in the anther. (c) Haploid anthers. (d) Haploid plantlet from anther. (e) Haploid plant. (f) Section through a leaf. (g) Isolated mesophyll cells. (h) Protoplasts from mesophyll cells. (i) Petri dish with protoplasts; mutagenesis at this stage or in the leaves before preparation of the protoplasts; agar medium with selecting material, e.g. a plant parasite toxin. (j) and (k) Only a few resistant colonies growing. (l) Regenerating plants. (m) and (n) Rooted plants diploidized with colchicine. (o) Amphidiploid mutant (in this case with the recessive gene for petiole winged leaves) should be tested for resistance to the toxin.

(After G. Melchers, Tübingen)

Diploid plants are flowered, subjected to anther or isolated microspore culture and haploid plants obtained. Such haploids may have numerous uses in plant breeding programmes (Nitzsche and Wenzel, 1977) and they can be made homozygous and fertile by diploidising with colchicine. In some plants high frequencies of haploids can be found amongst the progeny of sexual crosses. Indeed, even in *Brassica napus* spontaneous monoembryonic haploids can be found amongst populations of most varieties of spring and winter forms (Thompson, 1974). Using cultivar 'Oro' as starting material, Thompson selected haploids, diploidised them with colchicine and succeeded in producing a new variety, 'Maris Haplona', which gave consistently higher levels of oil than 'Oro'. If successful, anther or isolated microspore culture could lead to the production of very large numbers of haploid or homozygous individuals which would increase the possibility of uncovering beneficial gene combinations. Microspores (within cultured anthers) of *Brassica napus* are able to give plantlets of various ploidy levels, including haploids (Thomas and Wenzel, 1975; Wenzel et al. 1977) (Plate A 4,5). Keller and Armstrong (1977) were able to obtain high percentages of microspore plantlets using an anther incubation temperature of 30°C. Attempts have also been made to grow isolated microspores (Thomas et al. 1976$_a$). It was possible to obtain reproducibly up to 16-celled colonies but these failed to develop further under the conditions used (Plate A 2,3).

Theoretically, one of the great values of haploid plants lies in the possibility of mutant production, especially recessives, and if the mutation and selection were to be carried out at a single cell or protoplast level this would eliminate chimaera formation. To achieve this, mutagen treated single cells are cultured in the presence of a selective agent for the mutation being sought. Surviving cells are then allowed to form plantlets which are tested for resistance to the agent; resistant mutants can then be diploidised with colchicine and introduced into conventional sexual breeding procedures or in the case of non-sexually propagated crops, into vegetative propagation programmes. Where dominant mutations are sought, the use of haploid cells is not required. As will be discussed later, some of the mutants obtained could be of importance for studying particular pathways of metabolism, others may even be of economic significance. Much of their value however, could be in their use as selection markers in more ambitious programmes of genetic manipulation. These include the production of hybrid plants by protoplast fusion of sexually incompatible species or the introduction of defined genetic information from one species to another. Since the modification events generally occur at low frequencies powerful methods of selection are required to isolate the modified cells from those which are unchanged. The use of single cell culture and manipulation in plant genetics, biochemistry and breeding will only come about when large populations of single totipotent cells (somatic cells, protoplasts or microspores) of crop plants are readily available. Although such populations do exist for plants such as tobacco, they exist in few crops and even in these there are still barriers preventing their proper use.

At first sight, *Brassica napus* would appear to be a crop which adequately fits the ideal system depicted in Figure 1. We have already seen that large numbers of haploid plants can be obtained from microspores or from field populations. It is also possible to induce the growth of isolated protoplasts from both amphidiploid and amphihaploid protoplasts. In 1974, Kartha et al. succeeded in obtaining plants from protoplasts of greenhouse-grown amphidiploid rape but no details were provided of the number of plants produced. Subsequently, many thousands of calluses were regenerated from greenhouse-grown amphihaploid plants (Thomas et al., 1976) (Plate A, 8). However, less than ten of the calluses ever formed plants; the possible reasons for this and the way in which the problems may be overcome, have been previously discussed (Thomas et al., 1979$_b$). One way would be to use as a source of protoplasts, tissues containing cells of proven morphogenetic capacity. In rape, Kohlenbach (personal communication) has been using an embryogenic tissue culture derived from anther culture as a source of protoplasts and has obtained colonies again capable of embryogenesis. However, details of the procedure or the frequency of protoplast growth and plant formation are not yet available. Leaf mesophyll tissue gives high yields of viable protoplasts and their ease of handling would make them a more efficient system if the problem of shoot formation from the protoplast-derived calluses could be overcome. Recently, I have been focusing some effort on protoplasts isolated from leaves of *in vitro* propagated shoots (Plate A, 1) (Thomas, unpublished) and it is hoped that by alteration of the conditions under which the protoplasts are initially cultured it will be possible to alter the subsequent morphogenetic response of protoplast-derived calluses. Already thousands of callus colonies have been obtained and they are currently being tested for their morphogenetic capacity.

In the absence of an adequate protoplast system for genetic manipulation in rape, other systems have been examined. One such system is that of stem embryogenesis (Thomas et al., 1976) Most of the microspore-derived embryos formed in anther culture of rape do not give rise directly to plantlets which can be grown to maturity. Instead they form plantlets with abnormally swollen stems and these stems give rise to many hundreds of further embryos (Plate A, 6,7). The development of the secondary embryos is not synchronous; globular, torpedo and cotyledonary stages can all be found on a single stem. At the cotyledonary stage the embryos can be individually removed, transferred to fresh culture medium and the process of stem embryogenesis repeated. Eventually, after 4 - 6 transfers, plantlets develop without further stem embryos and these can be rooted in sand and grown to maturity in potting compost. Sections through embryo-bearing stems indicate that the embryos can arise from single cells of the epidermis (Plate A, 9, 10, 11, 12). This means that it should be possible to subject embryos to mutagenic agents and to isolate solid, mutant secondary embryos. Unfortunately, we do not really know how many of the cells are totipotent at any one time, how they interact or crossfeed under selection conditions and how routine mutagen and selection conditions can be quantified It is thus not clear whether such a system offers any real advantage over the whole plant selection systems at present used in plant breeding. Nevertheless, the system has been used not only for propagation of specific haploid and diploid homozygous genotypes but also for mutant production (Hoffmann, 1978). Attempts are being made to use stem embryogenesis methods for the propagation of synthetic rape produced by embryo culture of the products of the sexual cross *B. campestris* x *B. oleracea* (Gland and Röbbelen, personal communication). However, we do not really know whether the plants propagated by these methods are uniform, and this must be very carefully tested in view of the increasing evidence that the tissue culture process can <u>induce</u> a great deal of variation amongst the regenerated plants.

The most striking evidence of variability in tissue-derived populations is provided by Shepard and his colleagues (1979, 1980) in potato. He regenerated several thousand plants from mesophyll protoplasts of potato. Even after discarding 'wild-aberrants' resulting from gross chromosomal charges, large numbers of plants, possessing normal chromosome number, demonstrated a great deal of variation in agronomic characters. These included variation in flowering response, stage and size of plant and tuber, depth of tuber set and resistance to potato pathogens. This variation persisted through subsequent vegetative propagation. The message is clear: certain tissue culture systems may not be useful for propagation of special genotypes. On the other hand they may be extremely useful for inducing variation in a particular crop. However, it may be advisable to direct a selection pressure for a particular trait such as has been done in the case of eyespot, Fiji and downy mildew diseases of sugar-cane (Heinz et al., 1977).

The most obvious application of tissue cultures for mutant or variant selection in crop plants is positive selection for resistance to herbicides, fungicides, insecticides, nematocides and pathogen toxins. Flack and Collin (1978) have described the isolation of *B. napus* tissue cultures with increased tolerance to Anulam but it was not possible to regenerate plants from the cultures. Another extremely interesting possibility is to select for regulatory mutants causing increased or decreased production of a specific compound. The example so far most extensively worked out involves selection of plants resistant to toxic levels of amino acids of their analogues. Here it is envisaged that some variants will have alterations in feed-back sensitivity of key enzymes, resulting in massive overproduction and accumulation of corresponding amino acids. Such an accumulation <u>in seeds</u> has clearly been demonstrated in barley (Bright, personal communication) and maize (Green, personal communication). In barley the initial selection was performed on cultured barley embryos and in maize on shoot-forming tissue cultures.

Although it is not yet clear to what extent amino acid overproduction can alter protein quality in the seeds, it does seem that methods are being developed which can eventually alter the nutritional value of crop plants. More adventurous projects seek regulatory mutants which affect yield and others aim to select for tolerance to environmental stresses such as toxic salt levels, toxic heavy metal levels, drought and frost. Protoplasts of *Brassica napus* are currently being used as hosts for infection with Cauliflower Mosaic Virus; this could prove to be a useful vector for the transformation of plant cells (Hull, personal communication). Similarly, there is interest in transferring male-sterility from non-commercial rape types into important varieties (Primard and Pelletier, personal communication) using protoplast fusion techniques. Through similar techniques, attempts are being made to produce artificial rape by fusion of its progenitors (Schenk, personal communication). We are, however, at a very early stage in the development of single-cell technology and there are inevitably many gaps in our ability to work with crop plants at the single cell level. As we achieve a greater understanding of the physiology and biochemistry of plants, and the cells isolated from them, these gaps will be filled. Many years may elapse before the techniques become of use in plant breeding, but usefulness, even if indirect, is inevitable.

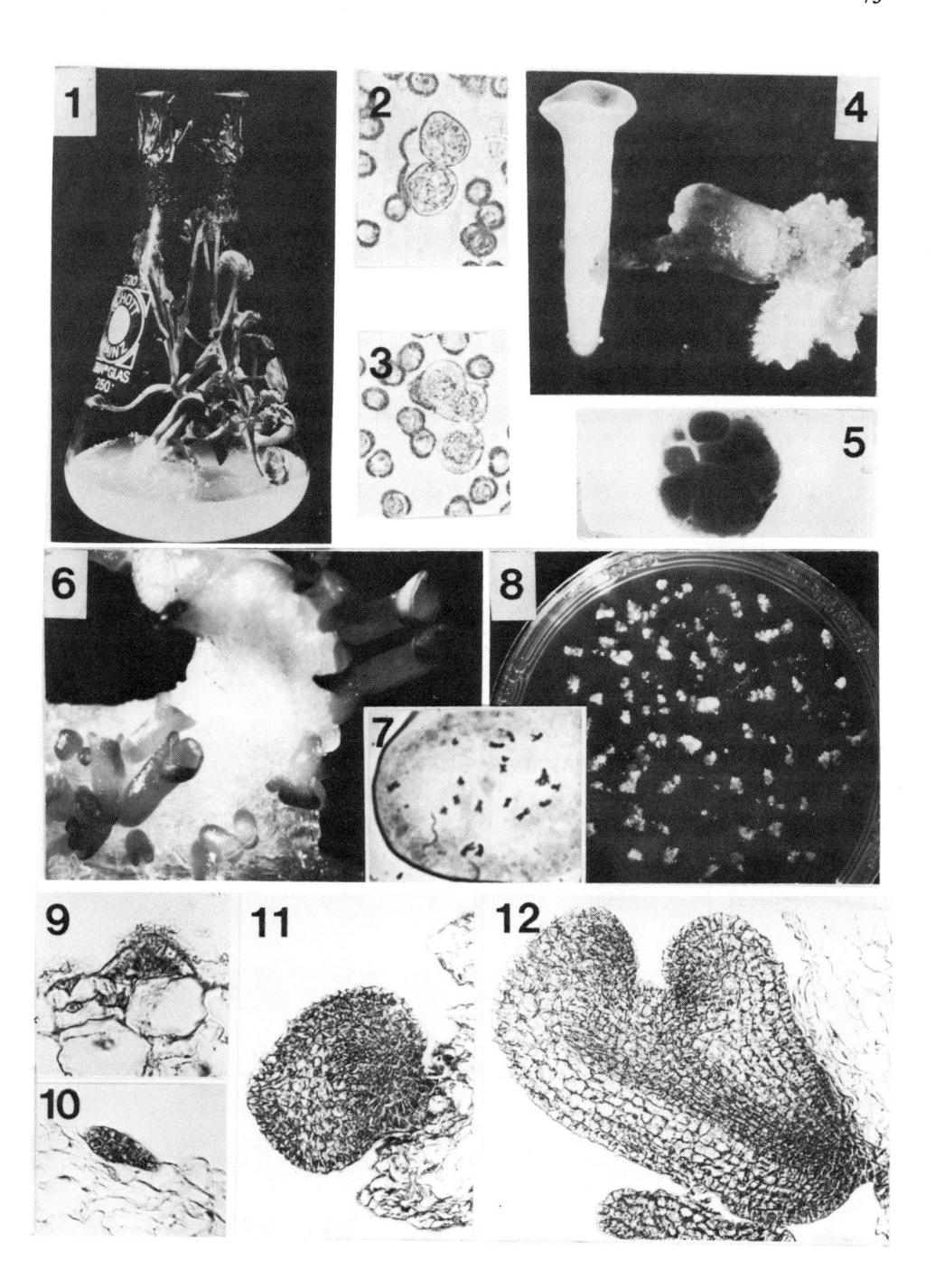

Plate A

Explanation of Plate A:

Aspects of plant regeneration from single cells, microspores and protoplast of rape.

1. Shoot derived from microspore embryoid. The shoots can also be used for protoplast isolation.

2/3. Divisions in populations of isolated microspores. 2) 2-celled stage; 3) 3-celled stage. Note the presence of the exine. After several divisions the microspores fail to develop.

4. Embryo derived from microspore in cultured anther.

5. Young embryo from anther culture; it is still enclosed within the exine.

6. Stem embryogenesis from microspore-derived plantlets. Different stages of embryogenesis can be observed.

7. Amphihaploid chromosome number of cells from the cotyledon of secondary embryo (n = 19).

8. Colonies derived from protoplasts.

9/10. Divisions within single cell of the epidermis.

11. Globular stem embryo in section.

12. Cotyledonary embryo. The embryos are very loosely attached to the original stem.

REFERENCES

Flack, J. and Collin, H.A., 1978. Selection of resistance to asulam in oil seed rape. Abstr. 4th Intern. Congr. Plant Tissue and Cell Culture, Calgary, No. 1744.

Heinz, D.J., Krishnamurthi, M., Nickell, L.G. and Maretzki, A., 1977. Cell, tissue and organ culture in sugarcane improvement. In: J. Reinert and Y.P.S. Bajaj (Eds.) Plant Cell, Tissue and Organ Culture. Springer-Verlag, Berlin.

Hoffmann, F., 1978. Mutation and selection of haploid cell culture system of rape and rye. In: Production of Natural Compounds by Cell Culture Methods, Ed. A.E. Altermann and E. Reinhard, Ges. Strahlen-und Umweltforschung, 319-326.

Kartha, K.K., Michayluk, M.R., Kao, K.N., Gamborg, O.L. and Constabel, F., 1974. Callus formation and plant regeneration from mesophyll protoplasts of rape plants (*Brassica napus* L. cv. Zephyr).

Keller, W.A. and Armstrong, K.C., 1977. Embryogenesis and plant regeneration in *Brassica napus* anther cultures. Canadian J. Bot. 55, 1383-1388.

Melchers, G., 1974. Haploids for breeding by mutation and recombination. International Atomic Energy Agency, Vienna. 221-231.

Nitzsche, W. and Wenzel, G., 1977. Haploids in Plant Breeding. Supplements to Z. Pflanzenzüchtg. 8, 1-101.

Shepard, J.F., 1979. Mutant selection and plant regeneration from potato mesophyll protoplasts. In: Emergent techniques for the genetic improvement of crops. Univ. Minnesota Press (in press).

Shepard, J.F., Bidney, D. and Shahim, E., 1980. Potato protoplasts in crop improvement. Science 208, 17-24.

Thomas, E. and Wenzel, G., 1975. Embryogenesis from microspores of *Brassica napus*. Z. Pflanzenzüchtg. 74, 77-81.

Thomas, E., Hoffmann, F., Potrykus, I. and Wenzel, G., 1976. Protoplast regeneration and stem embryogenesis of haploid androgenetic rape. Mol. Gen. Genet. 145, 245-247.

Thomas, E. and Wernicke, W., 1978. Morphogenesis in herbaceous crop plants. In: Frontiers of Plant Tissue Culture. Ed. T.A. Thorpe. University of Calgary press, 403-409.

Thomas, E., King, P.J. and Potrykus, I., 1979a. Improvement of crop plants via single cells *in vitro* - an assessment. Z. Pflanzenzüchtg. 82, 1-30.

Thomas, E., Harms, C.T., Hoffman, F., Lörz, H., Potrykus, I. and Wenzel, G., 1979. The production and utilisation of haploid plants *in vitro* for the improvement of commercially important crops. Proc. 2nd S. African Maize Breeding Symposium, 1976. Ed. H.O. Gevers.

Thomas, E., Brettell, R. and Wernicke, W., 1979. Problems of plant regeneration from protoplasts of important crops. Advances in protoplast research. Proc. 5th International Protoplast Symposium, Hungary, 269-273.

Thompson, K.F., 1974. Homozygous diploid lines from naturally occurring haploids. Fette, Seifen, Anstrichmittel, 76, Parts 7-8, Supplement p. 15.

Wenzel, G., Hoffman, F. and Thomas, E., 1977. Anther culture as a breeding tool in rape I. Ploidy level and phenotype of androgenetic plants. Z. Pflanzenzüchtg. 78, 149-155.

SHORT COMMUNICATION

F. Schenk
Institute of Agronomy and Plant Breeding,
Georg August University, Göttingen,
Federal Republic of Germany

As Dr. Thomas has pointed out, it is possible to fuse protoplasts from different species, for instance *Brassica campestris* with *Brassica oleracea*. We carried out these experiments in our Institute in Göttingen. We fused protoplasts of both species; these protoplasts divided and in the latest experiments we hope to be able to get some plants derived from hybrid protoplast material. We hope to get about ten plants rooted in solid agar. If this meeting had been held one week later I would have been able to tell you the results of these experiments. At the moment there is the possibility that the protoplast derived plants may be source plants, that is plants from either *Brassica oleracea* or *Brassica campestris* - plants which are not fused. This is possible because only a certain percentage of protoplasts will fuse together and theoretically it is possible that these particular plants will not be fused. However, we have had previous experiments to detect different isozyme patterns and we were able to show that hybrids had a different pattern. *Brassica campestris* had special bands and *Brassica oleracea* had other bands which are very characteristic for the species. The hybrid has all the bands more or less together in this pattern. So there is a good outlook at the moment for these fusion experiments.

SHORT COMMUNICATION

G. Röbbelen
Institute of Agronomy and Plant Breeding,
Georg August University, Göttingen,
Federal Republic of Germany

Perhaps I could add something to the story. We set the stage in Göttingen for a kind of race in that this specific cross was made by two people. One was a conventional cross, made by Dr. Gland, and the other unconventional, by Dr. Schenk. Indeed, the very same plants were used for both kinds of approach. So far the conventional technique, using embryo culture, is much more efficient but this does not mean that this must always be so. For example, with 6 000 pollinations, 300 hybrids of this type were found. Of course, we were interested in whether there was some correlation; some crosses are more difficult than others. The question is, is fusion by protoplast technique difficult in the same way, with difficult section combinations, and vice versa? This is not necessarily true, so it may be that what is difficult to arrive at by conventional methods may be more easily achieved by non-conventional techniques. These new techniques are well advanced but, as was pointed out by Dr. Thomas, we still need very efficient general plant breeding to isolate the superior genotypes required for practical production.

APPLICATION OF PROTOPLAST FUSION IN THE IMPROVEMENT OF RAPESEED YIELD

G. Pelletier
Laboratoire d'Amélioration des Plantes,
Université Paris XI
91405 Orsay Cedex, France

Productivity of rapeseed would be greatly enhanced by the creation of hybrid varieties. However, this requires the use of a good (stable) source of cytoplasmic male-sterility. The best one for this purpose has been obtained by Bannerot et al. by crossing *Brassica* spp. with an already cytoplasmic male-sterile radish found by Ogura.

The association of *Brassica* nucleus with *Raphanus* cytoplasm results in a very stable cytoplasmic-male-sterility, but also in chlorophyll deficiency at temperatures below $12^\circ C$.

From previous work in tobacco, and from unpublished analyses of chloroplastic and mitochondrial DNA, we have strong arguments to support the hypothesis that, for the cytoplasmic part of the interaction, the male-sterility resides in the mitochondria and the chlorophyll deficiency resides in the chloroplasts.

The aim of our work is to obtain, through protoplast fusion - the only possible method for this purpose - cytoplasmic reassociation of organelles and so develop a rapeseed plant possessing *Brassica napus* chloroplasts and *Raphanus sativa* mitochondria, which would be male-sterile but without chlorophyll deficiency.

In collaborative work, cytogenetic studies are being carried out at INRA, Rennes, to obtain restorer genes in rapeseed by interspecific crosses with radish.

We need to obtain a good protoplast culture system enabling us to reproducibly regenerate plants from them. This does not yet exist in *Brassica* spp. and we are therefore working in this direction.

DISCUSSION

M. Dambroth (FRG)

As you can see, ladies and gentlemen, we have some specialists in this field and you now have an opportunity for questions or comments.

D.G. Morgan (UK)

I have a question for Dr. Thomas. Speaking as someone who knows very little about this particular field, I looked at the medium he is using for culturing the Brassica calluses and see that he is having difficulty in getting a high proportion of shoots from them. To someone who is interested in plant growth substances, I noticed that your medium contains kinins and auxins in it. What affect have gibberellins? Gibberellins are very much involved in the differentiation of normal shoots and the apices of normal shoots.

E. Thomas (UK)

We have tried the gibberellins. Under the conditions tested so far, the only effect has been to induce root formation. There is no problem in getting roots. But with gibberellic acid, abscisic acid, anti-auxins, increases in osmotic pressure, and changes in salts, we were unable to produce many shoots in previous experiments. I am more confident about my recent experiments. The calluses look as if they are going to do something. But, like Dr. Schenk, I wish it was a week later; perhaps I could then have more results.

D.G. Morgan

You mentioned gibberellic acid, but often you find that a mixture of the gibberellins with the GA7s is more effective. Have you tried mixtures of the gibberellins at all?

E. Thomas

No, I have not. One of the problems in this field is that

there are so many things to test. I can give an example of my colleagues in Switzerland with protoplasts from cereals. They tested in the order to 80 000 different variations. But there is little point in trying variations if the cells themselves are not competent to respond. Perhaps this is the problem which we face. Perhaps it is the physiological condition of the protoplasts that we take at the beginning which is critical in determining their subsequent response in culture. However, it is a very good question.

Dr. Schenk *(FRG)*

Dr. Pelletier, what kind of selection system do you use?

G. Pelletier *(France)*

We have not tried to explore the possibilities of selective pressure. In this system it will be possible to use chlorophyll deficiency in selection, but we have not tried to do that *in vitro*.

E. Thomas

Can Dr. Schenk tell us what his selection system is?

Dr. Schenk

The fusion derived colonies had a much more vigorous growth than the original ones which were not fused. Other authors have also found this. For example, Dr. Schieder has used two of our species and has observed the same hybrid effect - a vigorous effect.

E. Thomas

On the contrary, following fusion in potato, there would appear to be no increase in vigour in the resulting products.

G. Robbelen *(FRG)*

No doubt because the potato is a herozygous type already,

whereas in these rapeseeds there is probably quite a bit of inbreeding depression. If you bring these together it is possible to get such an effect.

While I have the microphone I would like to present a question to Dr. Thomas. You talked about shoot embryos and, if I understood you correctly, you indicated that offspring from such shoot embryos were highly variable in some instances. What is the reason for this? They should be genetically uniform.

E. Thomas

The example that I referred to was the potato. You may know the recent results of Shepard in the States; he showed that different plants obtained from individual protoplasts showed a great deal of variation in agronomic characters. He cannot explain the reason for it. We do not know whether the variation is actually present within the different leaf cells, and, if it is, it puts a new light on cell differentiation, or whether the variation is induced during the tissue culture process. I think there is a lot of evidence coming out, especially from the Orsay laboratory, indicating that plants which come from tissue cultures - and you would expect them to be perfectly uniform - show a great deal of variability. There appear to be lots of chromosomal changes going on, and perhaps gene mutations which we cannot detect: there could be many reaons. My point is: if this occurs in potato, why should it not also occur in rapeseed? If we examine sufficient progeny from the stem embryogenesis system, we may find something which is of agronomic significance.

M. Dambroth

In this experiment I remember that they also found great variation in resistance against potato blight.

E. Thomas

That is correct.

G. Röbbelen

I think this is a big problem. If this is usual, I would question the merit of these techniques in general. Creation of variation is no problem for a plant breeder; getting it down to a homozygous stable type, or selecting for what he wants, is the real point.

Let me address a question to Dr. Mix. You mentioned the experiments of Dr. Sacristan. As I know, plants from these *Plasmodiophora* resistant selections are growing. However, I do not know whether these have been successfully tested for their *Plasmodiophora* resistance. Has this been done or is it something that is to be expected? I met Dr. Sacristan a few weeks ago but did not hear whether she had succeeded in proving this.

G. Mix (FRG)

I do not know if it has been proved.

G. Röbbelen

I would like to add that to have a full cycle of this fungus, in tissue culture, in cells *in vitro*, is a great achievement. No one has succeeded in doing it in this way before.

E. Thomas

Professor Röbbelen questions the merit of these techniques; so do I at the moment, but I would like to point out that we cannot be expected to do in five years what it takes the plant breeders a lifetime to do.

N.J. Mendham (UK)

I would like to ask why you are expecting the sort of variations which you have in potato, which is an auto-tetraploid and has a very unstable nature, to occur in rapeseed which is a fairly stable amphidiploid. Surely it is different?

G. Pelletier

There is a lot of chromosomal variation in the embryogenic strains cultivated in this way. We have looked at the chromosomes of numbers of plantlets obtained from strains, cultivated *in vitro*. The starting chromosome number level is haploid and you can find diploid plantlets and some with variation in the normal diploid number.

G. Röbbelen

I am not blaming the new techniques alone. I know that Miss Gland also got some instances with 36 chromosomes only, from her conventional crossing of *Brassica oleracea* and *Brassica campestris*. Apparently one pair of chromosomes was lost during the procedure. Such things can occur in many techniques, and may be more frequent under specific conditions.

One question is: what does agar do? Could you comment on new non-agar nutrient media? Have you had any experience of these?

E. Thomas

We use both agar and liquid and there does tend to be a difference in root production. There are experiments with legumes, especially from the laboratory of Professor Cocking in Nottingham, and it has been shown in alfalfa that if protoplasts are plated into agar or into liquid they do not grow very well. However, if the protoplasts are plated onto a filter paper in contact with either agar or liquid media, they grow and give more embryos.

L.G.M. van Soest *(FRG)*

I would like to comment on what Professor Röbbelen said about the work of Shepard. It *is* a problem to create variability in potato. Many of the older varieties are sterile. I could mention Sieglinde in Germany, King Edward in England and the Dutch variety Bintje. You cannot make any conventional crossings with these. Particularly in the field of *Phytophthora* resistance, it is very important to get variation in it by these methods. It is a real problem to create variability, because all these varieties have particular characteristics, mainly in protein, in cooking quality or in chip quality.

I would like to address another question to Dr. Thomas. What is the aim of your work? Are you doing the same work as others have done with potatoes, making haploids and doing the screening work (as you mentioned) in petri dishes and going back, later on, to the diploid level? Is that your intention?

E. Thomas

Unfortunately, at the moment, rapeseed is a very small part of my programme. All I have done is to try and provide a tool which the people at the John Innes Institute can use in their virus studies: to use viruses as vectors for the transfer of specific pieces of plant information. At the moment we are not doing any serious breeding. On the other hand, I would be extremely interested to go back into this, but it is a question of having enough people and enough finance to study these problems. Personally, I am interested in plant regeneration from protoplasts: I will not let this beat me because I am stubborn. I will keep working at the rapeseed regeneration and perhaps I could work together with Dr. Pelletier and he could transfer male sterility into my system. This would be the sort of approach I would anticipate - but first of all at a very scientific level. The more we can learn at the scientific level, then the more will come from the applied point of view in the future.

I have a question for Dr. Mix. Perhaps I misunderstood your paper, but you mentioned something about plant regeneration from soyabean and resistance to a herbicide. Could you enlarge on that for me, please? I ask this question because, throughout the years, soyabean has proved to be the most difficult legume to work with in tissue culture.

G. Mix

Yes, this was the work of Davis in 1977. He regenerated a soyabean plant tolerant to the herbicide chlorpropham.

E. Thomas

From tissue culture?

G. Mix

Yes, I can show you the paper.

E. Thomas

Thank you. I would be very interested - I must be out of date!

Dr. Schenk

Dr. Thomas, the enzyme Pectolyase Y23 that you used was very good for your protoplast isolation. Is it commercially available? Enzymes for isolating protoplasts are very important later on for getting divisions because a lot of enzymes are very toxic.

E. Thomas

The first rapeseed work that we did involved the use of an enzyme which we called P.A.T.E. and this was not available commercially. It was only under these conditions that the experiments would work. Pectolyase Y23 enzyme is very active and gives good protoplasts, but it is very expensive. It is available from Japan.

G. Röbbelen

I would add at this point that this seems to be a very critical situation. Since we are here at an EEC symposium it appears that we should consider how important these enzymes are for protoplast preparations. Some industries who supplied them in the past, do so no longer. It might be a serviceable function for the Commission to inform these companies of the importance of this protoplast work, and provide some support, so that when new enzymes that are suitable become available production can be maintained at a reasonable price. This has been done by individuals fairly frequently, but to no avail. It may be that the pectinases we have used so far are being replaced by better ones, but the problem is still there.

E. Thomas

I would agree completely. The enzymes that we use are extremely critical. In fact, at every stage during protoplast preparation and culture, we face extreme technical difficulties. This is why many scientists in this field get excited, even if they see a division.

M. Dambroth

Thank you. I think we can end this session here. I would like to thank all the speakers and the audience for their questions and comments.

SESSION II

GENETIC AND BREEDING ASPECTS OF RAPESEED,
SUNFLOWER AND SOYABEAN II

Chairman : B.O. Eggum

BREEDING FOR LOW CONTENT OF GLUCOSINOLATES IN RAPESEED

G. Röbbelen
Institute of Agronomy and Plant Breeding,
Georg August University, Göttingen,
Federal Republic of Germany.

Rapeseed meal, which remains after the oil has been extracted from the seed, contains about 40% protein in the dry matter with a well balanced amino acid composition (see, e.g., Miller et al., 1962). It should, therefore, be a valuable source of protein and energy in animal diets. However, it is not used as freely as its nutritive value and cost would suggest, especially in diets for pigs and poultry. These restrictions on its use are chiefly based on its content of glucosinolates, which yield toxic and goitrogenic cleavage products and which produce a pungent taste that lowers the palatability of the feed (see, e.g., Bowland et al., 1965).

Several methods have been considered for removing the glucosinolates from the meal by industrial treatment. However, processes involving destruction of glucosinolates by heat (Rutkowski, 1970), micro-organisms (Staron, 1970), or chemicals (Anderson et al., 1975) are either not economical or they result in reduced protein quality. Processes involving extraction of glucosinolates from defatted rapeseed meal (Ballester et al., 1970), ground rapeseed meal (Eapen et al., 1969) or from intact rapeseeds (Bhatty and Sosulski, 1972; Kozlowska et al., 1972; Sosulski et al., 1972) result in substantial losses of proteins when water or aqueous salt solutions are used. More promising results have been obtained by extraction with 70% aqueous acetone (Mukherjee et al., 1976), but in general, most of the methods so far tried have been too expensive and have resulted in protein losses, or reduced protein quality, or both (Bowland et al., 1965).

This situation has prompted plant breeders to study the possibilities of reducing the glucosinolate content to an innocuous level by plant breeding methods. It is this subject, which will be described in the following lecture.

A. CHARACTERISTICS AND OCCURRENCE OF GLUCOSINOLATES

Glucosinolates are a uniform class of naturally occurring anions, the general structure of which was first proposed and soon verified through synthesis by Ettlinger and Lundeen (1956; 1957). At present, more than 70 different glucosinolates are known and many of them are present in varying amounts in most of the cultivated *Brassica* species. Excellent reviews on the details of glucosinolate chemistry and biosynthesis have recently been written by Van Etten and Wolff (1973), Underhill et al. (1973) and Kjaer (1976). Glucosinolates are characterised by a central S-C=N group and an adhering side chain, which represents a great variety of complex structures including those with simple or branched alkyl, or akenyl, or heterocyclic groups (cf. Kjaer and Olesen Larsen, 1976).

Glucosinolates are considered non-toxic themselves; but physiologically active products are derived from them upon enzymatic hydrolysis. This cleavage is catalysed by a group of enzymes, called myrosinase, which are separately deposited but invariably accompany the glucosinolates in the living plant tissue (Ettlinger and Kjaer, 1968). They are released when the plant tissue is crushed or when autolysis within the plant occurs. When in contact with a glucosinolate, they induce detachment of glucose producing an aglucone. The aglucone under neutral conditions gives rise to sulphate and, by a Lossen-type of molecular rearrangement, yields an isothiocyanate (trivial name = mustard oil) as the first end product. Isothiocyanates exhibit great variations as to stability, volatility, pungency, or taste. On the other hand, myrosinase hydrolysis under weakly acidic conditions or in the presence of ferrous ions produces nitriles and elemental sulphur. Isothiocyanates carrying a hydroxyl group in the B-position spontaneously cyclise a yield substituted oxazolidine-2-thiones, while isothiocyanates derived from the indole glucosinolates are unstable in neutral or alkaline solutions and produce thiocyanate ions (Schlüter and Gmelin, 1972). Even other reactions may occur under certain conditions of autolysis, when moistened seeds are crushed or green plant parts are injured.

Glucosinolates appear to be diffusely present in the parenchymal tissues throughout the plant (Guignard, 1890a), while the concomitant myrosinases are only reported in particular cells, called idioblasts (Guignard, 1890b; Schweidler, 1910). Generally, a plant species contains more than one glucosinolate. In seed meal of the six main *Brassica* species the patterns of the major glucosinolate components display both qualitative and quantitative interspecific variation, shown in Table 1. Sinigrin is generally absent in *B. campestris* and *B. napus* (Josefsson, 1972), although it is a major component in *B. oleracea*. On the other hand, progoitrin, yielding specifically potent goitrogens upon hydrolysis, is absent in *B. nigra* and *B. juncea* (Namai et al., 1972). Obviously, the b genome of *Brassica nigra* (according to U 1935) includes the main factors for sinigrin synthesis (Hemingway et al., 1961), while progoitrin is chiefly controlled by genes located on the c genome of *B. oleracea*. In *B. campestris*, containing the a genome, gluconapin is the dominant gluosinolate (Josefsson, 1972; Anand, 1974), but higher values of glucobrassicanapin have also been reported (Youngs and Wetter, 1967; Krzymanski, 1970). In *B. juncea*, divergent results were obtained from samples of different origin (Hemingway et al., 1961; Vaughan et al., 1963). Samples from several countries contained only or mainly sinigrin, but those from India and Pakistan contained gluconapin, either as the only compound or in combination with sinigrin. Data like these, which are supported by independent analyses with other material (Josefsson, 1972; Anand, 1974), not only reflect events of evolutionary significance, but are also of considerable interest for possible broadening of the genetic base of rapeseed breeding, since genetic variation in glucosinolate content was found to be relatively small within the typical oilseed species.

A substantial amount of work on glucosinolates has been directed to seed constituents not only because of their significant economic interest, but also because development towards a mature seed is principally directed to arrive at a genetically highly determined end. The accumulation of glucosinolates in seeds proceeds continuously during their development, and maximum contents are not reached before their final maturity (Kondra and Downey, 1969; Byczyńska et al., 1970). A significant relationship between siliqua position on the plant and glucosinolate content of the seeds has been obtained, especially in plants grown in controlled environment rooms (Kondra and Downey, 1970). Seeds from siliquas developed at the base of the main inflorescence formed the highest amount of glucosinolates; those from the apical siliquas on the side branches the lowest. Of the different parts of a seed, the embryo contains most of the glucosinolates, while the seed coat has only a relatively small amount (Josefsson, 1970b). Removal of the seed coat, therefore, would not reduce the glucosinolate content of the meal.

Environmental variation in glucosinolate content of the rapeseed amounted to only ± 15% of the average value in analyses which included field grown material from 6 localities (ranging from Finland to Turkey) as well as seeds from soil-free cultures with 18 different combinations of nitrogen, phosphorus, and potassium fertilisation (Josefsson 1970a; Josefsson and Appelqvist 1968). Similar results were reported by other authors (Wetter and Craig, 1959; Finlayson et al., 1970). The only significant effect was an extremely low sulphur supply of the soil, which reduced glucosinolate content to approximately 35% of the average level.

B. METHODS FOR QUANTITATIVE DETERMINATION OF GLUCOSINOLATES

Prerequisites for any efficient study of glucosinolate biosynthesis are suitable analytical methods for their reliable quantitative and qualitative determination. This is particular true for the work of the plant breeder, which in its initial

TABLE 1

GLUCOSINOLATE CONTENT OF SEEDS OF CULTIVATED *Brassica* SPECIES. THE ANALYTICAL DATA ARE GIVEN IN µmol GLUCOSINOLATE/g DEFATTED SEED MEAL AFTER GAS LIQUID CHROMATOGRAPHY OF TMS DERIVATIVES (Thies, 1977). MINOR AMOUNTS ARE NOT LISTED. THE SAMPLES ARE SELECTED FROM UNPUBLISHED DATA TO REPRESENT THE RESPECTIVE SPECIES

Species	Genome	Form	Sin	Nap	Brn	Pro	Nal	Σ
B. campestris L.	aa	ssp. *trilocularis*	0.1	157.3	1.5	2.9	0.1	161.8
		ssp. *oleifera*	0.1	20.3	21.1	18.0	4.0	63.5
		ssp. *pekinensis*	0.1	22.9	2.4	2.0	0.3	27.7
B. nigra KOCH	bb	cv. Europe	150.3	0.1	0.1	0.1	0.1	150.7
		cv. Europe	82.1	0.1	0.1	0.1	0.1	82.5
B. oleracea L.	cc	ssp. *acephala*	104.6	23.0	0.9	81.9	3.7	214.1
		ssp. *acephala*	4.7	23.6	1.1	96.8	4.6	130.8
		ssp. *capitata*	44.7	2.3	0.1	1.9	0.1	49.1
		ssp. *gongylodes*	3.2	0.6	0.1	0.5	0.1	4.5
B. juncea COSSON	aabb	cv. Europe	113.6	0.1	0.1	0.1	0.1	114.0
		origin India	1.3	148.6	1.1	0.1	0.1	151.2
		origin India	42.8	173.9	17.9	2.3	0.1	237.0
B. napus L.	aacc	cv. 'Diamant'	0.1	33.3	8.2	109.4	5.2	156.2
		cv. 'Erglu'	0.1	5.5	1.0	8.3	0.4	15.3
		ssp. *napobrassica*	0.1	0.7	0.7	166.4	0.1	167.2
B. carinata BRAUN	bbcc	origin Abyssinia	52.3	5.6	15.5	0.1	1.3	74.8
		origin Egypt	120.4	12.0	2.2	6.2	0.1	140.9

Abbreviations: Sin: Sinigrin, Nap: Gluconapin, Brn: Glucobrassicanapin, Pro: Progoitrin, Nal: Napoleiferin, Σ: total content.

phase requires the screening of large series of samples each
derived from a single plant only. Thus, small amounts of
substance need to be analysed by rapid, cheap, and sensitive
methods of sufficiently high precision and reasonable accuracy.
Methods to measure the total amount of glucosinolates may be
adequate for early stages of the breeding programme; they may
even be preferred, if the occurrence of unknown glucosinolate
compounds cannot be excluded in genetically diverse plant
material. But determination of individual glucosinolates may
become crucial in other cases, e.g., if differences in
potential toxicity must be considered (for review of literature
see Lein, 1970).

The common method used in most of the early studies is
hydrolysis with myrosinase and assay of the resulting split
products. For determination of the volatile isothiocyanates
of rapeseed meal, Wetter (1955) used distillation at pH 4
into an ammonia solution, which converts the isothiocyanates
into substituted thioureas; by reaction with silver nitrate,
these produce silver sulphide, and the unreacted silver ions
were titrated with potassium thiocyanate. Progoitrin, however,
which is the major glucosinolate in *B-napus* rapeseed, is
hydrolysed enzymatically almost exclusively to the non-volatile
oxazolidine-2-thione (goitrin); this, therefore, requires a
separate determination based upon extraction from the
distillation residue by diethylether and photometric UV-
measurement (Wetter, 1957). Many other procedures have been
used, e.g. the determination of the volatile lipophilic
isothiocyanates by gas liquid chromatography (Youngs and
Wetter, 1967), or colorimetric assays of the thiocyanate ions
for estimation of indole glucosinolates (Johnston and Jones,
1966; Jürges, 1978; McGregor, 1978), although coloured
materials found in plant extracts often interfere with such
kinds of assay. Many of the methods require larger samples,
most are rather time consuming, and none allows the determination
of all the important glucosinolates of rapeseed in one
analysis. This requirement was finally met by a sufficiently

sensitive method formulated by Wetter and Youngs (1976), in which both the total substituted oxazolidinethiones and the alkali stable isothiocyanates are converted to thioureas and measured by their specific UV absorbance. Similarly, a conversion of the total glucosinolates to nitriles under acidic conditions was tried for a quantitative estimate of the potentially toxic compounds in rapeseed meal (Brak and Henkel, 1978).

Simple, quick, and reliable analytical methods for pre-screening in the first steps of plant breeding programmes cannot be expected, however, via quantitative determinations of the aglucones or their future cleavage products. Also sulphate as the second compound of the myrosinase hydrolysis does not lend itself to such assays, since the relatively easy titri-metrical or gravimetrical (McGhee et al., 1965) measurements require larger seed samples. The most important method for the plant breeder to assay for total glucosinolates, therefore, is based on quantitative or semi-quantitative determinations of the B-D-glucose liberated on enzymatic hydrolysis. This approach is suitable only for seed, which contain only traces of free glucose in intact tissues; leaves or other green parts of the plants contain significant quantities of glucose as a result of photosynthesis. With seed, enzymatic glucose determinations (Lein, 1970; Lein and Schön, 1969) using purified myrosinase on hot buffer inactivated material and the Glucose-UV-Test of Boehringer will give reliable estimates of the total glucosinolate content, even from a half-seed (Lein, 1970). Under certain conditions, the crude aqueous plant extracts should be passed through charcoal prior to assay in order to remove coloured and other interfering compounds (Van Etten et al., 1974).

By further simplification, Lein (1970) arrived at a most efficient and effective method for prescreening thousands of samples per day at almost no cost. After squashing a probe, which may be one seed only, by one hammer stroke and adding one drop of water to the pulp, glucosinolates are hydrolysed

by the endogenous myrosinase; after a few minutes the
concentration of the released glucose is read by a commercially
available test paper containing glucose oxidase peroxidase
and a chromogen turning green on high concentrations. Various
degrees of standardisation (accuracy of seed grinding, adjustment of seed weight to water volume, time of reaction, purification of the crude extract etc.) are possible to acquire the
optimal compromise between rapidity and accuracy of the test
under given conditions.

For the precise and reliable quantification of
particular glucosinolates or groups of glucosinolates, gas
liquid chromatography is the only available method. This
method can manage to separate and identify even complex mixtures
containing isothiocyanates, nitriles, organic thiocyanates, or
oxazolidine-2-thiones (Van Etten et al., 1976). Most
glucosinolates can be chromatographed without prior hydrolysis
with good resolution and quantification in the form of their
trimethylsilyl derivatives (Underhill and Kirkland, 1971;
Thies, 1976; 1977). For certain samples of plant tissues,
e.g., with a high amount of sucrose or other interfering
substances, or with a very low content of glucosinolates
(below 10 μmoles/g defatted dry matter), purification and
concentration of the extracts on ion exchange columns at pH
7 with K_2SO_4 as eluant was suggested by Thies (1977) for
microlitre scales prior to chromatography.

Essential improvement was attained by a simultaneous
enzymatic desulphatation on the same column using a sulpho-
hydrolase from the snail stomach, which hydrolyses the sulphate
ester bond of the glucosinolates without attack on the
analytically essential S-glycosidic bond (Thies, 1979). By
this means, the sulphate is removed before chromatographic
separation and no longer disturbs it, while all the rest of
the procedure, especially the derivatisation of the glucosinolates, remains unaltered (Thies, 1978). In this way, minute
amount of plant materials can be assayed accurately and
precisely in a highly automated procedure for any glucosinolate

compound (or mixture), including those which so far cannot be detected on chromatograms (indolylglucosinolates) or are, at least in part, destroyed (glucosinalbin). After examination of glucosinolates from 10 homologous series only the methylsulphinyl glucosinolate glucoiberin appeared on the chromatograms in the form of multiple peaks.

C. GENETICS OF GLUCOSINOLATE SYNTHESIS AND BREEDING FOR LOW CONTENTS

Genetic differences in glucosinolate content are relatively small within species of rapeseed (Josefsson and Jönsson, 1969; Lein, 1970; Namai et al., 1972; Namai and Hosoda, 1975); but considerably lower amounts were found in summer than in winter types of *B. napus* and *B. campestris* (Josefsson and Appelqvist, 1968). There was a broader distribution of glucosinolate contents between individual plants, especially in summer turnip rape (Josefsson and Jönsson, 1969; Lein, 1970). Initial efforts to select rapeseed with low glucosinolate content, begun in the early 1940s, were not successful (Schwarze, 1946), probably because of the inefficient analytical methods available at that time. The first genotypes with a reduced glucosinolate level were found in summer turnip rape, where continued selection succeeded in reducing the total glucosinolate content considerably (Downey et al., 1969). Lein (1970) indicated that also in winter turnip rape, i.e. the cultivar 'Lembkes Winterrübsen', the glucosinolate content in the seed could be definitely reduced by easy single plant selection.

Another major step in the same direction was the discovery in 1967 that the Polish *B. napus* cultivar 'Bronowski' contained a glucosinolate content almost as low as 10% (i.e. 12 µmol) of the conventional average (Kondra and Downey, 1969; Josefsson and Appelqvist, 1968, Krzymanski, 1970; Lein, 1970). It turned out that all the three major glucosinolates, i.e., progoitrin, gluconapin, and glucobrassicanapin, were simultaneously reduced (Downey et al., 1969; cf. the cv. 'Erglu' in Table 1). 'Bronowski', in addition, only exhibited 7 to 10% erucic acid in its seed oil (Krzymanski, 1970; Lein, 1970). From this exceptional gene source the two first high yielding 'double low' cultivars (low in erucic acid and in glucosinolate content) of summer rapeseed, named 'Tower' (Stefansson and Kondra, 1975) and 'Erglu' (Röbbelen, 1976), were developed by conventional backcross procedure and finally released for agricultural production in 1973/74 in Canada and Germany, respectively. Today, the 'Bronowski' gene, which so far remains the only gene source for low glucosinolate content, is widely used by most rapeseed breeders for the development of improved cultivars.

Irrespective of its usefulness, the genetic control of low glucosinolate content in 'Bronowski' is still only partially understood. The first results from progeny analyses after corssing were published by Krzymanski (1970), Lein (1970), and Kondra and Stefansson (1970), in this order. From reciprocal cross combinations, it turned out that the glucosinolate content was mainly determined in a matroclinous fashion, i.e., by the maternal genotype rather than by the genotype of the embryo. Correspondingly, after crossing 'Bronowski' on to the Canadian 'zero erucic' summer rape cultivar 'Oro', seeds from F_1 plants, which constitute the F_2 generation, uniformly contained the high glucosinolate content of 'Oro'; only from F_3 seeds an average of 1 recessive low type segregated out of about 60 individuals with higher glucosinolate values (Lein, 1970). For the single glucosinolates, the segregation ratios appeared rather complicated, since discrete genetic classes were difficult to distinguish. Kondra and Stefansson (1970) postulated from their data that three loci control gluconapin content with higher values being partially dominant to lower values. Glucobrassicanapin was determined by four or five loci with overdominance of the high values, while four partially dominant loci were found to condition high values of progoitrin (cf. also Krzymanski, 1970). The gene system controlling these three major glucosinolates of rapeseed did not segregate independently of each other. From eight possible phenotypes in a backcross population only four were present, indicating linkage or genes which act simultaneously on all three components; such genes most probably control early stages of the biosynthetic pathways. On the other hand, the three major glucosinolates were not altogether under an identical genetic control. Thus, it seems not to be impossible to select successfully within the total glucosinolates for a further reduction of, e.g., the progoitrin, which generates the most detrimental fission products (Krzymanski, 1970; Lein, 1972; Röbbelen, 1976).

Using reciprocal cross combinations of 'Bronowski' with the high glucosinolate but zero erucic cultivar 'Oro', Lein (1970) determined an additional genetic influence of the cytoplasm on glucosinolate synthesis. In correspondent progenies, the seeds from plants with 'Bronowski' cytoplasm had a mean glucosinolate content which was 8% lower than seeds from plants with 'Oro' cytoplasm. Thus, the breeder should select for low glucosinolate content in the 'Bronowski' cytoplasm, although possible adverse effects on other characters of economic importance, e.g., seed yield, oil content, or winter hardiness, may not yet have been sufficiently considered.

Because of the dominating economic importance of winter rapeseed, European breeders have extensively used the 'Bronowski' genotype for crosses with adapted winter cultivars. But, so far, lines originating from such crosses with the 'Bronowski' summer rape give lower yields than conventional winter rape cultivars, due to the lower yield of summer rape in general and also to poor winter hardiness (Jönsson et al., 1975; Röbbelen, 1976). Therefore, adapted winter forms of rapeseed with the 'double low' or 'OO'-quality, i.e., low in erucic acid content of their seed oil and in glucosinolate content of their meal as well, will not be available for agricultural production in the EEC until a few more years. Since the 'Bronowski' gene reduces the glucosinolate content to about one tenth of the conventional level, it has been agreed that it is economically desirable and feasible to licence as 00-cultivars only those which exhibit less than 30 µmoles of glucosinolate per g defatted dry meal determined at the breeders seed level (Röbbelen and Rakow, 1979). Cultivars with glucosinolate contents above this limit are likely to segregate and to be subject to genetic drift; therefore, it is not advisable to admit intermediate types to commercial rapeseed production.

During the evolution of *Brassica*, interspecific crosses have occurred widely in nature. For example, the generally low progoitrin content of *B. campestris* (see Table 1) was not found in the seed of various forage turnips (Namai et al., 1972), which are commonly supposed to have originated from spontaneous interspecific hybrids between rutabaga *(B. napus)* and turnip *(B. campestris)*. Although it will be very difficult to breed rapeseed cultivars essentially free from progoitrin and isothiocyanate forming glucosinolates, there should be an assured possibility of synthesising such genotypes by interspecific crosses between properly selected forms of the various *Brassica* species. For such strategies, recent methods of embryo and tissue culture, including *in vitro* fertilisation and protoplast fusion, may provide the breeder with exceptional new opportunities for efficient germ plasm transfer (cf. Reinert and Bajaj, 1977; Thomas et al., 1979).

Biologically, glucosinolates and in particular their volatile fission products are probably important in determining insect or nematode behaviour. Many of the isothiocyanates, nitriles, and thiocyanates, liberated by experimental hydrolysis are known to be highly toxic to several species of insects (Lichtenstein et al., 1964). Isothiocyanates are also powerful antibacterial and antifungal agents. On the other hand, glucosides themselves serve as a gustatory stimulant for insect pests of Cruciferous plants, while their isothiocyanate cleavage products are feeding and oviposition attractants for many insect species (see Beck and Reese, 1976). For wild vertebrates, such as deer or rabbit, glucosinolates act as repellents. Thus, we are finally left with the troublesome statement that we know little about the functions of glucosinolates in the plants which produce them, and about the ultimate biological effects which their elimination through plant breeding may have.

REFERENCES

Anand, I.J., 1974. Plant Biochem. J. 1, 26-31.

Anderson, G.H., Li, G.S.K., Jones, J.D. and Bender, F., 1975. J. Nutr. 105, 317-325.

Ballester, D., Rodrigo, R., Nakouzi, J., Chichester, C.O., Yanez, E. and Monckeberg, F., 1970. J. Sci. Food. Agr. 21, 143-144.

Beck, S.D. and Reese, J.C., 1976. Recent Adv. Phytochem. 10, 41-92.

Bhatty, R.S. and Sosulski, F.W., 1972. J. Amer. Oil. Chem. Soc. 49, 346-350.

Bowland, J.P., Clandinin, D.R. and Wetter, L.R., 1965. Publ. 1257, Canada Dept. of Agriculture, Ottawa.

Brak, B. and Henkel, H., 1978. Fette, Seifen, Anstrichmittel 80, 104-106.

Byczynska, Barbara, Krzymanski, J. and Wiazecka, Krystina, 1970. Hod. Rosl. Aklim. 14, 547-551.

Downey, R.K., Craig, B.M. and Youngs, C.G., 1969. J. Amer. Oil. Chem. Soc. 46, 121-123.

Eapen, K.E., Tape, N.W. and Sims, R.P.A., 1969. J. Amer. Oil Chem. Soc. 46, 52-55.

Ettlinger, M.G. and Kjaer, A., 1968. In 'Recent Advances in Phytochemistry'. T.J. Mabry, R.E. Alston and V.C. Runeckles, eds., pp. 59-144. North Holland Publ. Comp., Amsterdam.

Ettlinger, M.G. and Lundeen, A.J., 1956. J. Amer. Chem. Soc. 78, 4172-4173.

Ettlinger, M.G. and Lundeen, A.J., 1957. J. Amer. Chem. Soc. 79, 1764-1765.

Finlayson, A.J., Christ, C.M. and Downey, R.K., 1970. Canad. J. Plant Sci. 50, 705-709.

Guignard, L., 1890a. C.R. hebd. Séances Acad. Sci. 111, 249-252.

Guignard, L., 1890b. C.R. hebd. Séances Acad. Sci. 111, 920-923.

Hemingway, J.S., Schofield, H.J. and Vaughan, J.G., 1961. Nature 192, 993.

Jönsson, R., Josefsson, E. and Uppström, B., 1975. Sveriges Utsädesf. T. 85, 279-290.

Johnston, T.D. and Jones, D.I.H., 1966. J. Sci. Food Agr. 17, 70-71.

Josefsson, E., 1970a. Diss. Math. Nat. Fac., Univ. Lund, Sweden.

Josefsson, E., 1970b. J. Sci. Food Agr. 21, 94-97.

Josefsson, E., 1972. Z. Pflanzenzüchtg. 68, 113-123.

Josefsson, E. and Appelqvist, L.-Å, 1968. J. Sci. Food.Agr. 19, 564-570.

Josefsson, E. and Jönsson, R., 1969. Z. Pflanzenzüchtg. 62, 272-283.

Jürges, K., 1978. Proc. 5th Int. Rapeseed Conf., June 1978, Malmö, Sweden. Vol. 2, pp. 57-60.

Kjaer, A., 1976. In 'The Biology and Chemistry of the Cruciferae'. J.G. Vaughan, H.J. Macleod, and B.M. Jones, eds., pp. 207-219. Academic Press, London - New York - San Francisco.

Kjaer, A. and Olesen Larsen, P., 1976. In 'Biosynthesis'. T.A. Geisman, ed., vol. 5, pp. 179-203. Specialist Periodical Reports. Chem. Soc., London.

Kondra, Z.P. and Downey, R.K., 1969. Canad. J. Plant Sci. 49, 623-624.

Kondra, Z.P. and Downey, R.K., 1970. Crop Science 10, 54-56.

Kondra, Z.P. and Stefansson, B.R., 1970. Canad. J. Plant Sci. 50, 643-647.

Kozlowska, H., Sosulski, F.W. and Youngs, C.G., 1972. Canad. Inst. Food Sci., Technol. J. 5, 149-154.

Krzymanski, J., 1970. Hod. Rosl. Aklim. 14, 95-133.

Lein, K.-A., 1970. Z. Pflanzenzüchtg. 63, 137-154.

Lein, K.-A., 1972. Z. Pflanzenzüchtg. 67, 243-256.

Lein, K.-A., 1972b. Z. Pflanzenphysiol. 67, 333-342.

Lein, K.-A. and Schön, W.J., 1969. Angew. Bot. 43, 87-93.

Lichtenstein, E.P., Morgan, D.G. and Mueller, C.H., 1964. J. Agr. Food Chem. 12, 158-161.

McGhee, J.E., Kirk, L.D. and Mustakas, G.L., 1965. J. Amer. Oil Chem. Soc. 42, 889-891.

McGregor, D.I., 1978. Canad. J. Plant Sci. 58, 795-800.

Miller, R.W., Van Etten, C.H., McGrew, C., Wolff, I.A. and Jones, Q., 1962. J. Agr. Food Chem. 10, 426-430.

Mukherjee, K.D., Afzalpurkar, A.B. and El Nockrashy, A.S., 1976. Fette, Seifen, Anstrichmittel 78, 306-311.

Namai, H. and Hosoda, T., 1975. Japan. J. Genet. 50, 43-51.

Namai, H., Kaji, T. and Hosoda, T., 1972. Japan. J. Genet. 47, 319-327.

Reinert, J. and Bajaj, Y.P.S., 1977. 'Plant Cell, Tissue and Organ Culture'. Springer Verlag, Berlin - Heidelberg - New York.

Röbbelen, G., 1976. Fette, Seifen, Anstrichmittel 78, 10-17.

Röbbelen, G. and Rakow, G., 1979. Fette, Seifen, Anstrichmittel 81, 197-200.

Rutkowski, A., 1970. Proc. Int. Conf. Rapeseed, Sept. 1970, Sté-Adéle, Canada, pp 496-515.

Schlüter, M. and Gmelin, R., 1972. Phytochemistry 11, 3427-3431.

Schwarze, P., 1946. Züchter 17-18, 19-22.

Schweidler, J.H., 1910. Bot. Zbl. Beih. $\underline{1}$, 422-475.

Sosulski, F.W., Soliman, F.S. and Bhatty, R.S., 1972. Canad. Inst. Food Sci., Technol. J. $\underline{5}$, 101-104.

Staron, T., 1970. Proc. Int. Conf. Rapeseed, Sept. 1970. Sté-Adéle, Canada, pp 321-346.

Stefansson, B.R. and Kondra, Z.P., 1975. Canad. J. Plant Sci. $\underline{55}$, 343-344.

Thies, W., 1976. Fette, Seifen, Anstrichmittel $\underline{78}$, 231-234.

Theis, W., 1977. Z. Pflanzenzüchtg. $\underline{79}$, 331-335.

Thies, W., 1978. Proc. 5th Int. Rapeseed Conf., June 1978. Malmö, Sweden Vol. 1, pp. 136-139.

Thies, W., 1979. Naturwiss. $\underline{66}$, 364.

Thomas, E., King, P.J. and Potrykus, I., 1979. Z. Pflanzenzüchtg. $\underline{82}$, 1-30.

U, N., 1935. Japan. J. Bot. $\underline{7}$, 389-452.

Underhill, E.W. and Kirkland, D.F., 1971. J. Chromatogr. $\underline{57}$, 47-54.

Underhill, E.W., Wetter, L.R. and Chisholm, M.D., 1973. Biochem. Soc. Symp. $\underline{38}$, 303-326.

Van Etten, C.H., Daxenbichler, M.E., Williams, P.H. and Kwolek, W.K., 1976. J. Agr. Food Chem. $\underline{24}$, 452-455.

Van Etten, C.H., McGrew, C.E. and Daxenbichler, M.E., 1974. J. Agr. Food Chem. $\underline{22}$, 483-487.

Van Etten, C.H. and Wolff, I.A., 1973. In 'Toxicants Occurring Naturally in Foods'. 2nd ed., pp. 210-234. Committee on Food Protection, Nat. Acad. Sci., Wash. D.C.

Vaughan, J.G., Hemingway, J.S. and Schofield, H.J., 1963. J. Linn. Soc. (Bot.) $\underline{58}$, 435-447.

Wetter, L.R., 1955. Canad. J. Biochem. Physiol. $\underline{33}$, 980-984.

Wetter, L.R., 1957. Canad. J. Biochem. Physiol. $\underline{35}$, 293-297.

Wetter, L.R. and Craig, B., 1959. Canad. J. Plant Sci. $\underline{39}$, 395-399.

Wetter, L.R. and Youngs, C.G., 1976. J. Amer. Oil Chem. Soc. $\underline{53}$, 162-164.

Youngs, C.G. and Wetter, L.R., 1967. J. Amer. Oil Chem. Soc. $\underline{44}$, 551-554.

NEW METHODS OF QUANTITATIVE ANALYSIS OF GLUCOSINOLATES

H. Sørensen

Chemistry Department,
Royal Veterinary and Agricultural University,
Thorvaldsensvej 40, DK-1871 Copenhagen V, Denmark

Rapeseed production is quantitatively important in some parts of Europe and, especially, in Canada and India; FAO statistics on world production of oilseeds indicate that rapeseed currently ranks fifth in importance. The crop is primarily used as a source of vegetable oil (ca. 45% of dry wt.) and protein (ca. 40% of the meal). The rapeseed proteins are of high quality with a relatively well balanced amino acid composition.

The utilisation of rapeseed proteins is limited owing to the occurrence of glucosinolates in the meal. These compounds are the source of various manifestations of toxicity in animals and birds, ranging from depressed weight gain to enlargement of thyroid, kidney and liver, and even death of some experimental animals and birds. The new double-low rapeseed varieties are more promising than the previously used rapeseed varieties, but some nutritional problems still exist. A solution of the problems related to glucosinolates requires reliability in analytical methods, as previously used methods lack precision and reproducibility. Where it is necessary to analyse for very low contents of glucosinolates, methods of concentrations and/or clean-up steps are often required before final quantification. Thus, knowledge of the chemistry, stability and properties of intact glucosinolates are required.

The purpose of this presentation is to describe newly developed methods used in the isolation, identification, and quantitative determination of intact glucosinolates. The advantages and disadvantages of these methods in relation to previously used methods - which are based on estimation of glucosinolate degradation products - will be discussed briefly.

Glucosinolates are present in plants of families belonging to the order Capparales (Figure 1). In the last few years we have studied the quantitative determination of glucosinolates in some of these plants.

The investigations revealed the need for improvement in the experimental technique, as it was evident that some glucosinolates escaped detection by the methods traditionally used.

We consider that the problems concerning glucosinolates in food or feed are questions of the total amount of the types of glucosinolates present and of the degradation products produced from them. Therefore, investigations on these problems require consideration of the

(1) biochemistry and occurrence of glucosinolates

(2) chemistry and properties of both glucosinolates and products thereof

(3) methods used in the analysis of these compounds.

Glucosinolates are a well known group of natural products in which more than 90 different compounds are included. Only β-thio-D-glucopyranosyl is known as carbohydrate part of the functional group in glucosinolates, which, furthermore, contain a sulphate group. Both of these groups contribute to the hydrophilic properties of glucosinolates and the latter impart strong acidic properties to these compounds. These facts are important when we choose methods for studying glucosinolates.

The R groups are structurally related to the amino acids known from plants (more than 300) as derived from the fact that glucosinolates are biosynthetically derived from amino acids. In the biosynthesis, elongation of the side chain (R group) with ($-CH_2-$) units derived from acetate units often occurs. Furthermore, derivatisation of the side chain often occurs, leading to unsaturated, oxidised, hydroxylated, methylated or glycosated R groups (Table 1). These are important facts to consider in connection with plant breeding, where the purpose is the

All known glucosinolates have the structure shown in the following figure i.e. only differing in the R-group.

They are biosynthetically derived from amino acids and although about 350 amino acids are known (especially from plants) only about 90 different glucosinolates are known. The quantitatively dominating glucosinolates in rapeseeds are derived from methionine, phenylalanine and tryptophane.

a)

LIMNANTHACEAE; TROPAEOLACEAE; BRETSCHNEIDERACEAE*; SALVADORACEAE*; TOVARIACEAE; PENTADIPLANDRACEAE; BATACEAE*; GYROSTEMONACEAE*; MORINGACEAE; BRASSICACEAE; CAPPARACEAE; RESEDACEAE.

b)

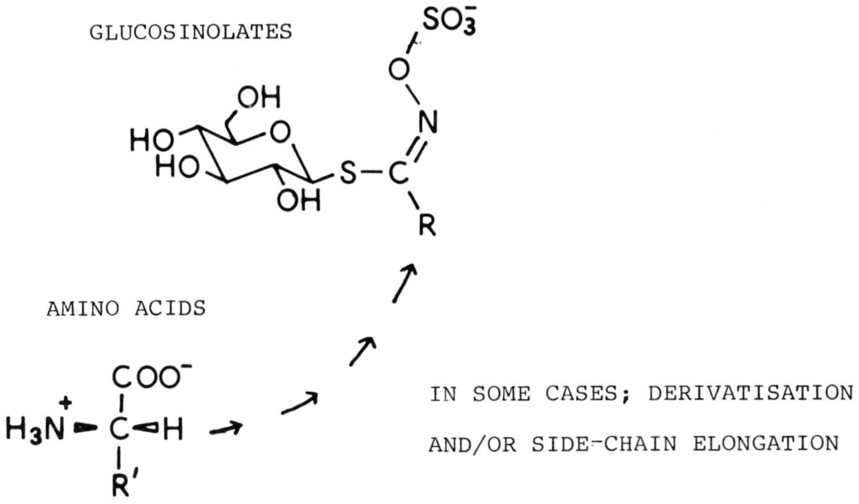

Fig.1. a) The distribution of glucosinolates in families of the order Capparales
  b) The general structure of glucosinolates

production of new low-glucosinolate varieties, and in connection with the choice of methods to study glucosinolate degradation products.

The β-hydroxylated, unsaturated and/or aromatic side chains are of special interest because, on degradation of the glucosinolates, they often lead to products which have drastic physiologic effects.

Degradation products of glucosinolates are shown in Figure 2. In acidic solution glucosinolates are transformed to carboxylic acids, but not quantitatively. In basic solution some glucosinolates are transformed to amino acids and/or other degradation products. Thus, in investigations of glucosinolates, strong acidic and basic conditions must be avoided.

The glucosinolate containing plants also contain thioglucosidases (myrosinases) EC 3.2.3.1, leading on autolysis or damage of the plant cells to formation of isothyiocyanates or other degradation products such as thiocyanates, thiocyanate ion, nitriles, amines, oxazolidinethiones and/or cyano-epithioalkanes. It seems most likely that some glucosinolate containing plants contain other enzymes or metabolites, beside the 'myrosinases', which modify or change the hydrolytic reactions in such a way that the products obtained are not isothiocyanates or oxazolidinethiones, but some other of the products shown in Figure 2.

The traditionally used methods in the analysis of glucosinolates are based on the myrosinase treatment of defatted plant material. Analysis is then performed on the degradation products, which are soluble in organic solvents. By use of gas chromatography (GLC) and mass spectroscopy (MS), it is possible to identify and quantitatively determine some of the individual glucosinolates. However, the procedures based on glucosinolate degradation products are generally suspect since enzymatic hydrolysis may not go to completion or may proceed by other routes.

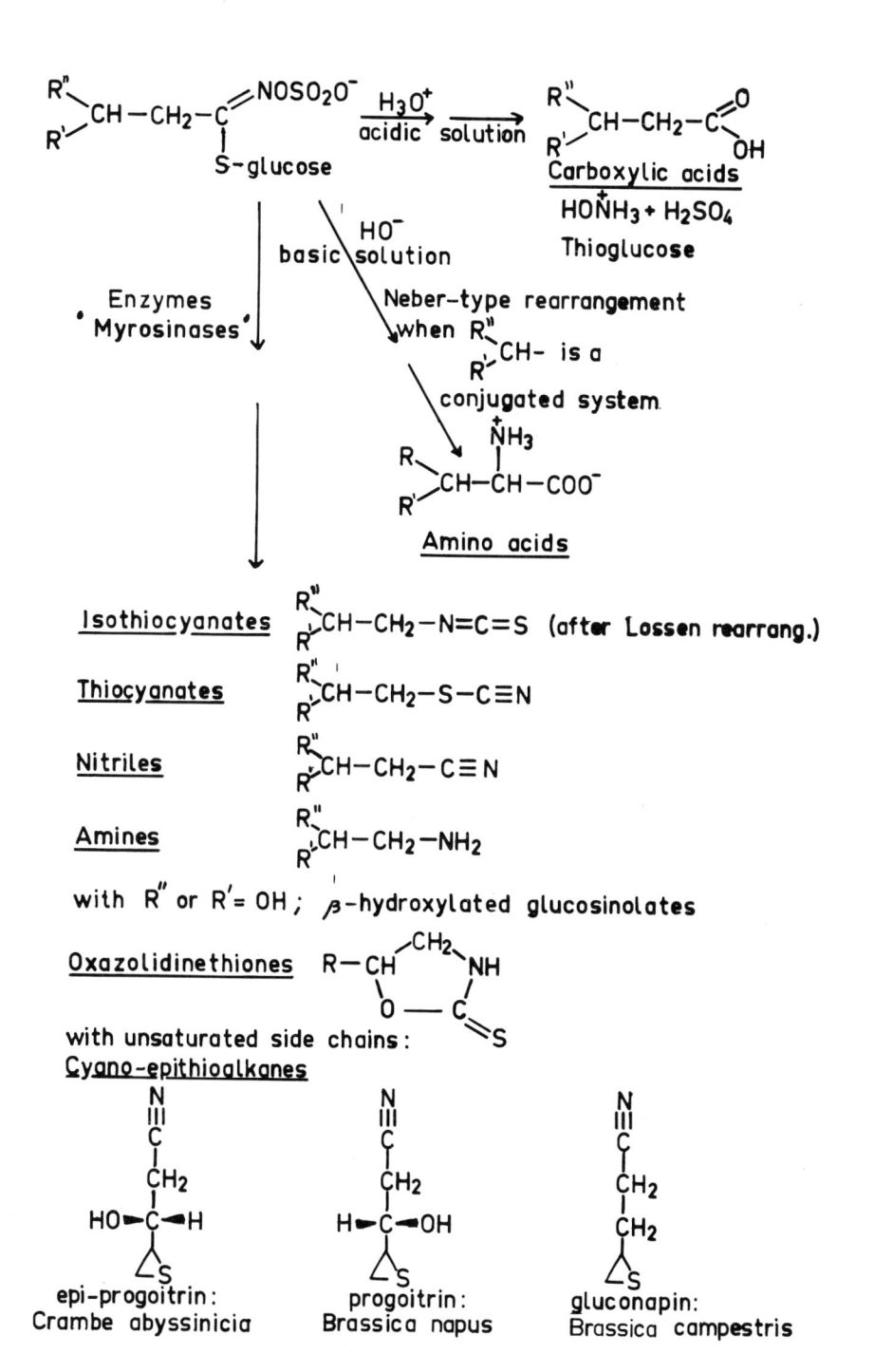

Fig.2. Degradation products of glucosinolates. Olsen and Sørensen, 1980. J. Agr. Food Chem. 28, 43-48.

TABLE 1

CHEMICAL STRUCTURES OF THE GLUCOSINOLATES SELECTED TO ILLUSTRATE THE REQUIREMENTS FOR METHODS OF GLUCOSINOLATE ANALYSIS

$$R-C\begin{matrix}S\text{-glucose}\\ \diagdown N OSO_2O^-\end{matrix}$$

#	R	Name	Trivial name
1	CH_3-	Methylglucosinolate	Glucocapparin
2	$CH_2=CH-CH_2-$	Allylglucosinolate	Sinigrin
3	$CH_3-CH_2-CH_2-$	3-Butenylglucosinolate	Gluconapin
4	$CH_2=CH-CH_2-CH_2-$	4-Pentenylglucosinolate	Glucobrassicanapin
5	$CH_2=CH-CH-CH_2-$ OH	2-Hydroxy-3-butenylglucosinolate	Progoitrin
6	$CH_2=CH-CH_2-CH-CH_2-$ OH	2-Hydroxy-4-pentenylglucosinolate	Napoleiferin
7	$CH_3-S-CH_2-CH_2-CH_2-$	3-Methylthiopropylglucosinolate	Glucoibervirin
8	$CH_3-S-CH_2-CH_2-CH_2-CH_2-$	4-Methylthiobutylglucosinolate	Glucoerucin
9	$CH_3-SO-CH_2-CH_2-CH_2-$	3-Methylsulfinylpropylglucosinolate	Glucoiberin
10	$CH_3-SO-CH_2-CH_2-CH_2-CH_2-$	4-Methylsulfinylbutylglucosinolate	Glucoraphanin
11	$CH_3-SO-CH_2-CH_2-CH_2-CH_2-CH_2-$	5-Methylsulfinylpentylglucosinolate	Glucoalyssin
12	$CH_3-SO_2-CH_2-CH_2-CH_2-$	3-Methylsulfonylpropylglucosinolate	Glucocheirolin
13	$CH_3-SO_2-CH_2-CH_2-CH_2-CH_2-$	4-Methylsulfonylbutylglucosinolate	Glucoerysolin
14	$C_6H_5-CH_2-$	Benzylglucosinolate	Glucotropaeolin
15	$C_6H_5-CH_2-CH_2-$	Phenethylglucosinolate	Gluconasturtiin
16	$C_6H_5-CH(OH)-CH_2-$	2-Hydroxy-2-phenylethylglucosinolate	Glucobarbarin
17	$(m\text{-}HO)C_6H_4-CH_2-$	m-Hydroxybenzylglucosinolate	Glucolepigramin
18	$(p\text{-}HO)C_6H_4-CH_2-$	p-Hydroxybenzylglucosinolate	Sinalbin
19	2-Arabinopyranosyloxy-2-phenylethyl-	2-Arabinopyranosyloxy-2-phenylethylglucosinolate	
20	o-Rhamnopyranosyloxybenzyl-	o-Rhamnopyranosylbenzylglucosinolate	
21	p-Rhamnopyranosyloxybenzyl-	p-Rhamnopyranosyloxybenzylglycosinolate	
22	indol-3-ylmethyl-	3-Indolylmethylglucosinolate	Glucobrassicin

The advantages and disadvantages of different glucosinolate analysis procedures can be discussed by consideration of the chemical structures of glucosinolates (Table 1), the steps involved in the different procedures (Figure 3), and by use of the information which is required and obtainable from the available literature.

The possibility of carrying out a satisfactory GLC-analysis based on glucosinolate degradation products is often hampered because a lot of glucosinolates contain R-groups which lead, on autolysis of myrosinase catalysed hydrolysis, to products which are hydrophilic and/or hardly volatile. If this is not the case, then the GLC-MS method is very efficient, i.e. for compounds of the structural type 1 - 4 and 14, 15 (Table 1).

The spectrophotometric UV determination of thiourea derivatives and oxazolidinethiones obtained from the corresponding isothiocyanates, the determination of liberated glucose, and the thiocyanate ion determination are in some cases used as fast and easy methods for determination of the total glucosinolate content. There are problems with all the above methods when they are used in quantitative glucosinolate analysis. In consequence, many reports are concerned with analytical methods for estimation of glucosinolates in plants or material produced therefrom. These methods have recently been reviewed elsewhere (Figure 3) but most of them require knowledge of which types of glucosinolates are present in the plant material.

Where the problem is to investigate glucosinolates in plant species not previously studied, it is important to know which types of compounds are present in the extract if the methods are based on glucosinolate degradation products. Our efforts in developing new methods of glucosinolate analysis are directed towards isolation, identification and determination of intact glucosinolates in order to avoid the above-mentioned problems.

The isolation of intact glucosinolates is performed by homogenisation in boiling methanol-water (thus arresting the

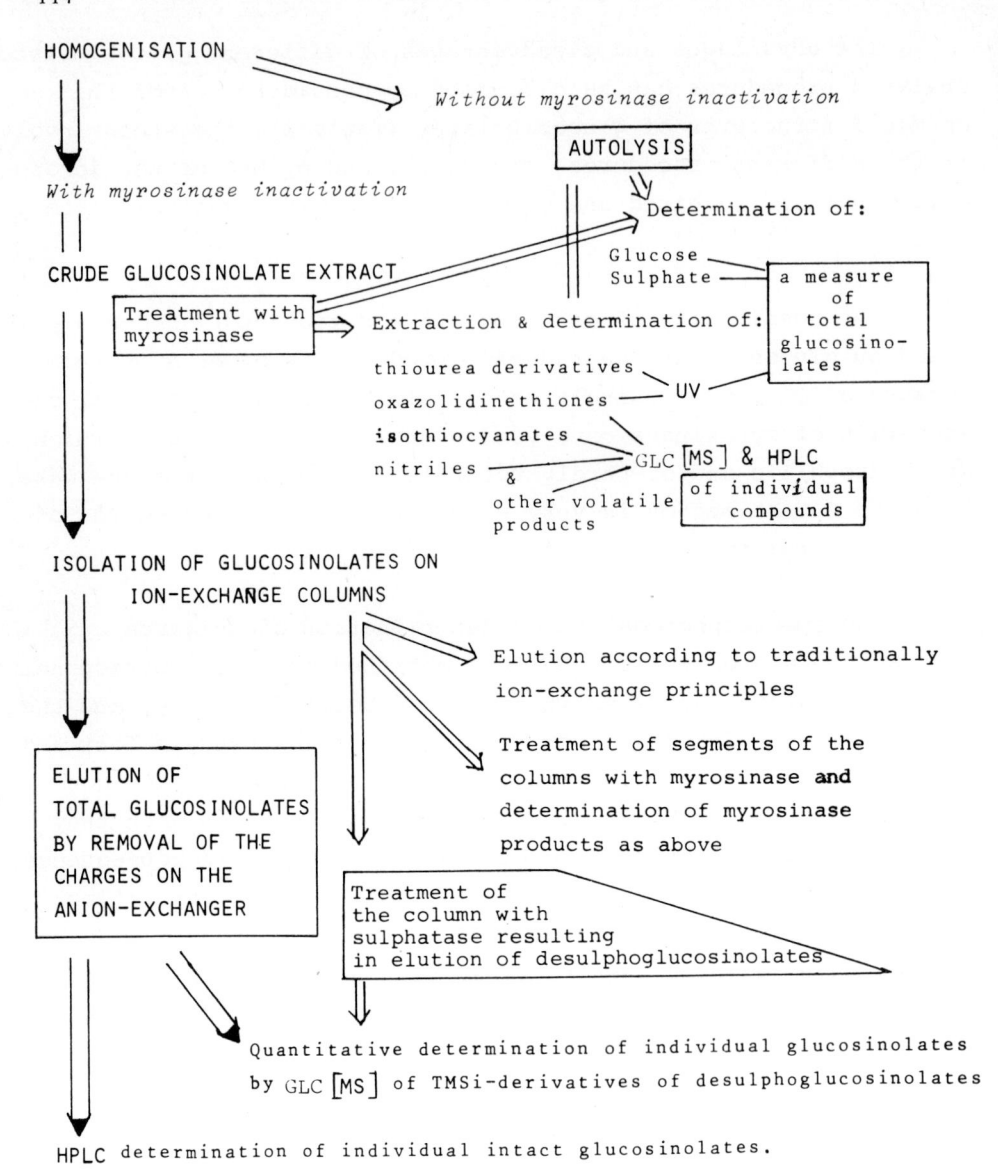

```
HOMOGENISATION
    |           \
    |            \──────▶ Without myrosinase inactivation
    ▼                              │
With myrosinase inactivation      ▼
    ‖                          AUTOLYSIS
    ▼                              ‖
CRUDE GLUCOSINOLATE EXTRACT        ▼
    │         ┌──────────────┐   Determination of:
    │         │ Treatment with│
    │         │ myrosinase    │──▶ Extraction & determination of:
    │         └──────────────┘
    │                    Glucose  ─────┐
    │                    Sulphate ─────┤  a measure
    │                                  │  of
    │                    thiourea derivatives │ total
    │                    oxazolidinethiones ── UV │ glucosinolates
    │                    isothiocyanates ────┐
    │                    nitriles ──── GLC [MS] & HPLC
    │                    & other volatile products │ of individual compounds
    ▼
ISOLATION OF GLUCOSINOLATES ON
    ION-EXCHANGE COLUMNS
              │
              ├──▶ Elution according to traditionally
              │    ion-exchange principles
              │
              ├──▶ Treatment of segments of the
              │    columns with myrosinase and
              │    determination of myrosinase
              │    products as above
              │
              └──▶ Treatment of
                   the column with
                   sulphatase resulting
                   in elution of desulphoglucosinolates
    │
    ▼
┌──────────────────────┐
│ ELUTION OF           │
│ TOTAL GLUCOSINOLATES │
│ BY REMOVAL OF THE    │
│ CHARGES ON THE       │
│ ANION-EXCHANGER      │
└──────────────────────┘
    │       │
    │       ▼
    │    Quantitative determination of individual glucosinolates
    │    by GLC [MS] of TMSi-derivatives of desulphoglucosinolates
    ▼
HPLC determination of individual intact glucosinolates.
```

Fig. 3. Different procedures for glucosinolate analysis. Olsen, O. and Sørensen, H. In: Recent Advances in Analysis of Glucosinolates. In press.

Amberlite IR-120 (H⁺)

(connected in serie;
to avoid strongly
acidic and basic
conditions).

Ole Olsen and Hilmer Sørensen.
Phytochem. <u>19</u> (1980)
1783-1787.

<u>ION-EXCHANGE CHROMATOGRAPHY</u>

Ecteola-cellulose (AcO⁻)

($pK_a \approx 7.5$)

Pyridine eluate with

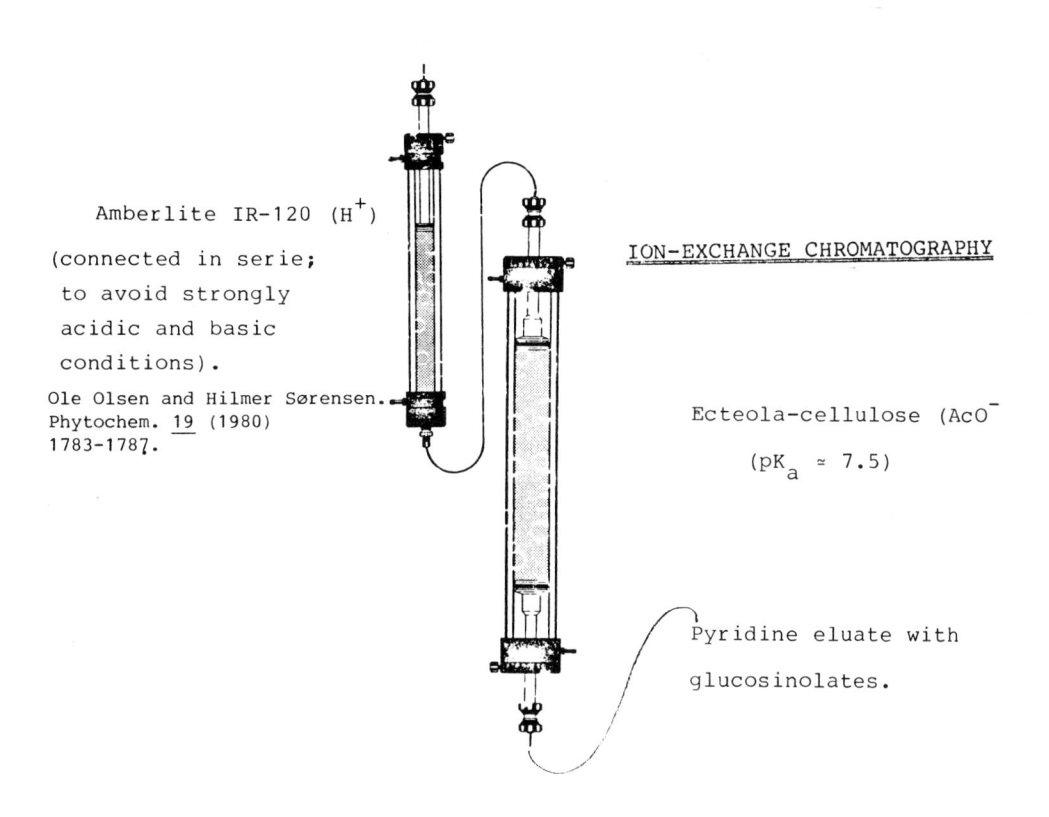

glucosinolates.

<u>Identification of intact glucosinolates</u> by HVE; PC;
^{1}H- and ^{13}C NMR-spectroscopy. O. Olsen and
Hilmer Sørensen. Phytochem. <u>19</u> (1980) 1783-1787.

<u>HPLC (High-Performance Liquid Chromatography)</u> using reversed
phase ion-pair liquid chromatography. P. Helboe, O. Olsen and
Hilmer Sørensen. J. Chromatogr. <u>197</u> (1980) 199-205.

<u>GC-analysis of the TMSi-derivatives</u> (desulphoglucosinolates).
O. Olsen and H. Sørensen. J. Agric. Food. Chem. <u>28</u> (1980) 43-48.

<u>Identification of degradation products</u> from acidic and/or
myrosinase catalyzed hydrolysis. O. Olsen and H. Sørensen,
Phytochem. <u>19</u> (1980) 1783-1787, ibid. <u>18</u> (1979) 1547-1552.

Fig.4. Isolation and determination of intact glucosinolates.

myrosinase-catalysed decomposition) followed by ion-exchange chromatography (Figures 3 and 4).

After drying of the pyridine eluate from the Ecteola ion-exchange column, the residue - containing the pyridinium salts o the glucosinolates - is suitable for silylation and quantitative determination of individual glucosinolates by GLC (Figure 5). This method is reliable for the analysis of complex glucosinolat mixtures but some compounds escape detection *(vide infra)*.

The identity of the glucosinolates used in our studies has been confirmed by PC, HVE, ^{1}H- and ^{13}C-NMR spectroscopy (Figure and Table 3). Furthermore, the identity of some of them has bee confirmed by identification of the degradation products obtained by acidic and/or myrosinase-catalysed hydrolysis (Figure 7).

It has been revealed from our studies of glucosinolates in rapeseed, for example, (Table 3), that some unsolved problems still exist with respect to GLC determination of the methylthio-alkyl-, the methylsulphinylalkyl- (e.g. 7 - 11; Table 1) and the methlysulphonylalkylglucosinolates (e.g. 12, 13; Table 1). The CH_3-S, CH_3-SO- and the CH_3-SO_2-groups are easy to detect by use of NMR-spectroscopy, and thereby it has been revealed that both of the methods based on GLC-analysis (Table 3) do not allow a reliable determination of glucosinolates containing these groups However, the results have shown that the GLC-analysis of the TMSi-derivatives of desulphoglucosinolates (Figure 5) isolated by the new ion-exchange method (Figure 4) leads to a much more reliable quantitative glucosinolate analysis than the previously applied GLC-analysis of isothiocyanates and oxazolidinethiones (Table 3).

Furthermore, some of the previously studied double-low rape varieties (Table 3) grown under different conditions, in differe years and on different soil types have been further investigated The results from this study show that the glucosinolate content of one and the same variety exhibited appreciable variation with respect to both relative and total glucosinolate content, which

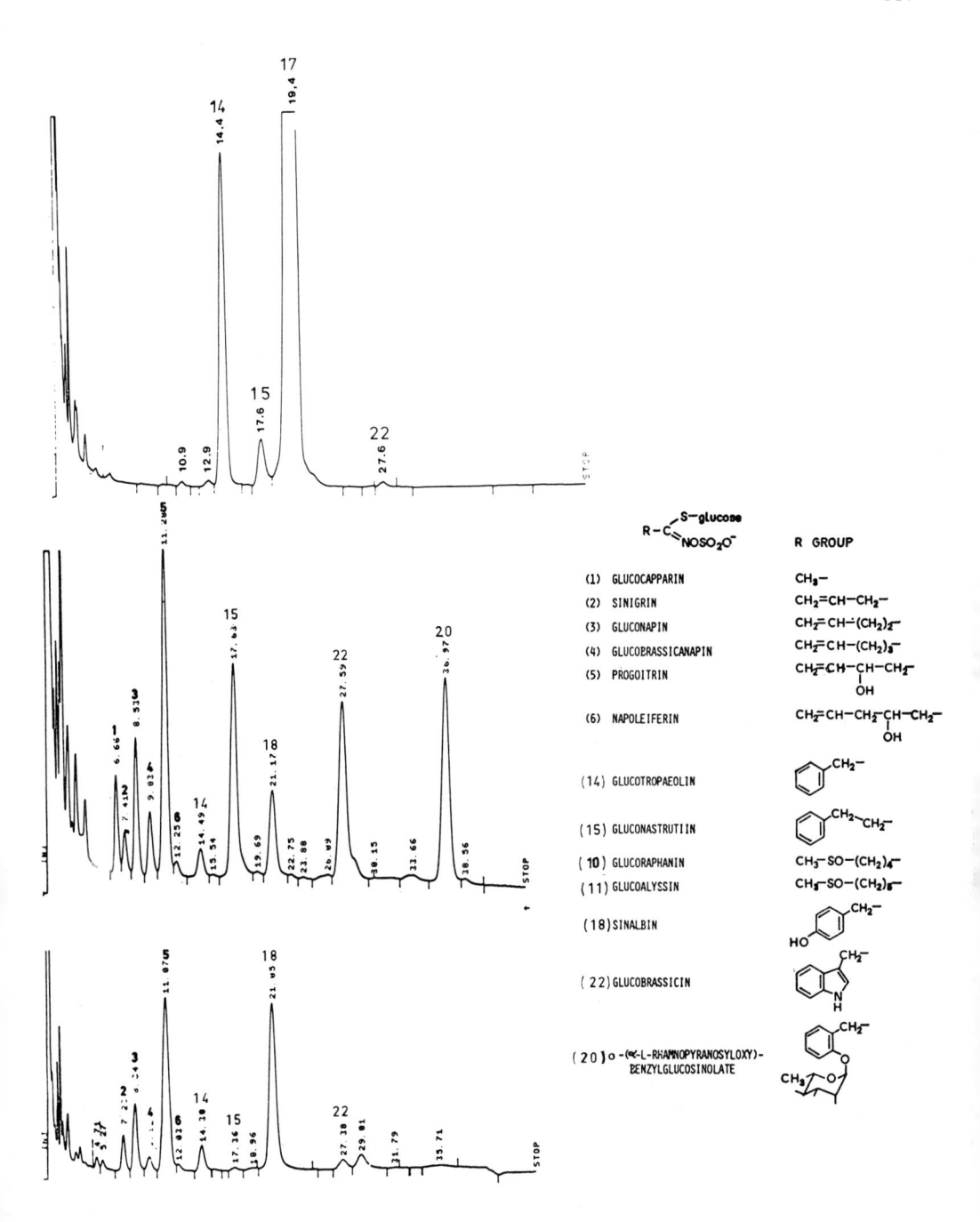

Fig.5. Separation and quantitative determination of individual glucosinolates by GLC of TMSi-derivatives of desulphoglucosinolates.
Olsen, O. and Sørensen, H. 1980. J. Agr. Food Chem. 28, 43-48.

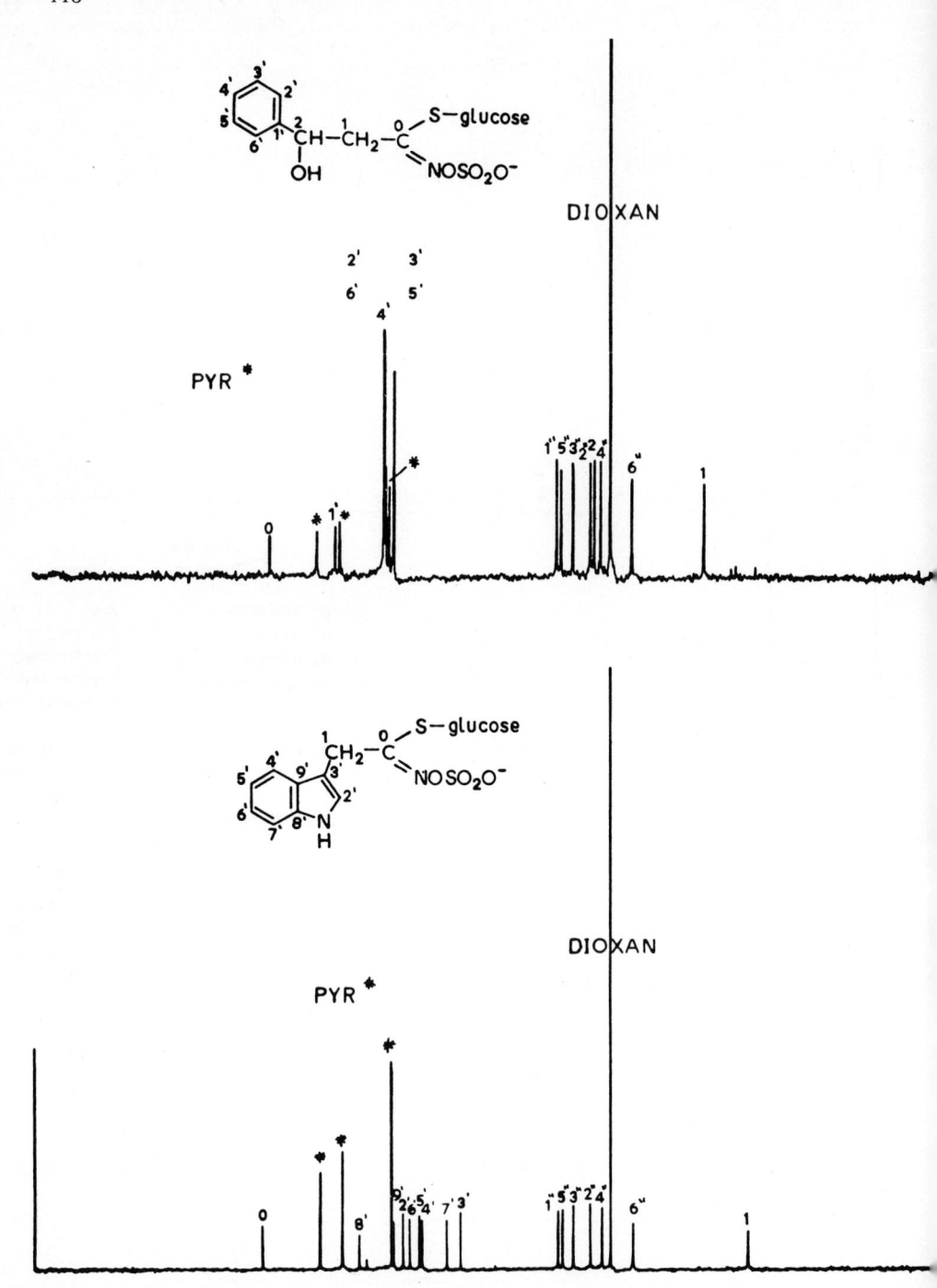

Fig.6. ^{13}C-NMR spectra of glucosinolates isolated as pyridinium salts by ion-exchange chromatography (Figure 4).
Olsen, O. and Sørensen, H. In press. In: Recent Advances in Analysis of Glucosinolates.

TABLE 2

Atom No.[a]		Compound										
		[b] Sinigrin	(8) 2-Hydroxy-2-methylpropyl-glucosinolate	[c] Glucotropaeolin	(9) 2-Hydroxy-2-phenylethylgluco-sinolate	(1) o-(α-L-rhamnopyranosyloxy)-benzylglucosinolate	[c] 1-Thio-β-D-glucopyranose, Na-salt	(4)[d] α-L-rhamnose	(7) o-Hydroxybenzylamine, HCl	p-Hydroxybenzylamine, HCl	(10) o-(α-L-rhamnopyranosyloxy)-benzylamine	p-(α-L-rhamnopyranosyloxy)-benzylamine	(6) N-(o-(α-L-rhamnopyranosyloxy)-benzyl)thiourea
Glucose moiety	1'	82.4 / 4.88	83.0 / 4.9	82.2 / 4.9	82.6 / 4.9	82.4 / 4.92	85.1						
	2'	72.9 / 3.20	72.9 / e)	72.7 / e)	72.7 / e)	72.7 / e)	77.9						
	3'	78.0 / 3.3	77.9 / e)	77.9 / e)	77.9 / e)	77.8 / e)	79.6						
	4'	70.1 / 3.3	70.1 / e)	69.7 / e)	69.9 / e)	69.5 / e)	71.4						
	5'	80.7 / 3.3	81.0 / e)	80.6 / e)	81.1 / e)	80.6 / e)	80.6						
	6'	61.6 / 3.39	61.5 / e)	61.2 / e)	61.5 / e)	61.1 / e)	62.3						
Rhamnose moiety	1'					98.4 / 5.62		95.1			98.6 / 5.61	5.61	5.61
	2'					71.0 / f)		71.9			71.5 / f)	f)	f)
	3'					70.4 / f)		71.1			71.1 / f)	f)	f)
	4'					72.8 / f)		73.3			73.1 / f)	f)	f)
	5'					70.4 / f)		69.4			70.7 / f)	f)	f)
	6'					17.7 / 1.28		17.9			17.9 / 1.24	1.24	1.24
Aromatic moiety	1*			136.0	143.0	120.8			119.1	125.3	118.3		
	2*			130.1 / g)	127.0 / g)	154.2			154.6	131.6 / 7.39	154.5	7.46	
	3*			128.9 / g)	127.0 / g)	115.2 / 7.25			115.2 / 7.01	116.8 / 6.98	115.4 / h)	7.21	h)
	4*			128.4 / g)	129.2 / g)	131.2 / 7.37			130.9 / 7.35	157.0	131.2 / h)		h)
	5*			128.9 / g)	127.0 / g)	123.5 / 7.13			120.4 / 7.01	116.8 / 6.98	123.8 / h)	7.21	h)
	6*			130.1 / g)	129.8 / g)	130.0 / 7.37			130.8 / 7.35	131.6 / 7.39	131.2 / h)	7.46	h)
Others	1	163.6	161.4	163.3	161.0	163.7							
	2	36.9 / 3.4	44.8 / 2.95	39.0 / 4.2	42.1 / 3.26	34.7 / 4.19			39.4 / 4.18	43.6 / 4.15	40.2 / 4.21	4.17	4.2
	3	132.9 / 6.0	71.7		71.9 / 5.36								
	4	119.1 / 5.24–5.11	29.5 & 28.8 / 1.35										

Olsen, O. and Sørensen, H. 1979. Phytochemistry 18, 1547-1552.

^{13}C and ^{1}H chemical shifts (δ) for glucosinolates and related compounds $\frac{\delta^{13}C}{\delta^{1}H}$ for the different atoms

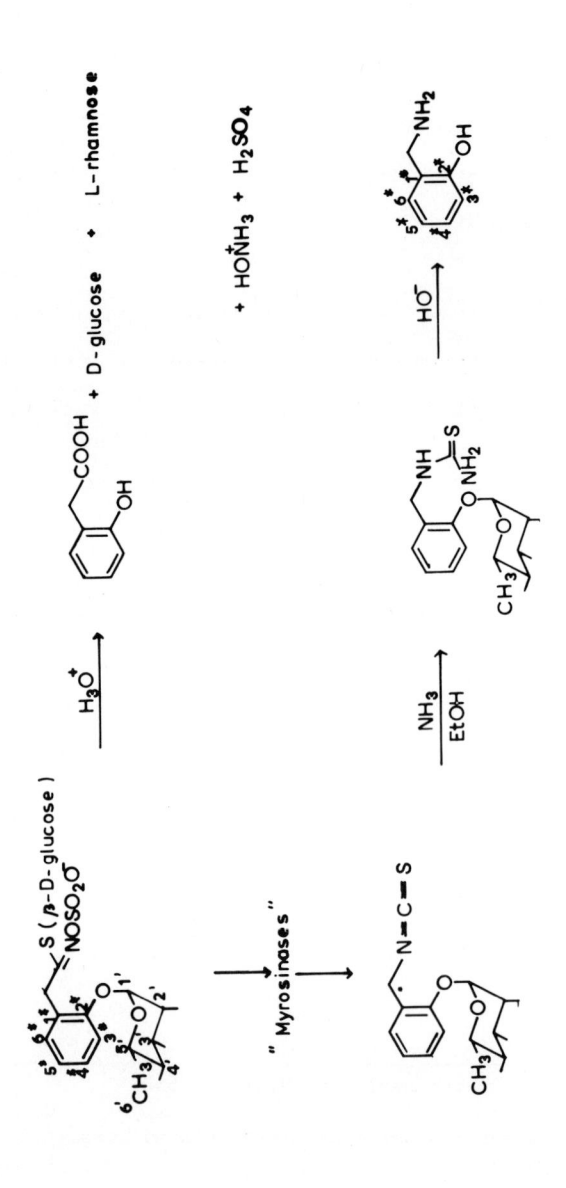

Fig.7. Degradation products of glucosinolates used in the identification of these compounds.
Olsen, O. and Sørensen, H. 1979. Phytochemistry 18, 1547-1552.
Olsen, O. and Sørensen, H. 1980. Phytochemistry 19, 1783-1787.

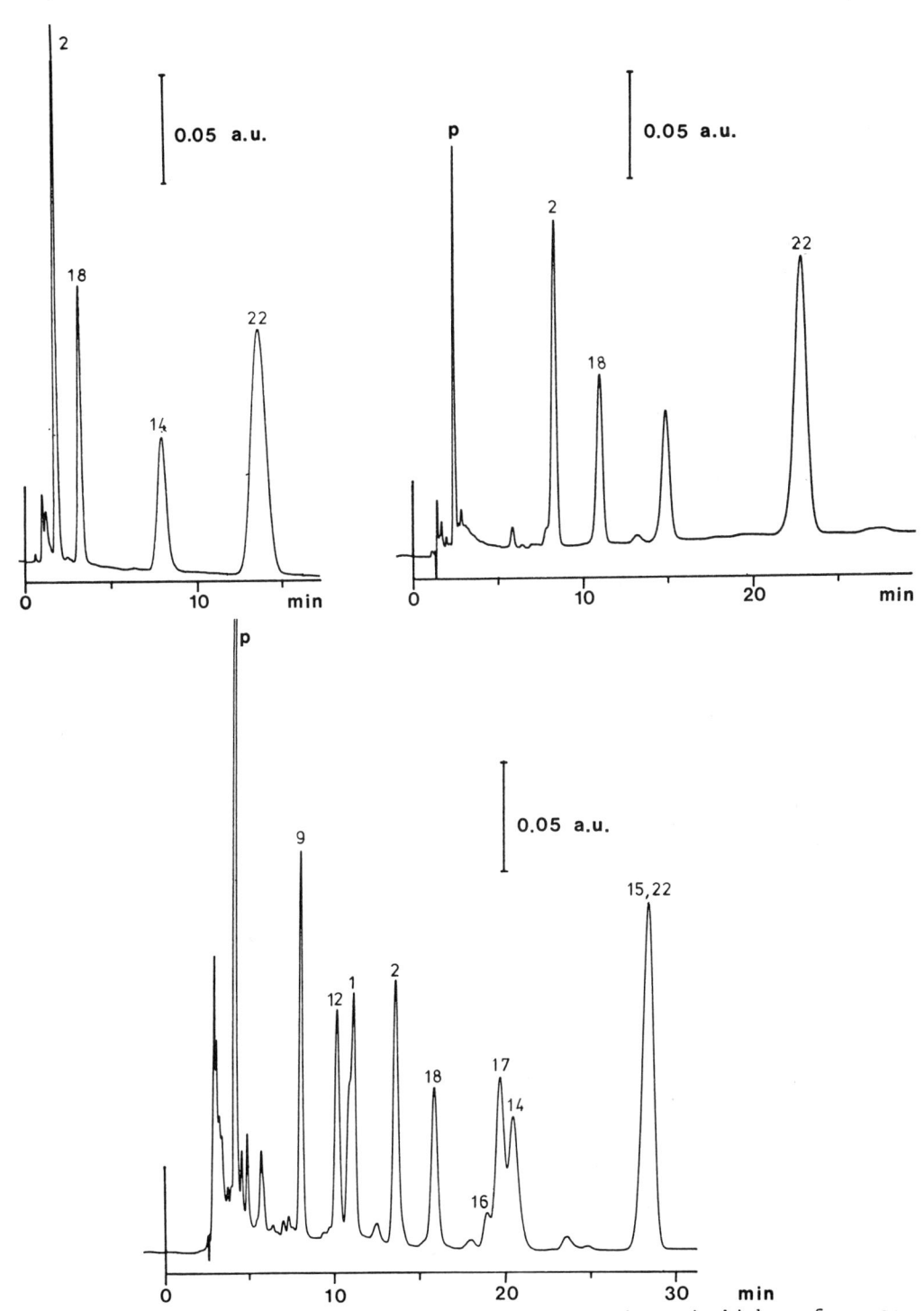

Fig.8. Separation of glucosinolates by reversed-phase ion-pair high-performance liquid chromatography.
Helboe, P., Olsen, O and Sørensen, H. 1980. J. Chromatog. 197, 199-205.

TABLE 3

CONCENTRATION OF THE QUANTITATIVELY IMPORTANT GLUCOSINOLATES IN RAPESEED. DETERMINED BY GLC-ANALYSIS OF TMSi-DERIVATIVES OF DESULPHOGLUCOSINOLATES AND BY GLC-ANALYSIS OF ISOTHIOCYANATES AND OXAZOLIDINETHIONES.

Glucosinolate	Retention time (r.t.) (min)	Gulle rape (B. napus L.)	Tower (B. napus L.)	Candle (B. campestris L.)	Erglu (B. napus L.)	DP-075 (B. napus L.)
Gluconapin	8.3	13.50	2.63	2.40	2.00	3.81
Glucobrassicanapin	9.7	5.78	0.64	1.72	1.09	0.51
Progoitrin	10.9	39.94	7.77	3.27	4.80	5.08
Napoleiferin	12.1	5.98	0.29	0.85	0.46	0.04
Gluconasturtiin	17.4	1.06	0.13	0.77	0.52	0.16
Sinalbin	20.9	*	9.46	0.26	3.18	0.11

GLC-analysis of the TMSi-derivatives of desulphoglucosinolates (Figure 5) isolated by ion-exchange chromatography (Figure 4).

Glucosinolate	Retention time (r.t.) (min)	Gulle rape (B. napus L.)	Tower (B. napus L.)	Candle (B. campestris L.)	Erglu (B. napus L.)	DP-075 (B. napus L.)
Gluconapin	8.7[b]	14.80	2.82	2.63	2.08	3.93
Glucobrassicanapin	10.2[b]	7.50	0.93	2.03	1.10	0.48
Progoitrin	10.0[c]	38.88	4.50	3.00	3.72	2.78
Napoleiferin	20.2[c]	7.74	0.78	0.90	0.30	0.02
Gluconastiin	16.7[b]	3.15	0.65	0.21	0.62	0.09

Corresponding results from GLC-analysis of isothiocyanates (b) and oxazolidinethiones (c).

Olsen, O. and Sørensen, H. 1980. J. Agr. Food Chem. 28, 43-48.

may be a result of genetic variation, growth conditions and/or developmental stage of the seeds. It is thus evident that the solution of the problems related to glucosinolates require methods which are able to detect all of the individual glucosinolates.

The separation of glucosinolates by reversed-phase ion-pair liquid chromatography is a new rapid and gentle method which leads to a simple, sensitive and fast analytical technique in combination with the new isolation procedure (Figures 3 and 4). The HPLC-method is not yet fully developed for quantitative analysis and some glucosinolates have not yet been studied, but the preliminary results are very promising (Figure 8).

In conclusion, our results have shown that quantitative glucosinolate analysis methods based on myrosinase- or acidic-catalysed degradation of the glucosinolates are unreliable for a lot of the known glucosinolates. The method based on isolation of intact glucosinolates by the described ion-exchange technique followed by GLC-analysis of the TMSi-derivatives of desulpho-glucosinolates is much more reliable for all of the glucosinolate types shown in Table 1, especially for those with β-hydroxylated aromatic and/or hydrophilic R-groups, but not for those with sulphur in the R-group. The latter seem, however, to be efficiently determined in the method based on HPLC-determination of individual intact glucosinolates.

DISCUSSION

G. Röbbelen *(FRG)*

I think there is agreement among all the people working in this field that we face a very complex situation as regards glucosinolates. There is total agreement in respect to the myrosinase: it needs to be inactivated. There have been tests using enzyme techniques for more or less complete degradation: for example, to find nitriles as the breakdown product and simply analyse on this level because almost all the glucosinolates are broken down in this way, but, as Dr. Sørensen said, this is not accurate enough. With the feeding experiments done so far there is still some risk of toxicity with the low glucosinolates. It was emphasised by Dr. Sørensen that the OO-types are a big step forward and their introduction into agriculture is a very positive aspect. But there is still this risk. It is different in different countries; some countries measure the quality of the carcass according to liver syndromes, etc.

There are two points of controversy. In his talk Dr. Sørensen said that all his previous experimental methods proved to be impossible, unreliable or unsafe. I cannot agree. It was emphasised that the glucose determination has some advantages, but I would say that, under certain conditions, all the others which we were given have some value. Under other conditions you may destroy what you have found: your data may be biased. This is due to sinalbin, for example. If you do desulphatation on a column, this is done at a pH of 6.2 or 6.5 so there is no acid degradation. If we use the technique of desulphatation, which was described as an error, we can measure accurately the sinalbin content in *Sinapis arvensis*. Therefore, there are other techniques available which, under certain conditions of accuracy, can also give good results.

The second point of concern is the amount of additional glucosinolates in the low, OO-varieties. There are about 15 mmols of glucosinolate known to be left in the OO-varieties and indications of about 9 mmols additional glucosinolate, which may be sinalbin. Therefore, the total content may not be 10 or 15 but nearer 24 mmols. Despite the fact that this is possible we could not trace it. We could trace the big amounts of sinalbin in *Sinapis arvensis*. When I say 'we', I mean Dr. Fenwick of the UK, Dr. Thies in our laboratory and Professor Gmelin, who was a student of Virtanen - who works with a different technique. He uses thin-layer chromatography and not gas chromatography. In an analytical seminar at which Dr. Sørensen and Dr. Thies were present, it was argued that in the thin-layer chromatography of Professor Gmelin aluminium oxide is used and this can also destroy the glucosinolates. This was taken up with Professor Gmelin; he wrote to me on the first of this month to say that he has discarded aluminium oxide and he has used a technique of cellulose thin-layer chromatography. Again he did not find high amounts of sinalbin in rapeseed, although they are easily detected in *Sinapis arvensis*. However, what was found was a compound on the same RF value with a rather high amount of something still undetermined (not sinalbin) and which could show a bias in analytical techniques. Therefore, I would say the question is still open in this regard. This is true for the analytical procedure, and also for the specific question of whether appreciable amounts of additional glucosinolate are still causing the problems with the present low varieties.

H. Sørensen *(Denmark)*

Our interest is in the methods used in connection with glucosinolate analyses. I agree it is necessary to have knowledge of different methods. Determination of glucose liberated during myrosinase action - as you described - is the basis for an easy and fast method which is certainly of great value for the plant breeders. However, it gives only <u>a measure</u> of the total glucosinolate content. There are a lot of interference possibilities and other methods are necessary in order to have reliable quantitative determinations of individual glucosinolates. I also want to mention that an important improvement of this method has been described by Dr. I. McGregor, Saskatoon.

Concerning the myrosinase degradation products and quantitative determinations of the products containing the R-group, it is obviously the isothiocyanates that are the initial products. However, owing to the reactivity and instability of some types of isothiocyanates, the final products obtained in this way are varying mixtures of a great number of different products. Therefore, procedures based on these products are generally suspect if we want a quantitative determination of all individual glucosinolates. If the purpose is qualitative analysis or identification of only some of the glucosinolates, it is possible to use GC-MS analyses of these myrosinase products with great advantage, e.g. glucosinolates derived from methionine (R corresponding to $CH_3-S-(CH_2)_m-$; $CH_3SO-(CH_2)_m-$; $CH_3SO_2-(CH_2)_m-$) apparently gives some problems in the methods based on GC analyses of TMSi derivatives of desulphoglucosinolates, whereas at least some of them are easily determined as isothiocyanates, as shown in the excellent paper of Dr. Daxenbiehler et al., J. Agric. Food Chem., 25 (1977) 12. The oxazolidinethiones are apparently undetermined in this procedure compared with the method based on GC of the TMSi derivatives of desulpho-glucosinolates, and some isothiocyanates are so unstable that it is nearly impossible to detect these compounds in methods based on isothiocyanates (UV of thiourea derivatives and GC-MS of isothiocyanates). The problem with quantitative determination of glucosinolates containing sulphur in the side chain, if we use the TMSi derivatives, is perhaps related to the silylation process. If this is the case, the problems may possibly be overcome if we use methods based on HPLC of intact glucosinolates as discussed in my paper. This method seems to be an important supplement to the method I described, based on GC of TMSi derivatives of desulpho-glucosinolates after purification by ion-exchange technique.

M. Dambroth *(FRG)*

Thank you. I am afraid there is no time for further questions. Perhaps there will be an opportunity later.

CONTENT AND PATTERN OF GLUCOSINOLATES IN RESYNTHESISED RAPESEED

Astrid Gland
Institute of Agronomy and Plant Breeding
Georg-August-University, Göttingen
Federal Republic of Germany

ABSTRACT

The utilisation of rapeseed (Brassica napus) *as a fodder plant is limited by its content of glucosinolates. This is true for the extracted meal when used as a protein concentrate as well as for the vegetative parts when fed in green or ensilaged condition.*

The five most important glucosinolates (sinigrin, gluconapin, glucobrassicanapin, progoitrin and gluconapoleiferin), which are present in all members of the Brassiceae, *were analysed by a gas chromatographic method developed by Dr. Thies (1977, 1978) in our laboratory.*

In vivo, *the glucosinolates are hydrolised by the endogenous enzyme myrosinase after cell lesion, in which case glucose and sulphate are released besides isothiocyanates or oxazolidinethiones. It has long been known that the oxazolidinethiones (progoitrin and gluconapoleiferin) have a greater inhibitory effect on thyroid gland function than the isothiocyanates have. In rapeseed* (Brassica napus) *about two-thirds of the total glucosinolate content are oxazolidinethiones; in seeds of turnip rape* (B. campestris) *these compounds amount to one-tenth only.*

Within the B. napus *species little variation is known in content and pattern of glucosinolates. Worldwide, breeders rely on one genotype only, the Polish variety 'Bronowski', for their breeding of low glucosinolate rapeseed. But closely related* Brassica *species, especially the diploid ancestors of* B. napus, B. oleracea *and* B. campestris, *show a great many differences not only in the total glucosinolate content but also in the proportion of individual glucosinolates.*

Therefore, it seemed possible to select special genotypes of these two species and to construct new rapeseed cultivars by interspecific crossing and polyploidisation of the amphihaploid hybrids. The following report will present some of our results in this direction (for details see Gland, 1980).

MATERIAL AND METHODS

Glucosinolate content and pattern were determined by gas chromatography for more than 700 genotypes available from the worldwide *Brassica* collection of the Institute in Göttingen. For hybridisation 102 different genotypes of *B. oleracea* and *B. campestris* were selected.

RESULTS AND DISCUSSION

Within the species *Brassica oleracea*, 367 varieties were analysed. Examples of the data are given in Table 1. It is evident that large differences exist, both in total content and in the pattern of glucosinolates present. Glucosinolate content ranged from 1 - 220 µmol/g defatted meal. Sinigrin and progoitrin contents also varied widely, as might be expected since cabbage has been selected by human taste as a vegetable for millenia.

Similarly complex variability was found within the 256 *B. campestris* forms which we analysed and which are also widely used as salad and vegetable crops in Eastern countries. Total glucosinolate content ranged from 8 to 200 µmol/g defatted meal. One important difference is the absence of sinigrin in *B. campestris* as compared to *B. oleracea* and *B. nigra*. Generally, gluconapin is the predominant compound in *B. campestris* genotypes.

In seeds of *B. napus*, glucosinolate content ranged from 10 to 250 µmol/g defatted meal. Progoitrin represents about two-thirds, and gluconapin about one-quarter of the glucosinolate content.

A few cultivars of swede showed an extremely high content of progoitrin, accounting for up to 85% of total glucosinolate content. Sinigrin was not found in any of the analysed samples; thus, *B. napus* is the second *Brassica* species to lack this glucosinolate.

Development work with most of the interspecific crosses between *B. oleracea* and *B. campestris* is still in progress, but the

TABLE 1
GLUCOSINOLATE PATTERN (CONTENT OF A SINGLE GLUCOSINOLATE IN % OF THE TOTAL CONTENT OF ANALYSED GLUCOSINOLATES) AND TOTAL GLUCOSINOLATES (IN µmol/g DEFATTED MEAL).

Collection No.	Cultivar	SIN	GNA	GBN	PRO	2-4	Total glucosinolates
1. Cabbage, *Brassica oleracea*							
525	White cabbage	92.5	2.9	0.7	3.7	0.2	45.6
575	Red cabbage	18.5	4.6	0.1	75.8	0.1	76.6
2184	Savoy cabbage	93.9	1.6	0.1	2.5	2.0	76.5
182	Brussels sprouts	36.0	8.6	0.2	54.9	0.2	45.5
2902	Cauliflower	95.8	1.5	0.2	2.4	0.2	61.8
2345	Broccoli	6.7	6.7	6.7	73.3	6.7	1.5
513	Broccoli	0.5	18.0	1.3	76.4	3.7	129.3
450	Kale turnip	71.1	13.3	2.2	11.1	2.2	4.5
2155	Curly kale	64.6	7.4	0.4	26.0	1.5	91.4
2243	Fodder borecole	60.0	9.3	0.7	27.7	2.3	150.8
2150	Brassica alboglabra	24.1	14.8	0.6	58.2	2.3	154.8
2. Turnip, *Brassica campestris*							
2473	Chinese cabbage	-	85.4	12.2	2.0	0.3	29.5
35	Chinese cabbage	-	90.2	8.5	1.3	0.1	151.1
849	Yellow sarson	-	97.8	2.1	0.1	0.1	205.9
2016	Turnip rape	-	34.1	25.0	37.5	2.3	8.8
76	Turnip rape	-	34.9	29.0	28.4	7.7	88.6
2130	Common turnip	-	65.5	6.2	26.5	1.8	98.5
3. Rapeseed, *Brassica napus*							
109	Rapeseed (Bronowski)	-	26.5	6.2	65.5	1.8	16.0
2499	Rapeseed (O559,Spain)	-	26.1	4.5	65.7	3.8	73.4
2534	Rapeseed (Diamant)	-	21.3	5.3	70.1	3.3	156.1
2682	Rutabaga (York)	-	0.4	0.4	99.1	0.1	167.9

SIN = Sinigrin GNA = Gluconapin GBN = Glucobrassicanapin
PRO = Progoitrin 2-4 = Gluconapoleiferin

first 35 combinations of resynthesised diploid rapeseed have just been analysed (Table 2). In these combinations the content of glucosinolates ranged from 66 to 196 µmol/g defatted meal and was as high as in any common rapeseed cultivar. With respect to the glucosinolate pattern the single components were also present in amounts similar to those from natural rapeseed. However, all the hybrids except 'H 474' contained sinigrin in amounts representing 7 - 15% of the total glucosinolate content.

TABLE 2

GLUCOSINOLATE PATTERN AND CONTENT IN SEEDS OF RESYNTHESISED RAPESEED FORMS

♀		♂		Hybrid plants					
Collection No.	Total glucosinolates	Collection No.	Total glucosinolates	No.	SIN	GNA	GBN	PRO	2-4	Total glucosinolates

Coll-ection No.	Total gluco-sino-lates	Coll-ection No.	Total gluco-sino-lates	No.	SIN	GNA	GBN	PRO	2-4	Total gluco-sino-lates
513	129.3	35	151.1	H474	–	29.1	3.4	66.1	1.5	143.0
525	45.6	2371	111.7	H 7	10.1	13.1	7.0	66.8	3.0	195.8
585	89.8	29	91.5	H485	14.6	13.1	6.5	63.0	2.7	66.3
613	25.1	170	86.1	H523	7.4	18.1	3.8	68.2	2.2	142.0
2152	31.4	2366	62.9	H865	10.5	33.4	7.8	46.7	1.8	73.4
2179	70.5	35	151.1	H829	14.0	14.2	6.0	63.3	2.5	176.2
2288	130.8	170	86.1	H878	9.7	21.3	2.9	63.9	2.2	120.4
2330	158.7	454	119.7	H 65	11.5	29.7	9.2	46.2	3.4	121.5

SIN = Sinigrin   GNA = Gluconapin   GBN = Glucobrassicanapin
PRO = Progoitrin   2-4 = Gluconapoleiferin

The glucosinolate patterns in two different hybrids, derived from the pollination of two *B. oleracea* by one *B. campestris* cultivar, are shown in Figure 1. The female plants were extremely different. No. '513' represents the progoitrin-type and No. '2179' the sinigrin-type. Nevertheless, both hybrid plants show a rather similar glucosinolate pattern.

Although we appear to have found a much greater variation in glucosinolate content and pattern in seeds of the *Brassica* species than has been reported before, most of the resynthesised rapeseed forms we have made so far show a glucosinolate pattern similar to any natural rapeseed cultivar. Whether the parents were of the pure sinigrin or the gluconapin type, and whether or not they also contained progoitrin, no similarity with these progenitors was detected in the new rapeseed hybrids. This indicates strong interactions between the two different parental genomes, which evidently contribute to the uniformity of the glucosinolate pattern typical of present rapeseed cultivars.

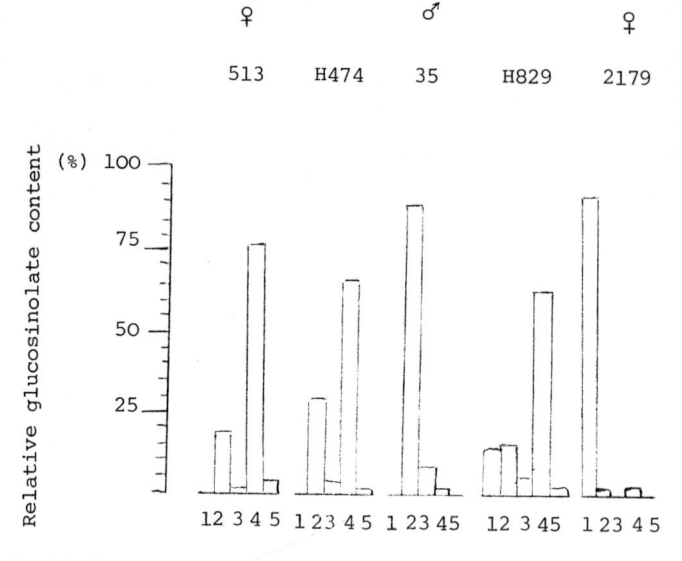

1 Sinigrin
2 Gluconapin
3 Glucobrassicanapin
4 Progoitrin
5 Gluconapoleiferin

Fig. 1   Glucosinolate pattern of resynthesised rapeseed forms as compared to their progenitors. The crosses involved two different *B. oleracea* females and one *B. campestris* cultivar as pollen donor.

REFERENCES

Gland, Astrid. 1980. Glucosinolatgehalt und -muster in den Samen resynthisierter Rapsformen. Diss. Landw. Fak., Göttingen

Thies, W. 1977. Analysis of glucosinolates in seeds of rapeseed *(Brassica napus* L.): Concentration of glucosinolates by ion exchange. Z. Pflanzenzüchtg. 79, 331-335

Thies, W. 1978. Quantitative analysis of glucosinolates after their enzymatic desulfatation on ion exchange columns. Proc. 5th International Rapeseed Conference, Malmö, Sweden, 1978. Vol. 1, 136-139

DISCUSSION

N.J. Mendham (UK)

Have you not used something like the broccoli, with very low glucosinolates, for crossing with the turnip which is also very low; what sort of result would you expect there?

A. Gland (FRG)

Yes, we tried to do it, but in the first year we were unsuccessful in getting embryoids from this combination. In 1979 we got some, but we have not analysed the seeds; this is still underway.

D.G. Morgan (UK)

I would like to ask a physiological question. There has been much talk about the glucosinolates this morning; they are there in large numbers and in different forms. What function do they have within the plant and in plant metabolism? Presumably they do have a role. And where are they formed within the plant?

A. Gland

Not much is known about their function within the plant. It is thought that they may protect it from insects or predators but the information is slight.

D.G. Morgan

What worries me is that you are selecting against them.

G. Röbbelen (FRG)

I think this is a question that affects any secondary plant metabolite. There are secondary effects which you sometimes get with respect to the glucosinolates. What Dr. Gland was showing was glucosinolate in the seed. Glucosinolate patterns in leaves may be different, as may be the case in roots as well. I know of some experiments with nematodes with respect to Plasmodiophora and it appears that there may be some effects but the data so far are contradictory. Some say they attract some of these pests; others say they repel them - it is debatable. One can only say that we are aware that there are effects.

Again, the question of accumulation and metabolism is open, and I would not be prepared to say where they are synthesised or at which organ or cell; we do not even know the biosynthetic steps from methionine. There is a very large gap in our knowledge in this story.

B.O. Eggum - Chairman *(Denmark)*

Thank you, Dr. Gland. Now we will proceed to the next paper.

BREEDING ASPECTS OF SUNFLOWER IN MIDDLE EUROPE

W. Schuster and I. Kübler

Institut für Pflanzenbau und Pflanzenzüchtung,
der Universität Giessen, Ludwigstrasse 23,
D-6300 Giessen, Fed. Republic of Germany.

ABSTRACT

The sunflower represents a new oil and protein crop for most parts of Europe. By breeding adapted genotypes, especially by developing hybrids on the basis of natural or chemically induced male sterility, highly productive varieties can be produced. For this purpose it is important to select material with only a slight short-day response (that is, either day-neutral or even long-day types), low temperature needs and good combining ability.

Furthermore, one must pay attention to disease resistance, or tolerance, to give yield stability. In addition to factors of yield, seed quality is significant; high contents of both oil and protein are wanted, the latter being important for improving the value of sunflower meal. Oil composition, which is most propitious under cooler conditions, at present, would be favourably influenced by the selection of types which have high linoleic acid content even when grown under warm conditions.

INTRODUCTION

In Europe interest in production of sunflower as an oilseed crop has increased substantially in recent years, because of its versatility as an oil and protein plant. Accordingly, cultivation has expanded and should continue to increase in the future. In 1978, the total acreage in Europe exceeded 1.9M ha, with an everage yield of 14.1 dt/ha and a production of 2.6M t (FAO Production Yearbook, 1978). This year, for the first time, sunflowers are being grown in Germany for oilseed production, and 50 ha have been sown.

The merits of sunflower are based on the high physiological value of the oil and the high biological value of the protein, both of which are important for use in human and animal nutrition (Robertson, 1972; Seibel, 1978). The hull percentage is a limiting factor in utilisation of the meal. In addition to selection for high yield, and increasing oil and protein content, it is important to select for low hull percentage (Schuster et al., 1980). Under European conditions highest yields and quality are attainable from types adapted to long daylength and relatively low temperature, but with sufficient drought resistance for south European conditions. On the basis of investigations at our Institute, it shall be demonstrated that it is possible to find new types adapted to different environments by means of existing material and breeding new varieties.

MATERIAL AND METHODS

Information about material and methods will be found in previous publications: Boye, 1970; Seibel, 1978; Schuster et al., 1980; Grauert, 1979.

RESULTS

In greenhouse trials (Boye, 1970) the development of 10 different sunflower varieties was studied under two temperature and two daylength regimes. Figure 1 depicts the accumulated temperature (sum of mean day temperature) at different stages of growth from germination to the beginning of flowering. (D-stage = flower-initiation; B_1 = formation of flower primordia; third stage = floral initiation). There is a clear difference between varieties in temperature needs.

The variety Vniimk 8931 has a distinct tendency to long-day reaction, showing a decrease in accumulated temperature requirement with increasing daylength, expecially in the D- and B_1-stages. On the other hand, temperature needs under long-day conditions were higher for other varieties, especially for Kiswardai, Borowski and the forage sunflower FW 436/59, which have pronounced short-day characteristics.

The course of growth and development as a function of temperature and photoperiod is presented in Figure 2. High temperature accelerates the rate of development and growth in both long-day and short-day conditions, while low temperature has a negative influence. Growth in height is also delayed under short-day conditions.

In addition to the greenhouse trials, Boye had field trials to study the effects of date of sowing at different locations. The temperature needs of 16 sunflower varieties during early growth from germination to D-stage in two years at two locations are reported in Figure 3. There are distinct interactions of the varieties Kort Russ and FW 436/59 in sowing time, years and locations, while Jupiter, Vniimk 8931 and Hesa have a better adaptability to changing environment. These investigations show that there is great variability within existing material in response to temperature and photoperiod, and this variation can be used to breed varieties adapted to differing environments.

In a worldwide variety test, numerous sunflower varieties were tested for yield and quality at different locations (Seibel, 1978). Seed yield in dt/ha is reported in Table 1.

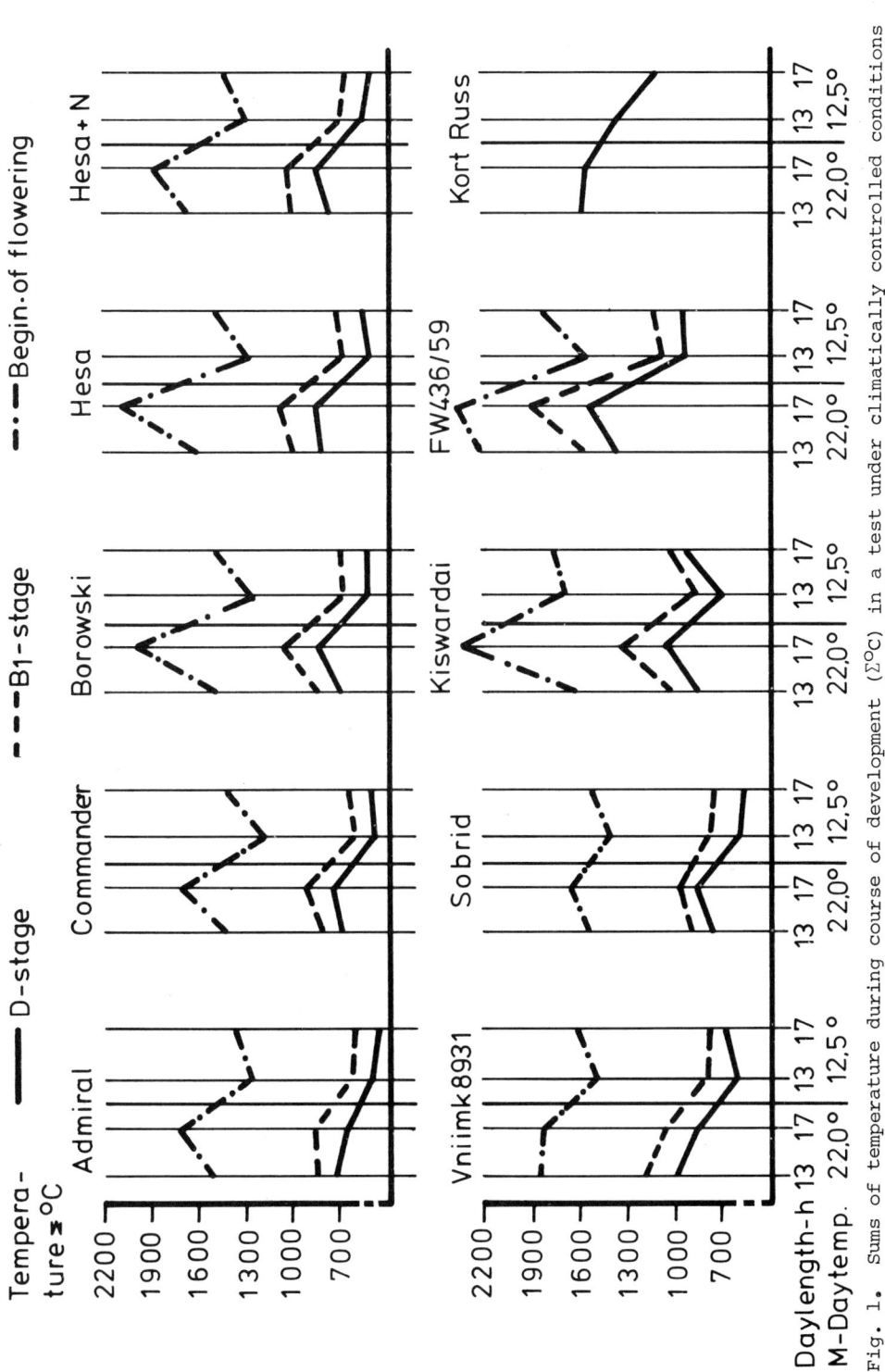

Fig. 1. Sums of temperature during course of development (Σ°C) in a test under climatically controlled conditions of 10 sunflower varieties at 2 temperatures (12.5°C, 22.5°C) and 2 daylengths (13h, 17h) (from Boye, 1970).

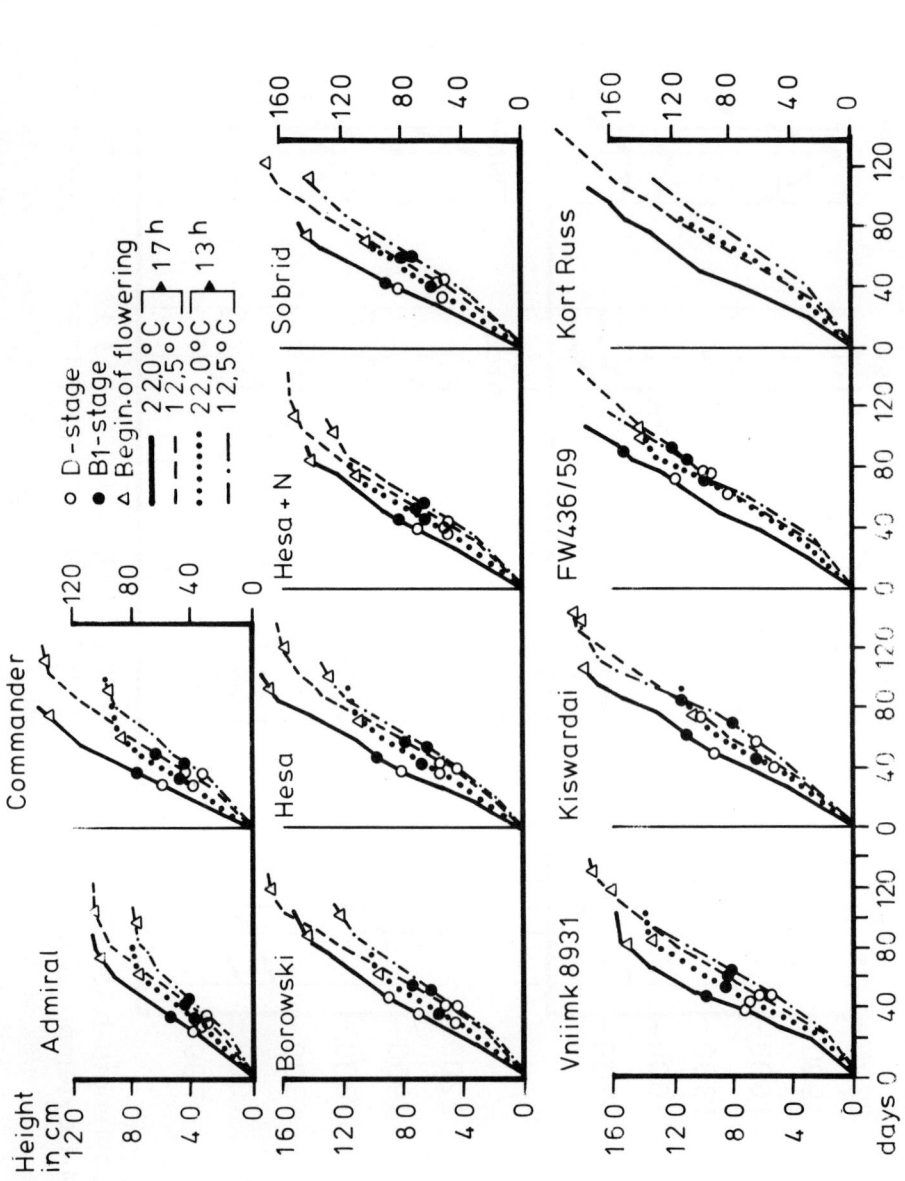

Fig. 2. Course of growth and development in test under climatically controlled conditions of 10 sunflower varieties at 2 temperatures (12.5°C, 22.5°C) and 2 daylengths (13h, 17h) (from Boye, 1970)

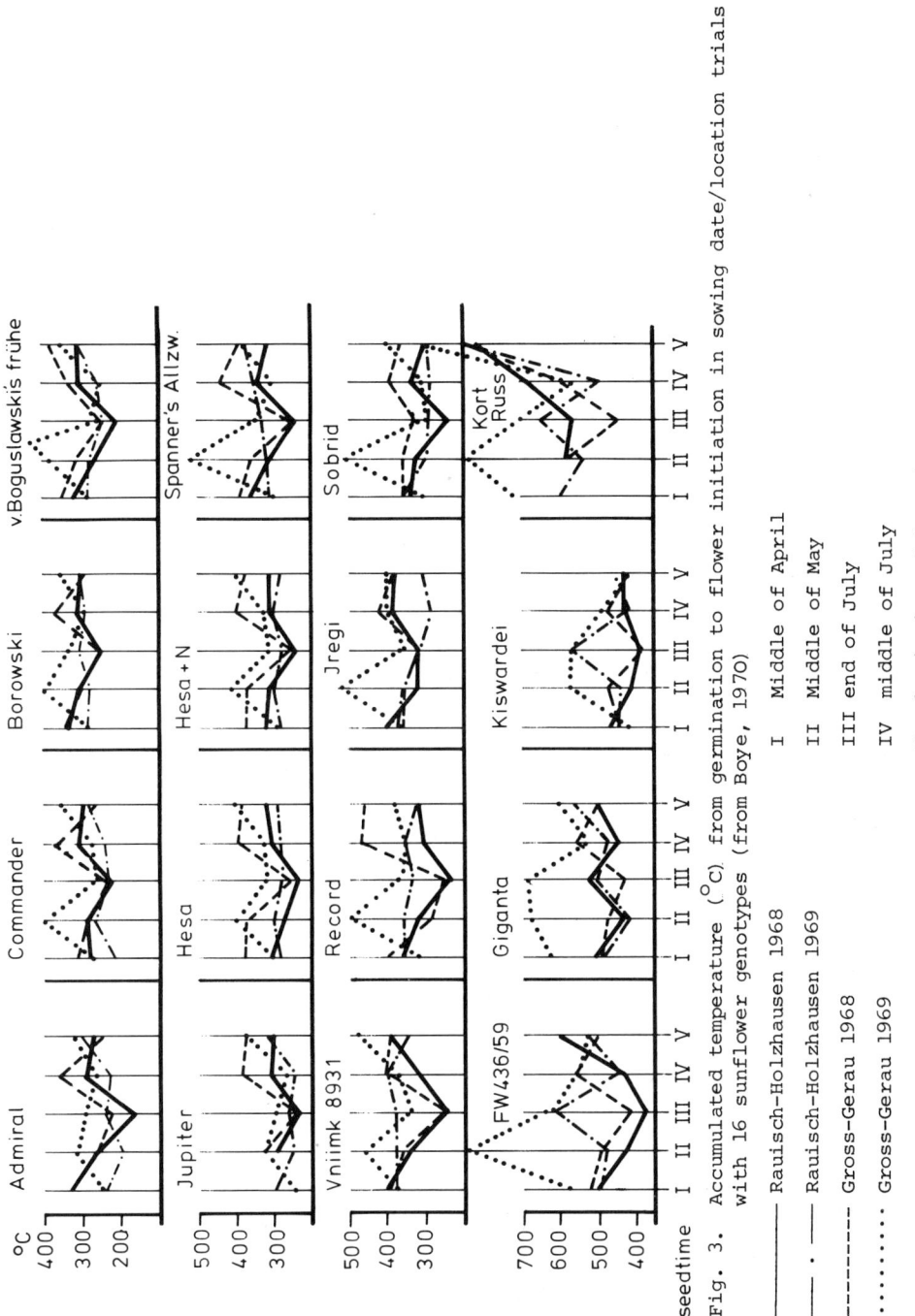

Fig. 3. Accumulated temperature (°C) from germination to flower initiation in sowing date/location trials with 16 sunflower genotypes (from Boye, 1970)

——— Rauisch-Holzhausen 1968
—·— Rauisch-Holzhausen 1969
------- Gross-Gerau 1968
········ Gross-Gerau 1969

I Middle of April
II Middle of May
III end of July
IV middle of July
V beginning of August

TABLE 1

SEED YIELD (dt/ha). WORLDWIDE VARIETY TESTS WITH SUNFLOWER 1973 AND 1974 (FROM SEIBEL, 1978)

Variety Origin Location	VNIIMK 8931 SU	Pere- dovik SU	cmsHA89x RHA266 USA	Valley CDN	HS 52 R	INRA 4701 F	INRA 6501 F	Sobrid D	Sorex D	Mean
Morden/CDN	18.4	18.0	47.1	20.5	42.1	15.6	16.2	21.3	30.4	25.5
Gross-Gerau/D	33.7	36.8	35.1	38.0	37.7	30.6	34.8	40.4	45.3	36.9
Novi-Sad/YU	42.3	40.1	39.3	30.6	38.4	33.5	35.9	36.1	30.6	36.3
Bornova/TR	19.0	18.6	7.0	15.8	16.6	6.9	6.5	13.2	17.8	13.5
Karadj/IR	32.1	31.5	29.9	31.8	33.0	31.1	30.6	29.8	30.1	31.1
Lincoln/NZ	20.1	23.6	28.0	26.4	27.5	20.2	24.6	22.8	25.4	24.3
Zevenaar/NR*	37.5	41.5	28.3	45.5	35.0	33.4	24.8	37.3	43.9	36.4
Rabat/MA*	16.3	14.0	11.6	8.5	17.2	5.0	10.4	14.5	12.7	12.2
Hayatnagar/IND*	8.3	9.1	8.5	3.4	6.8	5.6	4.6	4.5	4.5	6.4
Bako/ETH*	42.7	34.9	48.8	36.3	35.8	37.3	26.4	34.0	43.8	37.8
Mean	27.6	28.1	31.1	27.2	32.6	23.0	24.8	27.3	29.9	27.9
s %	44.64	43.29	53.45	52.49	40.17	58.56	43.25	47.38	48.50	

* Location not included in mean

The different ecotypes show different performances at the individual locations. The varieties Vniimk and Peredovik, like HS 52, are quite consistent in seed yield at all locations and are therefore very adaptable, but great variations in seed yield were obtained with the varieties Inra 4701 and Inra 6501. The interactions between varieties and locations were statistically significant.

Oil content of the seed (Table 2), as a mean over all locations, varied from 32.1% for Sobrid to 46.4% for Peredovik. Sobrid and Sorex are varieties bred for forage use in Central Europe and therefore have a low oil content. Great differences in oil content were found at the different locations. The highest oil contents were at locations with low temperatures, while low oil contents were found in the warm locations, Bornova and Karadj.

Crude protein content in seed, presented in Table 3, increases with higher temperature. As a mean over all locations, the difference between varieties was small, (19.1% Sobrid - 20.4% Inra 4701). The French varieties have highest protein contents, while the Russian varieties with high oil content have low protein content, relating to the negative correlation between oil and protein content.

In addition to yield, oil content and protein content, the quality of oil has also been investigated, and data are presented (Tables 4 and 5) on linoleic acid and tocopherol content of the oil in seeds from plants pollinated within the varieties. Varietal differences in linoleic acid content were small; as a mean of all locations it varied from 62.1% (Inra 6501) to 67.5% (Inra 4701). The effect of location was much greater. Under the cooler conditions in Morden and Gross-Gerau, linoleic acid contents of 70 - 80% were produced, while under warm cultivation conditions only 40 -50% linoleic acid was found.

TABLE 2

OIL CONTENT IN SEED IN % ADM 1973. WORLDWIDE VARIETY TESTS WITH SUNFLOWER 1973 AND 1974 (FROM SEIBEL, 1978).

Variety Origin Location	VNIIMK 8931 SU	Pere- dovik SU	cmsHA89x RHA266 USA	Valley CDN	HS 52 R	INRA 4701 F	INRA 6501 F	Sobrid D	Sorex D	Mean
Morden/CDN	42.5	45.6	45.7	39.0	42.8	37.1	30.7	30.7	34.0	40.0
Gross-Gerau/D	47.0	47.4	52.1	42.3	50.2	48.7	47.0	36.4	42.0	45.9
Novi-Sad/YU	48.4	50.3	43.5	31.0	51.0	39.8	44.6	32.0	38.9	42.2
Bornova/TR	41.0	40.2	43.4	32.8	38.5	41.4	35.2	27.1	31.9	36.8
Karadj/IR	41.3	42.7	46.8	39.4	42.7	36.0	40.3	29.7	37.0	39.3
Lincoln/NZ	43.5	46.5	48.8	42.4	47.4	43.2	40.8	33.3	38.1	42.7
Zevenaar/NL*	47.5	48.3	48.5	41.4	47.2	45.6	42.1	33.4	36.5	43.4
Rabat/MA*	-	-	-	-	48.5	45.1	45.0	35.1	37.6	-
Hayatnagar/IND*	46.5	49.2	45.5	36.9	45.5	45.4	41.5	35.0	41.7	43.0
Bako/ETH*	49.1	50.5	48.1	29.2	46.9	44.0	43.8	34.0	39.8	42.8
Mean	45.0	46.4	47.1	37.2	45.8	42.6	41.4	32.1	37.3	41.7

* Location not included in mean

TABLE 3

CRUDE PROTEIN CONTENT IN % ADM IN UNEXTRACTED SEEDS 1973. WORLDWIDE VARIETY TESTS WITH SUNFLOWER 1973 AND 1974 (FROM SEIBEL, 1978)

Variety Origin Location	VNIIMK 8931 SU	Pere- dovik SU	cmsHA89x RHA266 USA	Valley CDN	HS 52 R	INRA 4701 F	INRA 6501 F	Sobrid D	Sorex D	Mean
Morden/CDN	18.6	20.8	20.3	22.3	19.0	20.3	21.6	19.9	21.9	20.5
Gross-Gerau/D	18.7	19.0	20.9	16.3	18.8	18.1	18.9	16.7	16.2	18.2
Novi-Sad/YU	19.2	19.2	20.6	20.5	18.7	22.8	19.6	18.0	18.0	19.6
Bornova/TR	22.8	21.4	23.1	22.4	23.4	21.7	22.3	21.7	20.8	22.2
Karadj/IR	21.5	20.8	19.9	21.3	21.4	22.1	21.2	21.0	21.6	21.2
Lincoln/NZ	18.3	17.9	16.2	18.6	17.9	17.4	18.2	17.4	17.6	17.7
Zevenaar/NL*	19.9	19.5	20.2	20.7	19.3	19.5	20.1	18.9	19.4	19.7
Rabat/MA*	-	-	-	-	20.2	21.0	20.7	19.4	19.8	-
Hayatnagar/IND*	24.2	24.1	23.5	22.3	23.2	23.4	23.7	20.0	19.9	22.7
Bako/ETH*	20.7	18.6	19.1	22.4	19.7	19.6	20.1	16.5	17.3	19.3
Mean of ortho- gonal locations	19.9	19.9	20.2	20.2	19.9	20.4	20.3	19.1	19.4	19.9
(Mean of all locations)	(20.0)	(19.7)	(20.0)	(20.6)	(19.8)	(20.2)	(20.3)	(18.8)	(19.1)	(19.8)

* Location not included in mean

TABLE 4

LINOLEIC ACID AS % OF TOTAL FATTY ACIDS IN OIL FROM SEEDS OF ISOLATED POLLINATED PLANTS. MEAN OF 2 YEARS AT 6 LOCATIONS IN WORLDWIDE VARIETY TESTS WITH SUNFLOWER 1973 AND 1974 (FROM SEIBEL, 1978)

Variety	VNIIMK 8931	Pere-dovik	cmsHA89x RHA266	Valley	HS 52	INRA 4701	INRA 6501	Sobrid	Sorex	Mean
Origin	SU	SU	USA	CDN	R	F	F	D	D	
Location										
Morden/CDN	77.5	78.5	78.2	76.4	77.6	79.2	73.1	76.1	77.8	77.2
Gross-Gerau/D	73.1	74.5	72.2	72.6	73.3	76.4	69.6	73.4	73.4	73.2
Novi-Sad/YU	64.4	67.1	64.5	66.5	69.2	70.0	63.7	65.5	68.3	66.6
Bornova/TR	44.6	45.9	47.0	46.1	42.2	47.1	47.3	45.5	50.4	46.2
Karadj/IR	52.4	53.9	57.1	52.4	54.1	57.5	51.1	54.2	53.1	54.0
Lincoln/NZ	72.7	74.4	75.6	70.9	73.7	74.8	67.8	72.6	73.6	72.9
Mean	64.1	65.7	65.8	64.2	65.0	67.5	62.1	64.6	66.1	65.0

TABLE 5

TOCOPHEROL CONTENT AS mg/kg OIL IN SEEDS OF ISOLATED POLLINATED PLANTS. MEAN OF 2 YEARS AT 6 LOCATIONS IN WORLDWIDE VARIETY TESTS WITH SUNFLOWER 1973 AND 1974 (FROM SEIBEL, 1978)

Variety Origin Location	VNIIMK 8931 SU	Pere- dovik SU	cmsHA89x RHA266 USA	Valley CDN	HS 52 R	INRA 4701 F	INRA 6501 F	Sobrid D	Sorex D	Mean
Morden/CDN	766	747	1030	765	729	777	616	693	742	763
Gross-Gerau/D	537	495	373	334	241	428	487	419	355	408
Novi-Sad/YU	650	702	810	663	594	668	665	593	642	665
Bornova/TR	1000	972	1048	905	781	805	845	687	754	867
Karadj/IR	633	619	845	598	564	656	637	441	651	627
Lincoln/NZ	621	624	722	684	732	618	588	471	580	627
Mean	701	693	805	658	607	659	640	551	621	659

Besides fatty acid composition, the tocopherol content (vitamin-E-content) is an important quality characteristic of sunflower oil. The influence of temperature is again higher than the influence of varieties. In a cold location, Gross-Gerau, all varieties had a low tocopherol content of about 400 mg/kg oil, while tocopherol content increased up to 1 000 mg/kg oil in the warm location, Bornova.

Results from these investigations show that in yield, and especially in quality characteristics, there are considerable differences between sunflower varieties and between locations. But there is also a large variability within varieties and lines (Schuster et al., 1980). Figure 4 depicts the variation in linoleic acid content of individual seeds of different inbred lines of the variety Vniimk. The Figure shows that, as far as linoleic acid is concerned, the differences between inbred lines from the same population are sufficient to be used in selection.

Disease resistance of sunflower is very important for cultivation in Europe. Such diseases as *Plasmopara halstedi*, *Puccinia helianthi*, *Peronospora helianthi* and *Alternaria helianthi* are not, as yet, important, but immense loss of yield is caused by *Sclerotinia sclerotiorum de Bary* and *Botrytis cinerea*. Two stages of infection are of importance; the first is infection of the stem during germination, and the other is infection of the head during the long period from flowering to maturity.

Infestation values of field and greenhouse tests of 10 varieties and lines are reported in Table 6 (Grauert, 1979). It is interesting to note that some inbred lines showed better tolerance, as represented by lower infestation. Infestation of the line Sf 83/73 was 12.17% in the field trial and 45.5% in the greenhouse trial, that of line PK 104/75 only 6.37% in the field trial and none in the greenhouse, while varieties Inra 7702 Sorex and Hs 301 showed high rates of infestation in both trials.

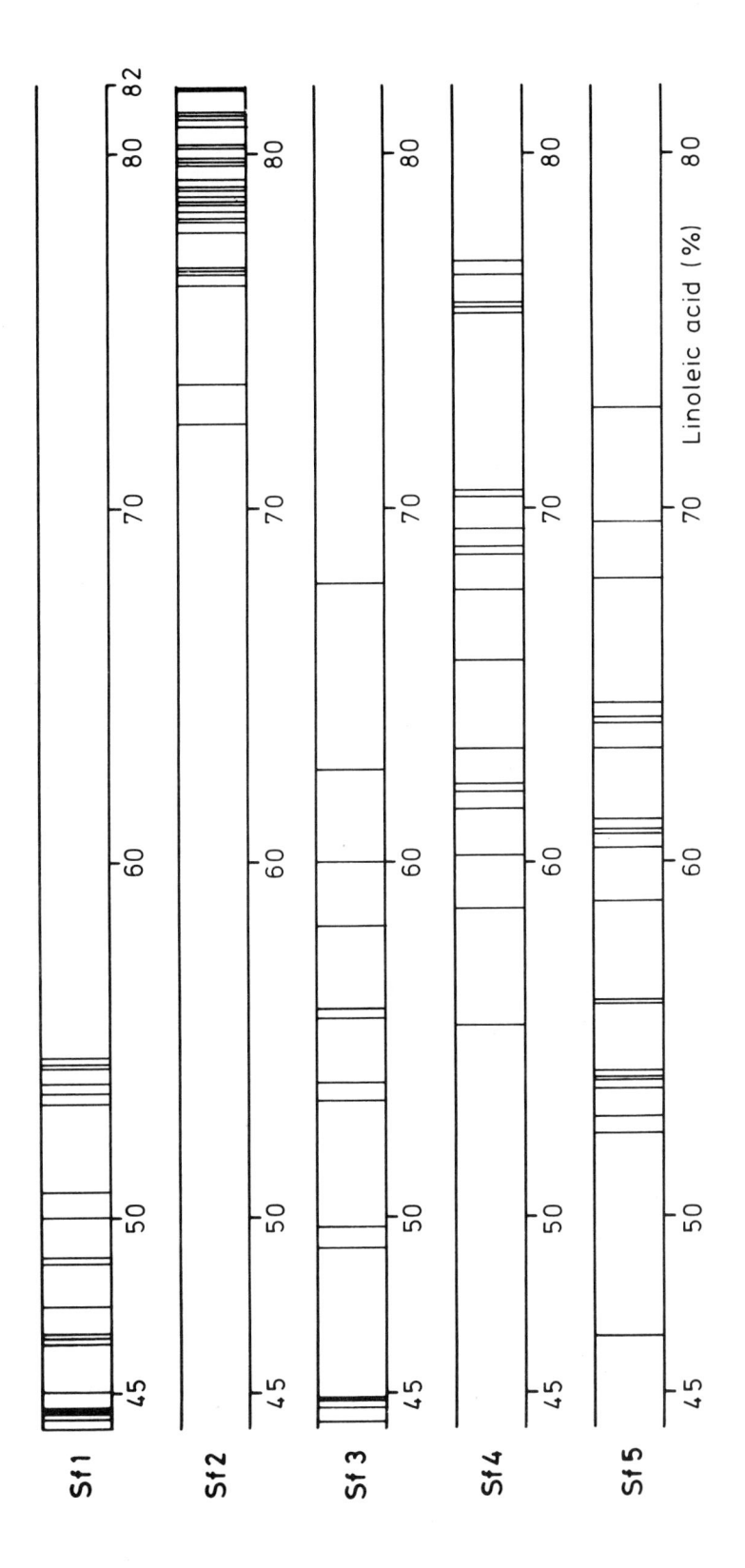

Fig. 4. Variability of linoleic acid within selfings I_2 and I_3 of Vniimk 8931.

TABLE 6

RESPONSE OF 10 VARIETIES AND LINES TO HIGH SCLEROTINIA INFECTION PRESSURE (FROM GRAUERT, 1979)

Variety/line	Infestation Field test $\bar{x}$	Greenhouse test % 1976/77*)	rel. to Inra 7702 1977
Inra 7702	39.70	100.0	100.0
Sobrid	25.97	85.4	65.4
Sf 83/73	12.17	45.5	30.7
Sorex	33.33	93.3	84.0
PG 3/72	19.80	7.1	49.9
PK 104/75	6.37	0.0	16.1
Peredovik	24.00	93.3	60.5
Airelle	20.83	58.5	52.5
Sf 302/73	19.33	64.3	48.7
HS 301	62.82	93.3	158.2
$\bar{x}$	26.30		
s	15.94		

$LSD_{5\%}$ = 14.41% $LSD_{1\%}$ = 19.76% $LSD_{0.1\%}$ = 26.89%

*) = In climatic-controlled greenhouse

REFERENCES

Boye, R., 1970. Unterschiedliche Reaktion von verschiedenen Sonnenblumensorten *(Helianthus annuus L.)* auf Photoperiode und Temperatur. Diss. Giessen.

FAO Production Yearbook, 1978. Vol. 32, 130.

Grauert, P., 1979. Untersuchungen zum Resistenzverhalten der Sonnenblume *(Helianthus annuus L.)* gegenüber *Sclerotinia sclerotiorum* (Lib.) de *Bary*. Diss. Giessen.

Howell, R.W., 1971. Breeding for Improved Oilseeds. JAOCS 48, 492-494.

Robertson, J.A., 1972. Sunflowers: America's Neglected Crop. JAOCS 49, 239-244.

Schuster, W., Kübler, I. and Marquard, R., 1980. Die Variabilität des Protein- und Fettgehaltes sowie der Fettsäurezusammensetzung einzelner Sonnenblumenfrüchte innerhalb von Sorten und Linien. Fette, Seifen, Anstrichmittel (in press).

Seibel, K.-H., 1978. Die Veränderung der Qualtität von Sonnenblumenfrüchten und -Samen durch unterschiedliche genetische und ökologische Einflüsse. Diss. Giessen.

DISCUSSION

J.M.F. Martinez *(Spain)*

I would like to ask about the lines with different content of linoleic acid. Did you cross those lines?

I. Kübler *(FRG)*

These are selfings of the I2 and I3 generations. I did not make crosses.

J.M.F. Martinez

I have another question about daylength reaction. You found long-day plants and short-day plants in different populations. Did you find any day-neutral plants?

I. Kübler

There was the variety Vniimk 8931. It may be a day-neutral variety.

J.M.F. Martinez

Your experience with the influence of temperature on fatty acids is similar to our own. I am interested because in some crosses we have found evidence of dominance of the high linoleic types when crossed with the high oleic types.

I. Kübler

We investigated the fatty acid composition only with selfings because it is influenced by the embryo. We wanted to look at these problems and not at crosses.

J.M.F. Martinez

Did you make single seed analyses?

I. Kübler

Yes, the individual seeds were investigated.

J.M.F. Martinez

Did you get an insight into the influence of the embryo and the influence of the sporophyte and the whole maternal influence?

I. Kübler

This is what we are trying to do; we are continuing with the investigations.

J.M.F. Martinez

In our experience both genotypes, the maternal and the embryo, influence the composition of fatty acids.

J.D. Gimenez (Spain)

In your table relating to infection you did not mention what the parasite was. Was it Sclerotinia?

I. Kübler

Yes, Sclerotinia.

N.J. Mendham (UK)

What kind of yields do you get from adapted varieties in this rather 'sunless' part of the world? Is it really feasible to grow sunflowers in this latitude?

I. Kübler

I would say that the sunflower is a crop suited to our region. The yield is about 30 dt/ha under our conditions.

N.J. Mendham

What sort of fluctuations would you have with different seasons. There are changes from season to season and you can have cold years and warm years. This must be on the edge of the sunflower growing areas. Surely, three years out of five must produce a fairly poor crop. Is that correct?

I. Kübler

There are variations from year to year.

C. Calet *(France)*

Sunflower protein is very good except for its lysine content. I am not a geneticist but I would like to hear your opinion on the importance or interest of selection for lysine content in sunflowers.

I. Kübler

Prospects are not encouraging. It is important for the utilisation of the meal to have a high protein content, but I think you must get the lysine from other plants. The biological value of the protein is about 60%.

G. Röbbelen *(FRG)*

I think lysine could be improved if we really worked on it. Results with other crops show this to be so; it is a matter of going into it properly. It is also an economic problem, although I agree that at present there is little variation available.

I. Kübler

There is a negative correlation between protein content and lysine content.

J. Greco *(Italy)*

There is a negative correlation between oil and protein, and a negative correlation between protein and lysine, so that there is a positive correlation between oil and lysine. When we breed for oil content there is an increase also in the lysine content.

E.S. Bunting *(UK)*

You have not mentioned the possible cultivation of the

sunflower as a forage crop or (in the light of the activities of this Institute) for possible biomass production. The conditions in the UK are even less favourable, I should imagine, for sunflowers than they are here. However, in the UK there can be very high total crop yields from sunflower, especially from late sowings (June or July sowings). Has any work on that been done in Germany?

I. Kübler

My first investigations were with Boye. He worked on the forage use of sunflower too. I did not mention it because I was talking about seed production. If you are interested I can give you a copy of the publication.

G. Röbbelen

The comments on the correlation between high oil and good lysine quality brings me back to my question to the physiologists. Why do the oil producing plants have this high protein amount and high quality protein, and the starch producing plants have this negative effect on protein amount and also protein quality? This must mean something. I think we should go into it.

E. Thomas (UK)

I would like to go back to the problem of lysine. I thought you might like to know that, using tissue culture techniques in both maize and barley, it is possible to obtain a great amount of amino acid production in the seeds simply by planting out embryos or tissue cultures on culture media containing toxic levels of amino acids or their analogues. Dr. Green from the USA reported that he can obtain something like 300 times the normal levels of amino acids in the seeds. It is a bigger problem to get free amino acids into the protein, but if sufficient mutants are available I see no reason why this should not be possible. It may be feasible also in sunflower: just dissect the embryos, plate them onto media with amino acid analogues and select for amino acid overproducers.

B.O. Eggum *(Denmark)*

Does this mean that the free amino acids are in the kernels?

E. Thomas

That is correct, yes; a very high level.

B.O. Eggum

Nutritionally this would be satisfactory.

E. Thomas

The particular amino acid that I am talking about in corn is threonine, which is not too interesting, but I think that for others it is just a matter of time and selecting out. In barley there are proline accumulators: this is the work of Dr. Simon Bright at Rothamsted. They are also finding new enzymes in amino acid synthesis; it seems to be quite an interesting approach.

G. Röbbelen

I must confess that whatever we found in the physiology of protein production it was never limited by the sources of amino acids. In barley, at the late stage when you get the low lysine content of the protein, you will get an accumulation of lysine at the very same moment. Therefore, it is not a matter of the precursor of the amino acids (the essential amino acids) not being there at the time of the production of the protein, it is a matter of pulling the amino acids into protein production. Though your information is very interesting, I have doubts as to its real importance.

E. Thomas

At least it is a scientific approach to the problem.

B.O. Eggum

Protein is very important as regards nutrition, but do you have any data on variations in the crude fibre?

I. Kübler

Yes, we have found that there are great variations in crude fibre among different varieties. We have some data on this.

B.O. Eggum

I think this is quite important. We know it is a problem in rapeseed and also in sunflower.

G. Röbbelen

Do you mean the dehulled material is varying in the varieties, or do you mean percent of hull?

I. Kübler

I mean the percentage hull in the complete achene. We may use a technical method for dehulling, but there is also genetical variation in single seeds.

B.O. Eggum

Thank you very much.

EXPERIENCE IN SOYABEAN BREEDING IN MIDDLE EUROPE

W. Schuster and J. Böhm
Institut für Pflanzenbau und Pflanzenzüchtung,
der Universität Giessen, Ludwigstrasse 23,
D-6300 Giessen, Fed. Republic of Germany.

The soyabean is one of the most ancient crop plants. It is presently grown to some extent in most parts of the world and is a primary source of vegetable and oil protein. With the continuously increasing demand for protein it has been necessary to produce protein not only in greater quantity, but also of higher nutritive value. Soyabean is the crop that best meets the requirements of human as well as animal nutrition and, in the USA for example, the area under soyabean cultivation has increased from less than 2M ha in 1939 to nearly 30M ha in 1979.

The soyabean originated in East Asia, a region with short days. In general, therefore, the soyabean shows its best development under short-day conditions. Less time is needed to come to flowering, and from flowering to ripening, under short-day conditions than under long-day conditions. Nevertheless, there are genotypic differences in response to photoperiodic reaction (Rudorf, 1935).

Since the end of the 19th century, breeders have tried to develop adapted varieties for planting under middle European conditions (Haberland, 1878; Riede, 1938; Sessous, 1938, 1942; Oberdorf, 1947; Sachs, 1949). The varieties developed, however had a very low yield and were not able to compete with other crops. In addition, there were sufficient import possibilities in the 1950s, and people lost interest in breeding and planting soyabeans. At that time there were only two places in Germany working on soyabean breeding: the Max Planck Institute at Köln-Vogelsang and the Institute of Crop Production and Plant Breeding at the University of Giessen. These two Institutes worked on an international basis, using material of different origins; by working with combinations of gene pools they were able to develop material that was better adapted to middle European climatical conditions, including a higher yield level (Rudorf, 1959). Results of this research work were the varieties Praemata and Adepta (Max Planck Institute) and Caloria (Giessen). Since the 1960s the Institute at Giessen is the only one in the Federal Republic of Germany still working on soyabeans, and it has developed the varieties Gieso and Olima. As a result of the short-day character of the soyabean all the earlier introductions showed delayed development under our conditions. Thus, the first and most important breeding aim was to select types with a low response to daylength, or even a long-day tendency.

In the years 1972 and 1973, Schuster and Jobehdar-Honarnejad tested 10 soyabean varieties from different regions of the world, investigating their reaction to temperature and photoperiod under controlled conditions in a phytotron. Figure 1 shows the development of these varieties under different climatical conditions.

Two daylengths were tested (12 and 18 h/day) under high temperatures (22°C day and 15°C night, average 19.5°C) as well as low temperature (15°C day and 8°C night, average 12.5°C). The tests in the phytotron showed clearly that development is quantitatively influenced by photoperiod and temperature. Also, there were clear differences between the varieties in response. All development stages were accelerated by higher temperatures, and slowed down by low temperatures, but there were different levels of this extended vegetation period between the different varieties.

Most of the varieties showed distinct photoperiodic responses. The varieties Caloria, Gieso, Altona, Merit, Corsoy and Fiskeby showed a long-day tendency under high temperatures as well as under low temperatures, and a daylength of 18 h shortened the period from emergence to beginning of flowering. The variety Geduld, however, had a distinct short-day reaction.

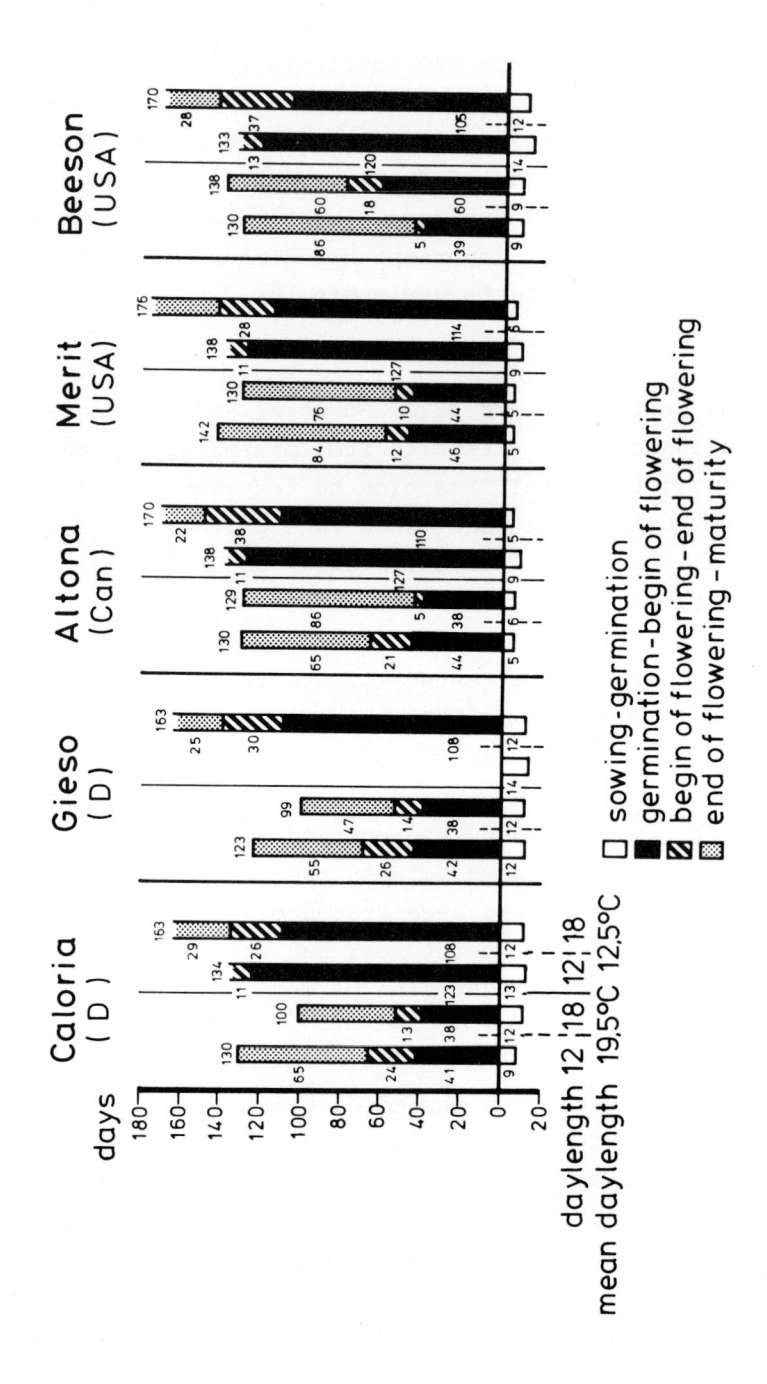

Fig. 1 Development of 10 soyabean varieties grown under controlled conditions at 2 temperatures and 2 daylengths (Schuster, 1973).

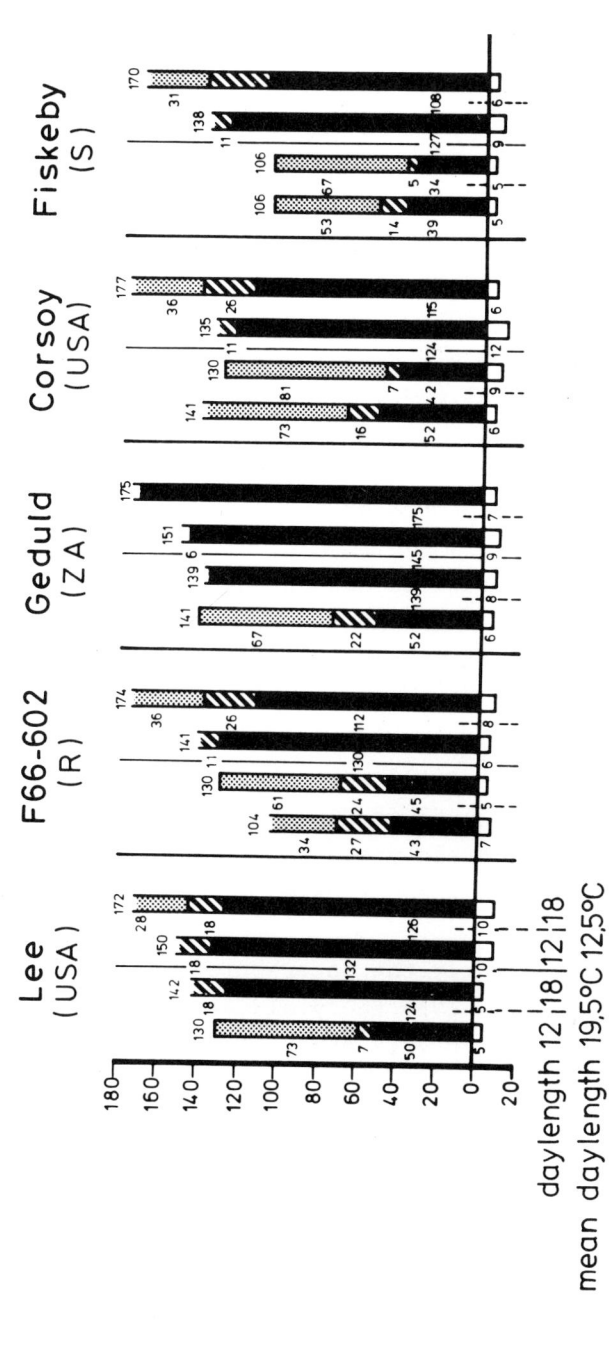

Fig. 1. continued. For legend see part one.

The reaction of the varieties Beeson, F 66/602 and Lee was different under changed temperatures. High temperatures caused a short-day reaction, whereas low temperatures shortened the vegetative stage under long-day conditions. These results show clearly that there are interactions between temperature and daylength, as well as between temperature and varieties, affecting photoperiodic reaction. The same quantitative reactions between daylength and temperature could be observed during the stage from the end of flowering to ripening. Distinct differences between the varieties were expressed by such characteristics as plant height, number of leaves per plant, ramification per plant, number of pods per plant, as well as growth rate and yield of green-matter. In all cases, under high temperatures and long days, plant height and growth rate was increased. Under short-day conditions and high temperatures a reduction of plant height could be observed. Finally, grain yield was influenced very much by photoperiod and temperature (Table 1).

Field experiments, carried out at our Institute, show large variations in yield over the years. The performance of the variety Caloria at the breeding stations Gross-Gerau and Giessen is given in Table 2. At these two locations Caloria gave nearly the same average grain yield of 13.1 and 13.2 dt/ha, respectively. The highest yield was obtained in 1953 at Gross-Gerau (24.0 dt/ha) and in 1958 at Giessen (23.2 dt/ha). In some years, however, the grain yield was less than 10 dt/ha, and in 1961 only 1.7 dt/ha was harvested at Giessen.

The Table shows clearly the interactions between location and year. For example, in 1955 the yield at Gross-Gerau was only 5.3 dt/ha compared with 22.7 dt/ha at Giessen, while in the year 1972, 17.3 dt/ha was harvested at Gross-Gerau and 13.0 dt/h at Giessen.

The next Table (Table 3) shows the performance of several varieties, originating from different countries under different climatical conditions. The grain yields are given of 13 or 14 varieties from 4 different regions, tested at 4 locations

TABLE 1

RESULTS WITH 10 VARIETIES GROWN UNDER CONTROLLED CONDITIONS: DATA GIVEN FOR 2 DAYLENGTHS AT THE HIGH TEMPERATURE. (SEE TEXT). NUMBER OF KERNELS PER PLANT, NUMBER OF KERNELS PER POD, AND 1 000 KERNEL WEIGHT (TKW) IN g. (JOBEHDAR-HONARNEJAD, 1974).

Variety Daylength variant		Caloria	Gieso	Altona	Merit	Beeson	Lee	F66/602	Geduld	Corsoy	Fiskeby
Grains per plant	12^h	1.6	2.3	0.8	0.16	7.6	6.3	(-)	12.8	(-)	3.3
	18^h	8.0	6.2	3.6	9.1	0*)	(-)	6.0	(-)	9.3	3.6
Grains per pod	12^h	0.7	0.8	1.2	0.5	1.5	1.0	(-)	1.5	(-)	0.9
	18^h	1.3	1.0	1.1	1.0	0*)	(-)	0.9	(-)	1.5	0.8
TKW g**)	12^h	160	171	160	150	169	101	100	186	(-)	75
	18^h	102	111	159	130	(-)	(-)	130	(-)	150	100

(-) No pod formation
*) Pods without kernels
**) Approximate values, because of the small quantity of grains

TABLE 2

EFFECT OF YEARS (1950 - 1972) AND LOCATIONS (GROSS-GERAU*, GIESSEN**) ON SEED YIELD AND CONTENTS OF OIL AND PROTEIN IN SOYABEAN, VAR. COLORIA. (SCHUSTER, 1973)

Year	Seed yield (dt/ha)		Crude protein content in % of ADM		Crude oil conent in % of ADM	
	Gross-Gerau	Giessen	Gross-Gerau	Giessen	Gross-Gerau	Giessen
1950	16.3					
1951	20.0					
1952	9.2					
1953	24.0					
1954	8.3	8.5				
1955	5.3	22.7				
1956	5.3	4.3				
1957	16.4	18.4				
1958	12.4	23.2				
1959	12.8	18.1				
1960	21.8	20.9				
1961	9.7	1.7				
1962	9.2	7.4				
1963	15.7	15.6				
1964	11.2	14.9	37.9	37.7	19.2	18.7
1965	12.7	3.4	30.6	32.1	16.6	17.3
1966	5.4	9.5	37.5	36.3	16.2	16.8
1967	10.2	6.7	37.7	38.0	16.7	17.2
1968	16.7	10.9	33.6	34.5	19.2	17.7
1969	20.1	18.9	37.7	35.7	21.9	16.4
1970	13.5	14.7	39.6	38.2	18.2	16.0
1971	8.6	17.1	35.9	36.5	19.5	17.8
1972	17.3	13.0	33.9	34.2	19.0	16.2
Mean	13.1	13.2	36.0	35.9	18.5	17.1

* Humous sandy soil (number of plates: 22 - 25) rainfall 550 mm
  mean temperature of year: 9.6°C

** Loess loamy soil (number of plates: 64 - 68) rainfall 600 mm
  mean temperature of year: 8.8°C

(Germany, France, Italy and Turkey). In 1969, a year with high temperatures, all locations showed higher yields than in 1970, and the highest grain yield (32.2 dt/ha) was recorded that year at Agna/Italy. The variety giving the highest yield in 1969, as an average of trials at all locations, was Merit (25.6 dt/ha) in 1970 it was Gieso (17.3 dt/ha). The lowest yields were obtained from the Japanese varieties, Okuhara (14.6 dt/ha) in 1969 and Akita (6.2 dt/ha) in 1970.

TABLE 3

GRAIN YIELDS (dt/ha) FROM DIFFERENT SOYABEAN VARIETIES GROWN AT 4 LOCATIONS IN 1969 AND 1970. (SCHUSTER, 1973)

Location year Variety	Gross-Gerau/D 1969	1970	M	Agna/I 1969	1970	M	Adapazari/TR 1969	1970	M
Lincoln/USA	8.1	3.0	5.6	26.1	9.3	17.7	26.0	10.1	18.1
Mandschur/USA	16.4	2.2	9.3	21.1	12.0	16.6	20.0	9.4	14.7
Adams/USA	14.9	3.2	9.1	24.9	12.5	18.7	21.0	9.7	15.4
Olak/USA	10.3	2.5	6.4	25.2	13.4	19.3	28.0	9.6	18.8
Merit/CA	23.8	12.8	18.3	27.6	15.9	21.8	32.0	11.5	21.8
Portage/CA	12.5	8.4	10.5	20.2	15.9	18.1	17.0	9.9	13.5
Clay/USA	21.6	12.1	16.9	28.8	17.7	23.3	27.0	11.0	19.0
Okuhara/J	15.3	3.3	9.3	13.1	9.2	11.2	15.0	6.1	10.6
Akita/J	17.7	1.6	9.7	19.9	8.3	14.1	15.0	9.3	12.2
Caloria/D	19.1	13.5	16.3	32.2	17.7	25.0	23.0	11.8	17.4
Gieso/D	19.9	14.2	17.1	27.9	17.6	22.8	25.0	12.0	18.5
St. 464/64/D	14.2	11.0	12.6	21.4	14.6	18.0	21.0	10.8	15.9
St. 454/64/D	19.6	11.1	15.4	26.7	14.6	20.7	24.0	13.3	18.7
(Altona CA)*		15.3		23.0	15.5		25.0	11.1	
M location/year	16.4	7.6		24.2	13.7		22.6	10.3	
M location			12.0			19.0			16.5
M year									

* Not included in mean

(Cont/d)

TABLE 3. (Cont/d)

GRAIN YIELDS (dt/ha) FROM DIFFERENT SOYABEAN VARIETIES GROWN AT 4 LOCATIONS IN 1969 AND 1970. (SCHUSTER, 1973)

Location year Variety	Soulaires/F 1969	1970	M	Mean year 1969	1970	Variety M
Lincoln/USA	12.5	10.8	11.7	18.2	8.3	13.2
Mandschur/USA	13.1	13.4	13.3	17.7	9.3	13.5
Adams/USA	13.6	12.6	13.1	18.6	9.5	14.0
Olak/USA	–	9.6	9.6	21.2	8.8	15.0
Merit/CA	18.9	25.2	22.1	25.6	16.4	21.0
Portage/CA	18.7	14.9	16.8	17.1	12.3	14.7
Clay/USA	19.2	15.7	17.5	24.2	14.1	19.1
Okuhara/J	15.0	11.6	13.3	14.6	7.8	11.2
Akita/J	13.8	5.3	9.6	16.6	6.1	11.3
Caloria/D	14.4	21.7	18.1	22.2	16.2	19.2
Gieso/D	15.4	25.2	20.3	22.1	17.3	19.7
St. 464/64/D	13.2	22.6	17.9	17.5	14.8	16.1
St. 454/64/D	17.0	25.3	21.2	21.8	16.1	18.9
(Altona/CA)*						
M location/year M location	15.4	16.4	15.9			
M year				19.8	12.1	15.9 total mean

* Not included in mean

These results show that the most important necessity in soyabean breeding is to develop varieties with a high and stable yield, which means a better adaptability for different locations and planting conditions. In particular, a better tolerance to cold temperatures during flowering time is required.

As Table 4 shows, the soyabean needs high temperatures at the time of flowering, when the average daily temperature should be within the range of 19 - 25°C (Enken, 1959). The optimum temperatures at this stage, 22 - 25°C, are very seldom reached under middle European conditions. This shows the necessity of breeding cold-insensitive varieties to ensure satisfactory yields.

The quality characteristics of most importance in soyabean grain are the contents of crude fat and crude protein. In the literature you can find protein contents of 30 - 50%; in comparison, the fat content is lower, with a range of 14 - 25%. The amino acid composition of soyabean protein gives it a high nutritive value, which is little influenced by environmental effects. This is demonstrated by experiments of Marquard, Schuster and Jobehdar-Honarnejad (1980). The composition of oil, and especially of the unsaturated fatty acids of the C_{18}-group, shows a greater dependence on genotype and environment.

Table 5 shows the content of the fatty acids oleic, linoleic and linolenic, in soyabean oil of 5 different varieties grown at 4 locations.

For oleic acid content there are significant differences only according to the different locations. The oil from seed produced at the cool locations, Gross-Gerau and Wien, gave contents of oleic acid 6.5% to 8.5% lower than that produced at the warmer locations, Bornova and Ghardinaou. The content of linoleic acid also showed significant differences only between locations. Distinctly higher contents of linoleic acid were found in the material produced at the cool locations Gross-Gerau

TABLE 4

THE TEMPERATURE NEEDS OF SOYABEAN AT DIFFERENT STAGES OF DEVELOPMENT (ENKEN, 1959)

Phase of development	Biological minimum	Sufficient temperature	Optimum temperature
Germination	6 - 7	12 - 14	20 - 22
Emergence	8 - 10	15 - 18	20 - 22
Formation of reproductive organs	16 - 17	18 - 19	21 - 23
Flowering	17 - 18	19 - 20	22 - 25
Formation of grains	13 - 14	18 - 19	21 - 23
Maturity	8 - 9	14 - 16	19 - 20

and Wien. There seems to be the same negative correlation between oleic acid and linoleic acid contents as already found in other oil crops, where we also have similar interactions between these two fatty acids and temperature. In this experiment the difference in linoleic acid contents, as an average of all varieties, at Gross-Gerau and Bornova was more than 10%.

The average content of linolenic acid was lowest at Ghardinaou (5.5%) and highest at Wien (10.7%). This is a remarkable difference considering that the contents of linoleic acid were quite similar at these two locations.

Several investigations made at our Institute have shown that there is little genotypic variation in the content of crude fat or in fatty acid composition. This means that there cannot be expected to be much success in changing these characteristics by plant breeding methods. An improvement of protein content seems to be more likely because of the higher variability of this characteristic.

TABLE 5

THE CONTENT OF OLEIC, LINOLEIC AND LINOLENIC ACID (AS % OF TOTAL FATTY ACIDS) IN THE OIL OF 5 SOYABEAN VARIETIES PRODUCED AT 4 LOCATIONS. MEAN DATA FOR THE 2 YEARS (MARQUARD AND SCHUSTER, 1980)

Variety Location		Caloria	Gieso	Altona	Merit	Prota	Mean
Gross-Gerau D	Oleic acid	25.8	23.9	28.6	22.3	21.8	24.7
	Linoleic acid	53.4	54.2	52.6	59.1	58.0	55.5
	Linolenic acid	10.5	9.7	7.1	8.2	8.9	8.9
Wien A	Oleic acid	24.6	23.2	27.1	25.6	22.8	24.6
	Linoleic acid	52.6	53.7	51.9	52.0	54.6	53.0
	Linolenic acid	11.6	10.9	9.6	10.2	11.2	10.7
Bornova TR	Oleic acid	30.4	31.4	35.4	34.3	33.9	33.1
	Linoleic acid	46.6	44.9	44.1	45.8	44.1	45.1
	Linolenic acid	6.6	6.4	5.8	5.3	5.7	6.0
Ghardinaou TN	Oleic acid	35.0	29.4	31.4	31.3	29.0	31.2
	Linoleic acid	46.9	52.5	52.6	50.8	49.9	50.5
	Linolenic acid	5.4	4.9	5.6	5.3	6.4	5.5
Mean	Oleic acid	29.0	27.0	30.6	28.4	26.8	28.4
	Linoleic acid	49.9	51.3	50.3	51.9	51.7	51.0
	Linolenic acid	8.5	8.0	7.0	7.3	8.1	7.8

Several other breeding aims should be mentioned: good standability, pods set higher from the ground to decrease harvest loss, and sufficient resistance to diseases and pests.

Weichsel (1955) already recognised in 1955 that the use of polyploidy would not be successful for a strongly autogamous plant like the soyabean. A selection in developed varieties from different origins also failed to give promising results (Oberdorf, 1947). Combination breeding, by crossing and combining different genetical material from all available sources, seems to be more successful (Schuster, 1973). This process seems to be the best way of developing successful varieties for middle European conditions.

Testing varieties and selected inbred lines for cold resistance and long day reaction, under controlled conditions in a phytotron, is a satisfactory method of getting fundamental information to determine suitable parent material for crossing. Subsequently, the pedigree method seems to bring the best results.

In the USA they are working to develop male-sterile inbred lines which could be the beginning of producing real hybrids. As reported by Schuster (1973), however, hybrid breeding in soyabeans is not as promising as it is in allogamous plants. The use of male-sterile inbred lines, however, could be an important advantage for getting new combinations; up to now all crosses have to be made by hand, which is very expensive because of the low rate of success (varying from 0 to 30%, with an average of 10%) and the low seed setting. By using male sterility it would be much easier and more successful to make new combinations of material for getting a higher variability.

REFERENCES

Enken, V.B., 1959. Soja, Moskau

Jobehdar-Honarnejad, R., 1974. Physiologische Reaktion verschiedener Sojabohnensorten auf tageslänge und Temperatur sowie die Ertragsleistung unter verschiedenen Umweltbedingungen. Dissertation Giessen.

Marquard, R. and Schuster, W., 1980. Protein- und Fettgehalte des Kornes sowie Fettsäuremuster und Tocopherolgehalte des Öles bei Sojabohnensorten von stark differenzierten Standorten. Fette, Seifen, Anstrichmittel 82; 137.

Marquard, R., Schuster, W. and Jobehdar-Honarnejad, R., 1980. Produktivität, Öl- und Eiweissgehalt von 6 Sojabohnensorten in Anbauversuchen auf 2 deutschen Standorten. Fette, Seifen, Anstrichmittel, 82, 89.

Oberdorf, F., 1947. Einiges über Sojabohnenzüchtung und -Anbau. Züchter 18, 411-413.

Riede, W., 1938. Die wichtigsten Voraussetzungen für einen erfolgreichen Sojaanbau in Deutschland. Phosphorsäure 7, 251-262.

Rudorf, W., 1935. Untersuchungen über den Einfluss veränderter Tageslängen auf Sorten von Sojabohnen und Buschbohnen. Z.f.Pflanzenzüchtg. 20, 251-267.

Rudorf, W., 1959. Dreissig Jahre Züchtungsforschung. Gustav Fischer Verlag, Stuttgart.

Sachs, E., 1949. Bedeutung, langjährige Versuchsergebnisse Anbaumöglichkeiten der Sojabohne. Landw. Jahrb. Bayern 26, 46-64.

Schuster, W., 1973. Die Züchtung von Sojabohnen für europäische Anbauverhältnisse. Bericht anl. d. Arbeitstagung 1973 der Arbeitsgemeinschaft der Saatzuchtleiter in Gumpenstein vom 4. - 6 Dezember.

Sessous, G., 1938. Züchterische Arbeiten und Kulturversuche mit der Sojabohne. Forschungsdienst, Sonderheft 8, 297-300.

Sessous, G., 1942. Stand und Ziel von Anbau und Züchtung der Soja. Forschungsdienst, Sonderheft 16, 400-403.

Weichsel, G., 1955. Polyploidy bei der Sojabohne. Wiss.Z.Karl Marx-Univ. Leipzig, Nat. R. 4, 31.

DISCUSSION

W. Hondelmann *(FRG)*

   Where do your original gene pools come from?

J. Böhm *(FRG)*

   From all over - wherever we can get them. Our first material came from America, but now we are working mainly with material from Sweden and Canada; these are the most promising gene sources. Material from Japan is not as promising as these.

D.G. Morgan *(UK)*

   The variations in yield that you get at this latitude with the soyabean; are they linked more with pod numbers, seeds per pod or mean seed weight?

J. Böhm

   It is pods per plant. The soyabean is very sensitive to rainy, cold periods during the flowering stage. If there is a rainy period during the flowering stage, one part of the plant is without any pods and therefore without any seeds. The soyabean sheds flowers if there is cold and rainy weather during the flowering stage.

D.G. Morgan

   My second question is related to the use of the phytrotron for the varietal comparisons. I see you had two day-lengths and two temperatures. I am a little concerned about the light intensity, the light quality and the radiation income that you had there. Normally the light intensity and the radiation income in a phytrotron are relatively low in comparison with the outside environment; these things could influence the rate of growth and the rate to a certain stage of development. I would be very surprised if they did not interact with day-length and with temperature. How do you think this might affect the pattern of results that you were getting in sorting out your varieties?

J. Böhm

I do not have the information with me, and so I cannot answer your question.

L.G.M. van Soest

I would like to ask something that does not have much to do with breeding, but is about nitrogen fixation. As you are introducing soyabeans to a new environment, is it necessary to introduce also rhizobium strains?

J. Böhm

Yes, it is; but we are fertilising soyabeans in our experiments and we have not used rhizobia.

L.G.M. van Soest

It is said that the optimum growth of the plant is when the nitrogen fixation is optimum.

J. Böhm

They are working on this problem at Hohenheim, but I believe thay have not had much success to date with rhizobia treatments.

G. Röbbelen (FRG)

This means that there is plenty of rhizobia in the soil. Only this year we obtained some soyabeans from Poland. We had never grown soyabeans in our area, but we found plenty of nodules in the roots. Therefore, I doubt whether rhizobia infection is really necessary because generally the rhizobium is in the soil even if this crop has not been grown for ten years.

J. Böhm

We made tests at our locations, and we could not work without giving nitrogen to soyabeans; without it there was not a very good yield. We never had many nodules in the roots.

Perhaps it depends on the soil; we have a very light, sandy soil at Giessen.

M. Dambroth (FRG)

I agree with you, Dr. Böhm, that we can dress the soyabeans with nitrogen under our conditions. However, Dr. Röbbelen, I do not think that the amount of nodulation is necessarily an indication of nitrogen efficiency. We know from American work, and that of scientists elsewhere, that the efficiency of the different strains is more important. The question from Dr. van Soest is very important for the future progress of legume growing in our area. This is not only for soyabeans but also for the lupin and *Vicia faba*, etc. In the past we always looked for nodulation and the amount of rhizobia, but these figures did not relate to the amount of nitrogen fixation. In future we must look for efficient strains of rhizobia to give increased nitrogen fixation.

L. Toniolo (Italy)

We have been cultivating soyabeans for about 60 years in Italy. There is a future for soyabean if it nodulates and produces its own nitrogen; otherwise it is uneconomical. The first step is to have nodulation; the second step is to have correct rhizobial strains for the different varieties. They are studying this in France.

G.J. de Jong (The Netherlands)

In the past we have obtained results on the effectiveness of strains of rhizobia in the new Polders in Holland. To begin with, rhizobia were completely lacking over there and they had to be introduced. We developed a method of inoculating seed just prior to drilling with a mixture of rhizobia with wet peat. It is an effective means of treating the seed and introducing rhizobia into fields where they are lacking. We found no problems at all. However, there are few legumes being grown in that area at present and we have stopped the work, but I believe there are centres with a good collection of different strains, and knowledge about the effectiveness of different strains in relation to different crops.

J. Böhm

Did you compare the yield with and without rhizobia infected seed? What were the differences?

G.J. de Jong

We did it with peas, alfalfa and other legumes, but not with soyabeans.

J. Böhm

This is the problem. In Europe it has been done with all other plants, but not with soyabeans.

L. Toniolo

There is quite a lot of information about seed treatment, and about the type of inoculant: there is peat inoculum and granular also. It is not a problem to inoculate seeds; it can be done in different ways. Sometimes, when soyabeans are first grown in a new area, there are insufficient nodules, but inoculating again in the second year can bring results.

G. Röbbelen

One aspect is that rhizobium strains have differences in efficiency of nitrogen accumulation, and also in competition between each other and the microbia of the soil. This has led to the proposal that there should be, for each generation of soyabean, a new infection every year of the rhizobium strain with the highest efficiency, with eventual elimination of other rhizobia from the soil - through competition, etc. Then, in the following years, the only effect will be that of the strain with the highest efficiency in nitrogen fixation.

J. Böhm

In Germany, there could be a problem in getting support for work with rhizobia. Each project must be financed. We are the only Institute in Germany working with soyabeans and I cannot see us going into this at present; we must do the other things first, day-length, adaptability, etc. - before we start on that problem.

B.O. Eggum *(Denmark)*

It has been interesting to hear from people concentrating on sunflower and soyabean. I think we all feel that the yields are still quite low. On the other hand we must remember that there are almost twice as many calories in fats as there are in carbohydrates. We should bear this in mind when we think of the number of calories produced per hectare and we double up the calories in the fat fraction. I feel that we did not cover all the analytical problems with rapeseed this morning. I hope we can find a reliable method for glucosinolates; we would prefer it to be fast, but perhaps that is too much to ask for. One thing that could be done is the exchange of samples between institutes. We could then get together and discuss the methods and results. This has been done with amino-acid analysis and has helped a great deal; perhaps it could work with glucosinolates. It is just a suggestion from your Chairman.

Thanks are due to all the speakers, and to the audience for the comments and questions. I hereby close the session.

SESSION III

AGRICULTURAL ASPECTS OF RAPESEED, SUNFLOWER AND SOYABEAN

Chairman : E.S. Bunting

REGULATION OF POD AND SEED NUMBERS IN OILSEED RAPE

D.G. Morgan
Department of Applied Biology, University of Cambridge,
England

Differences in seed yield between varieties of oilseed rape *(Brassica napus)* and between plants receiving different nitrogen treatments have been shown to be based on differences in pod numbers and seed numbers/pod (Allen and Morgan, 1972; Allen and Morgan, 1975). These are, therefore, important yield determining characters in this species and the paper discusses some of the factors which have been found to regulate them in spring rape.

GENOTYPE

A systematic comparison of Maris Haplona and Bronowski in the field in 1975 and 1976 showed that Haplona produced more pods than Bronowski, largely because it developed more flowers on the terminal and axillary inflorescences and a larger percentage of these formed pods which were retained to maturity (McPherson, 1980, personal communication) (Figure 1). However, and in general, pods from comparable inflorescences and positions on them carried more seeds in Bronowski than Haplona; this was largely because the ovaries contained more ovules and not because a higher percentage of them set seeds which did not abort (Figure 2). Present studies of seed development in rape are being greatly facilitated by the use of X-ray radiography (Pechan et al., 1980).

SUPPLY OF CARBON ASSIMILATES

Higher pod numbers and seed numbers/pod in the field were correlated with larger Leaf Area Indices at and after anthesis (Allen and Morgan, 1972; Allen and Morgan, 1975). This suggested that the supply of carbon assimilates is important in regulating these characters and this hypothesis was tested in

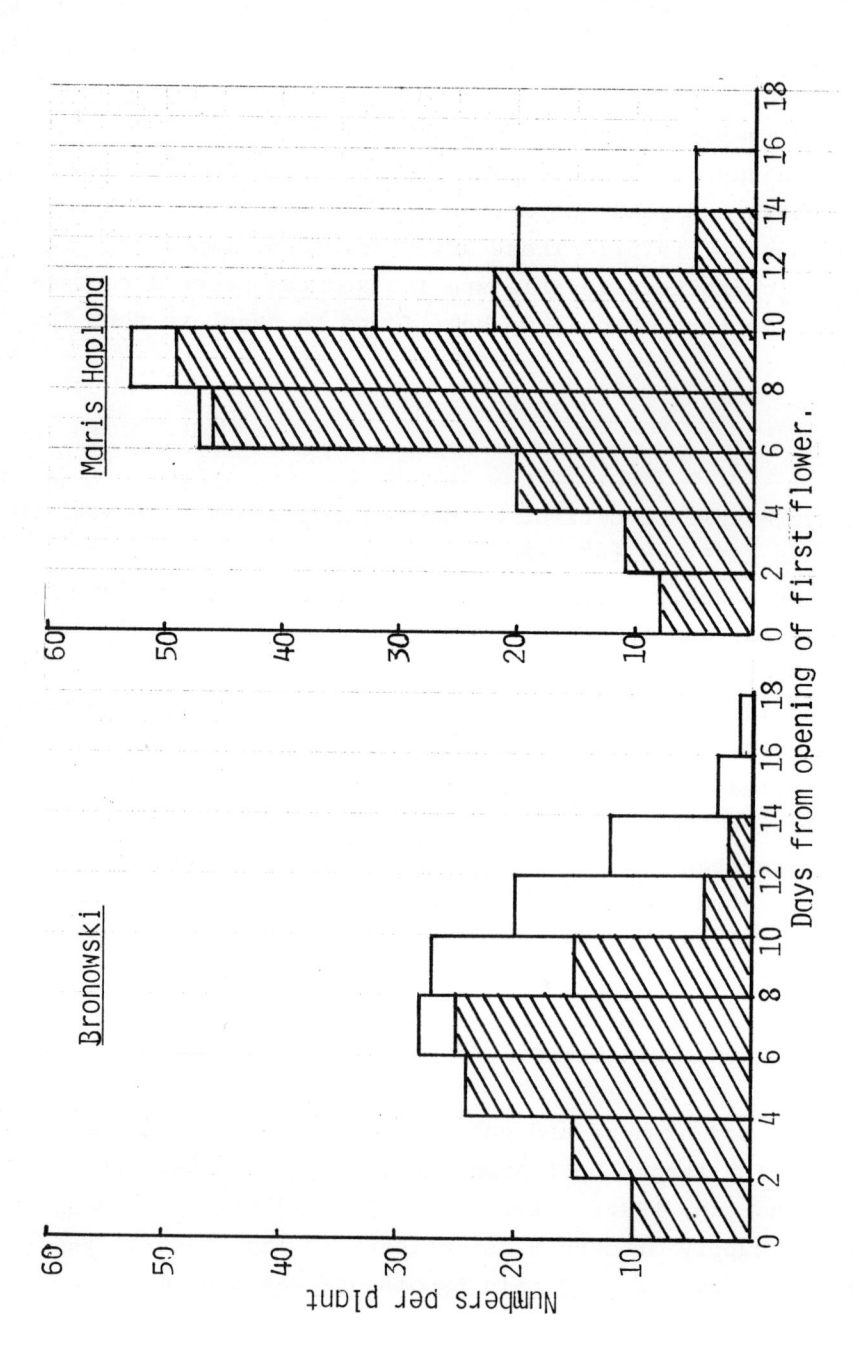

Fig. 1. Flower and pod production by Bronowski and Haplona

Open bars represent the number of flowers opened during two day intervals from the opening of the first flower on the terminal inflorescence. Hatched bars represent pods ultimately formed from the flowers.

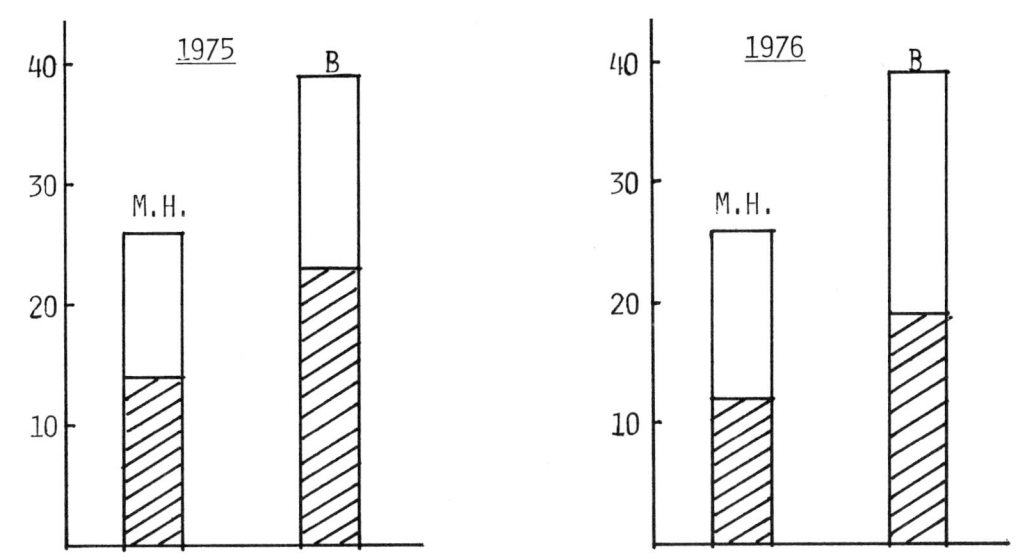

Fig. 2a. Seed numbers/pod and ovule numbers/ovary in Maris Haplona (MH) and Bronowski (B) in 1975 and 1976.

Open bars represent the mean number of ovules/ovary in the lowermost 10 flowers on the terminal inflorescence. Matched bars represent the mean number of seeds/pod in the lowermost 10 pods on the terminal inflorescence.

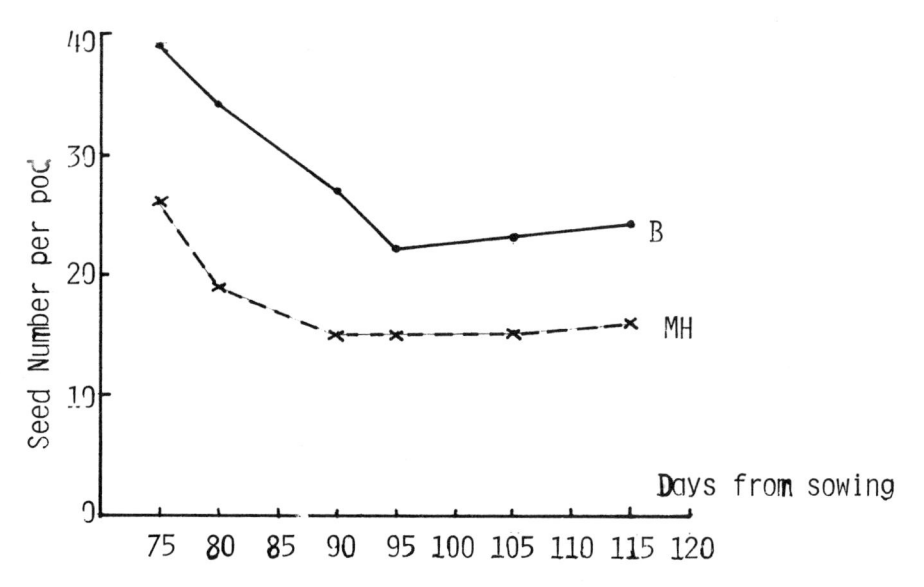

Fig. 2b. Establishment of seed number/pod over time in Maris Haplona (MH) and Bronowski (B).

Data represent mean seed number/pod for the 10 lowermost pods on the terminal inflorescence.

controlled experiments in which the supply was reduced by shading or removing leaves at different stages of plant and, therefore, inflorescence development. Irrespective of the time of treatment, shading or removal of leaves reduced pod numbers and especially on the axillary branches, where the effects were linked with reductions in the number of open flowers (Tayo and Morgan, 1979). Leaf removal, probably because it led to a greater stress in the supply of assimilates, led to the greater reduction in pod numbers. Seed number/pod, calculated on a per plant basis, also decreased slightly following the treatments at anthesis and thereafter but not significantly so.

SUPPLY OF NITROGEN

The effects of nitrogen on yield components in oilseed rape have been studied not only in the field but also in carefully controlled pot experiments in which plants were given extra nitrogen throughout or for two week periods from different stages of development (Prynne, 1980; personal communication). Inflorescences were restricted to the terminal on which the developmental changes in the pods were followed systematically in plants sampled at regular and frequent intervals. Irrespective of treatment, extra nitrogen increased pod numbers because more flowers opened each day to give rise to pods and the production of open flowers was more prolonged. More nitrogen also increased seed numbers in pods formed after the extra nitrogen was given.

Supplementing the supply of nitrogen increased leaf area and, in this way, probably increased the supply of carbon assimilates to the inflorescences. The promotion of pod and seed development by the extra nitrogen could therefore have been achieved directly or indirectly.

HORMONAL CONTROL

4-chlorophenoxy acetic acid (4-CPA) has been found to increase pod numbers and seed yield in *Brassica campestris* but its effects were inconsistent (Saxena et al., 1965). In a pot experiment it has been found that this substance was effective on *Brassica napus* when applied two weeks after anthesis but not at anthesis (Figure 3).

The results of various field and greenhouse experiments have shown that in spring rape pod numbers and seed numbers/pod are determined in the 3 - 4 weeks after anthesis. During this time genotype, the supplies of carbon assimilates and nitrogen and probably hormonal factors interact to regulate them. In order to achieve high seed yields it is essential to select varieties with many large seeds and to provide good growing conditions (nitrogen and water supply) at important developmental stages to enable the potential to be realised. It is likely that in these circumstances the supply of carbon assimilates to the developing flowers and pods will be favourable throughout development because the photosynthetic area (leaves, stems and pods) will be adequate and operating at a high photosynthetic efficiency (Figure 4).

The paper deals primarily with those factors which control pod and seed numbers in the weeks immediately following anthesis. However, considerable losses of seeds can take place shortly before harvest when pods are prone to shatter. Such losses are particularly important since there is then little or no scope for compensatory development within the plant (Figure 4). This stage of development and the phenomenon of pod shattering is being investigated in several genotypes.

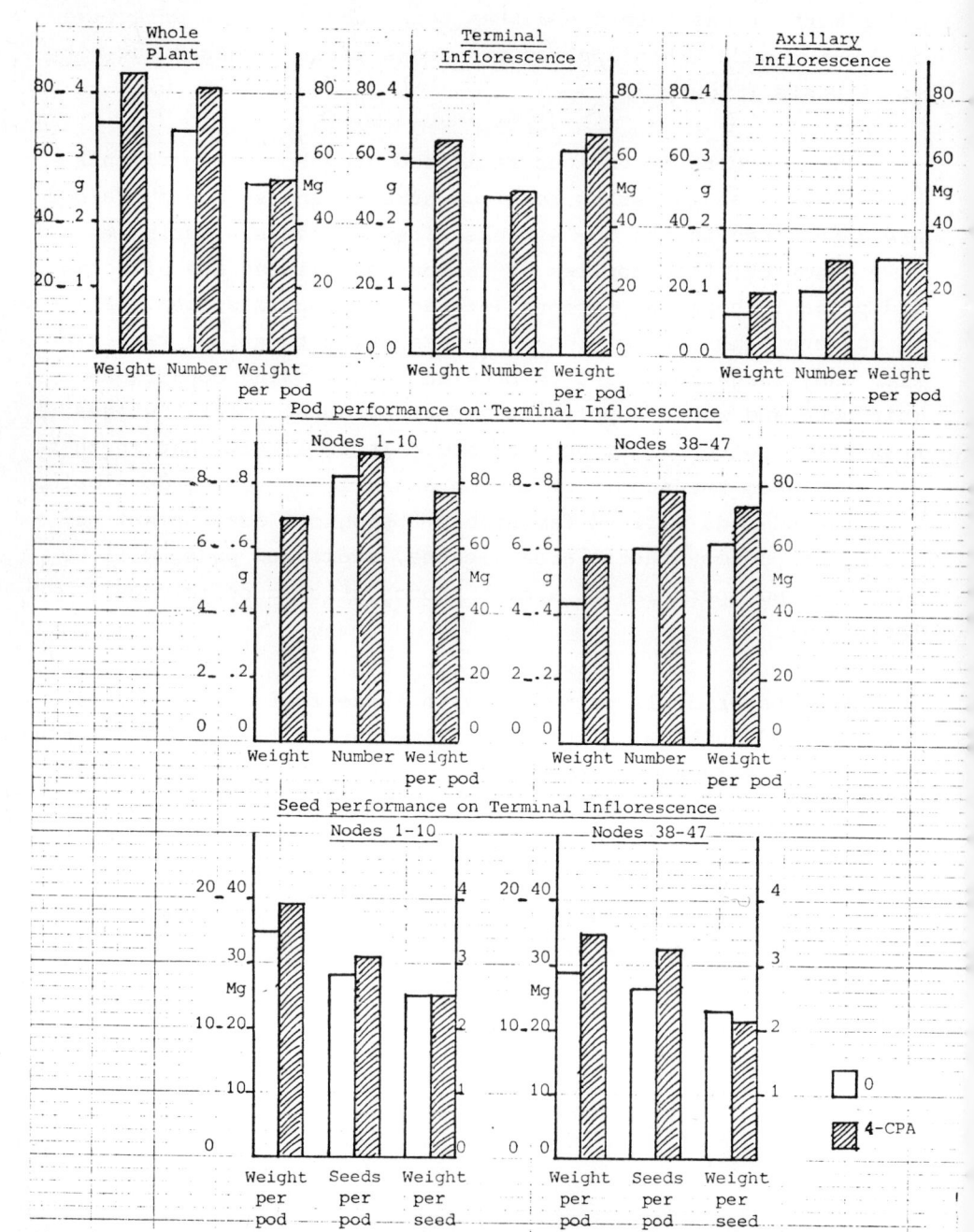

Fig. 3. Pod performance over whole plant and on terminal and axillary inflorescences.

Comparison of yield components in controls and in single plants of oilseed rape (*Brassica napus* var. Zollerngold) treated with 1 ppm 4-Chlorophenoxyacetic acid (4-CPA) 2 weeks after anthesis. Treatments replicated 6x.

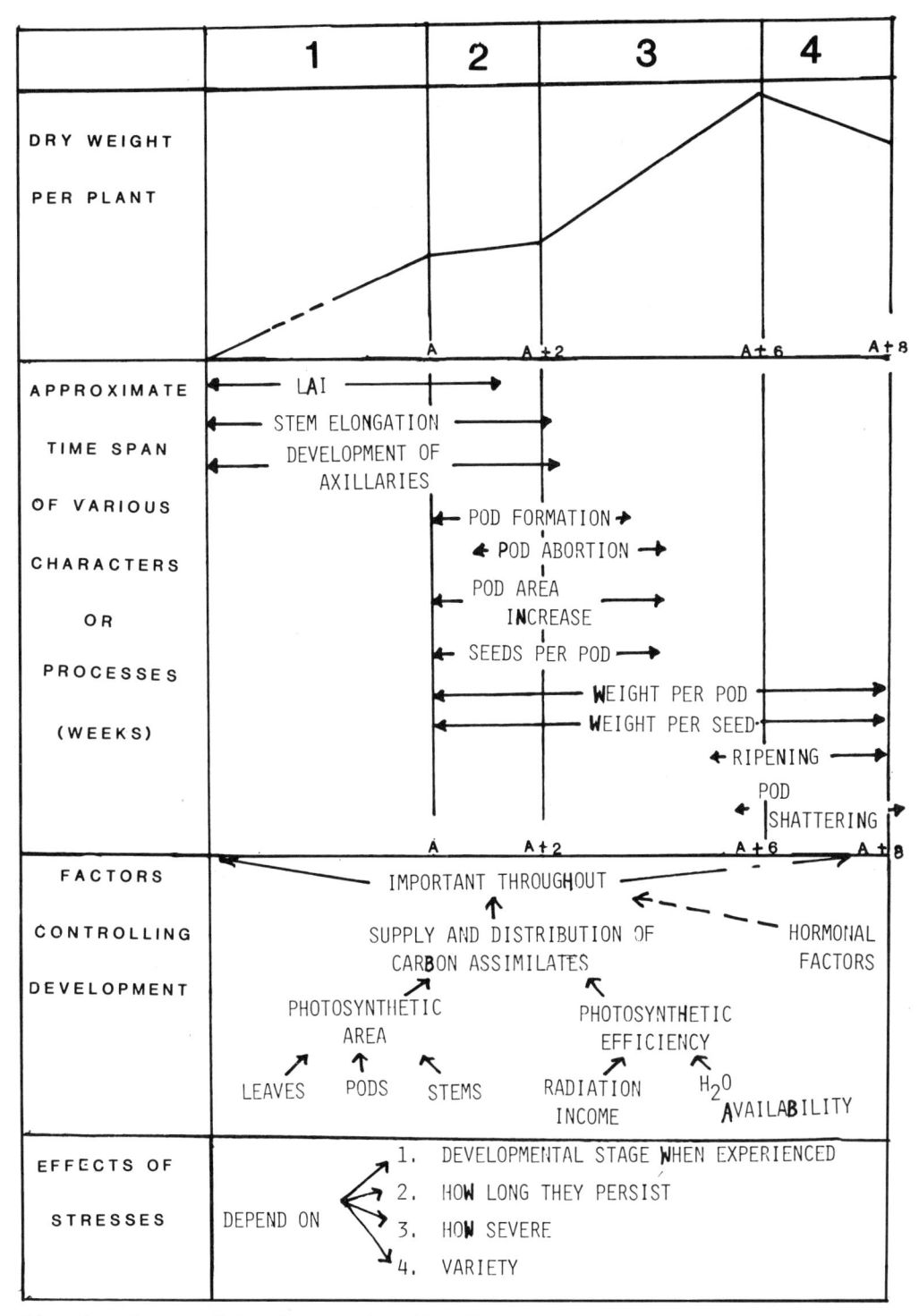

Fig. 4. Phases of development in oilseed rape

REFERENCES

Allen, E.J. and Morgan, D.G., 1972. J. Agric. Sci. (Camb.), 78: 315-324.
Allen, E.J. and Morgan, D.G., 1975. J. Agric. Sci. (Camb.), 85: 159-174.
McPherson, N.J., 1980. Personal communication.
Pechan, P.M., Bashford, J.A. and Morgan, D.G., 1980. J. Agric. Sci. (Camb). 95: 25-27.
Prynne, A.O., 1980. Personal communication.
Saxena, H.K., Watal, R.N. and Singh, S.P., 1965. Indian Oilseeds Journal, 9: 40-46.
Tayo, T.O. and Morgan, D.G., 1979. J. Agric. Sci. (Camb.), 92: 363-373.

DISCUSSION

N.J. Mendham *(UK)*

One point I would like to take up is the statement that the number of pods relates closely to the yield of the crop. Yesterday I said that, under many conditions in our experiment, it appeared that crops with less pods were capable of giving yields that break through an apparent barrier because of the better distribution of light over those pods. Mr. Morgan said rightly that, in most cases, the number of pods is closely related to the leaf area of the crop; in other words, the more vigorous the crop in terms of leaf, the more pods it produces. But in some cases where that correlation is broken it is possible to get very high yields. We were able to show that a late-sown crop with relatively few leaves at initiation, relatively few primary branches, and rather few pods, could - when it grew well in the spring - produce a lot of leaf but few pods, each of which was able to intercept quite a bit of light and maintain most of its 30 ovules through to final harvest. These are the crops which I feel have the potential for a breakthrough in yield which is badly needed. It may seem a conflicting statement to say that fewer pods can give higher yields, but the potential is there; if we can control the development of the crop to make it grow well and restrain itself in pod production, that is where we might get the yield breakthrough.

D.G. Morgan

We carried out some analyses in the field with Jet Neuf, comparing it with Quinta. In this case Jet Neuf produced a large number of pods on the terminal and upper axillary branches. The pods were also large; they had large numbers of seeds per pod and the seeds were also relatively large. I would be the last person to say that there are not many ways to get to the same high yield point. I would still maintain, **it may** be asking too much, that you want a plant producing as many pods as possible, a high ovule number and a high percentage of those set, and you want them to develop to a good seed size. I know I am asking for everything, but I cannot see why not if there is the right carbon assimilate economy and distribution, the right light distribution in the crop, and the right water availability. There is a tremendous wastage in most of the oil-seed rape crops, both in terms of flowers, pods and ovules.

N.J. Mendham

It is very clear that there is enormous wastage. Do you have any estimates of the numbers of pods per m^2, not per plant, produced by the likes of Jet Neuf compared to older varieties? As far as I can see you are quite right; Jet Neuf seems to set a lot of pods on the main stem and the first couple of branches, but rather less from then on. Jet Neuf may be getting closer to the plant-type I was talking about, with fewer pods per m^2 and better light distribution amongst them. From our observations on the older varieties, with 10 000 to 15 000 pods/m^2, the lower 5 000 to 10 000 are almost in shade and, as a result, almost all of their seeds abort. With 4 000 to 6 000 pods/m^2, the number of seeds per pod can be maintained at an acceptable level and there may be a higher yield than from those with the larger number of pods.

D.G. Morgan

Yes, I think we are getting to the same point along different roads. I am thinking more in terms of the architecture of the plant and even the canopy. One thing I would agree with is that there is no point in the pods being produced down at the bottom of the plant, for the very reasons which you stated yesterday: they are shaded, they are the later ones to develop and, from all the experiments we have done, they are the most susceptible to any competition - they are the least competitive of all the pods. I know it is easy to be wise after the event but, with Jet Neuf and the derivative from it, it strikes me that it has a sensible structure; I do not think the canopy is as shaded as with Victor which you showed us yesterday. With Jet Neuf there is a much better distribution of light into the crop; it produces large numbers of pods on the terminal inflorescences. It also has the capacity for setting large numbers of seeds per pod and those seeds get to a pretty good size as well. If you are saying that in rapeseed there are too many pods at the bottom of the plant, I would agree with you - we do not want them.

N.J. Mendham

Dr. Thompson of the Plant Breeding Institute, Cambridge, first brought my attention to the fact that Jet Neuf was bred for high seed weight per pod, which involves both mean seed weight and seed number per pod. If you do that, you are almost automatically selecting for fewer pods; perhaps this has led to this determinate habit even if it was not selected for.

D.G. Morgan

I would question the word 'automatic'. Because there is a large number of seeds per pod, why should you automatically select for a small number of pods? I cannot see that.

N.J. Mendham

There is almost always a strong negative correlation between those components.

D.G. Morgan

What has been in the past need not necessarily be the case in the future.

E.S. Bunting (UK)

There, I am afraid, we must end the discussion. Thank you very much indeed.

PLANT-WATER RELATIONS OF RAPESEED

A. Bramm

Institut für Pflanzenbau und Pflanzenzüchtung
der Bundesforschungsanstalt für Landwirtschaft,
Braunschweig-Völkenrode (FAL), Bundesallee 50,
D-3300 Braunschweig, Federal Republic of Germany.

ABSTRACT

Rape seems to be a plant which is only suitable for heavy soils. However, cultivation on light soils might be possible with adequate water- and nutrient supply. Pot experiments have been carried out to determine the water balance of rape. In these experiments a varying water supply of 30%, 50% and 70% of the effective water holding capacity was given to the rape during the development stages of shooting until flowering, of flowering until green maturity, and of green maturity to maturity.

The results show that the critical growing phase of rape with its highest water requirement is the period between flowering and green maturity. Before flowering the water requirement of rape is small, therefore an irrigation during this period seems not to be necessary. During the critical growing phase the water content of soils should not decline below 50% of the effective water holding capacity. The water requirement during the period from green maturity to maturity is being investigated at present, but it can be assumed that during this period rape should only be irrigated in exceptional cases.

INTRODUCTION

Rapeseed is said to be a cultural species that needs soils rich in nutrients and with good water storage for optimum development. At present in the Federal Republic of Germany rape is being cultivated on about 114 000 ha (Figure 1), 62.5% of which are in Schleswig-Holstein where, due to its maritime climate and to favourable soil conditions, the highest rape yields in the world - 30 dt/ha on average - can be achieved.

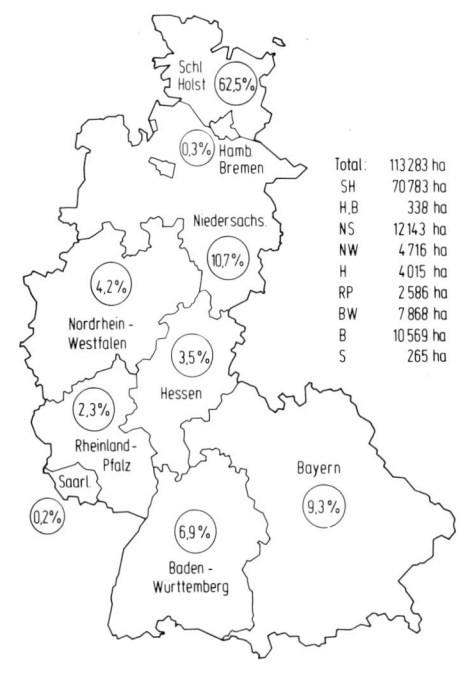

Fig.1. Cultivation area of rapeseed in Germany 1978.

The fact that the EEC is only 13% self-sufficient in plant and animal oils and fats indicates the need for a further expansion of rape cultivation. The high yield level of about 25 dt/ha in the southeast of Lower Saxony proves that economically efficient cultivation is not confined to Schleswig-Holstein.

Theoretically it should be possible to cultivate rape also in areas where the climate and soil conditions are rather less favourable than present centres of cultivation. There are, for example, the sandy soils of the 'North German Lowlands', where rape is a suitable alternative to sugar beet and potatoes when water and nutrient supply are assured.

So far few investigations have been made in the field of rape irrigation in the Federal Republic of Germany, because at present rape is only cultivated in areas where irrigation is not necessary. With the probable expansion of the crop to drier areas it seemed to be necessary to initiate basic investigations on the water requirement of rape.

METHODS

Experimental investigations into the water balance of rape are carried out at the Institute's pot station, where experiments in 'Mitscherlich' pots have been performed since 1978. To determine the optimum water supply for all developmental stages of rape it is necessary to supply different quantities of water during the various development stages by keeping the water content of soils at defined levels.

Since rape in its early stage needs little water, the water content in the soil is sufficient; the investigations start from the beginning of stem extension - the time of 'shooting'. We then differentiate three development phases until harvest; these development stages and the levels of water supply considered are as follows:

Three stages of development:

    time of shooting until flowering
    flowering   until green maturity
    green maturity   until maturity

Three variants of water supply are offered:

    30% effective water holding capacity
    50% effective water holding capacity
    70% effective water holding capacity

The result is a great number of possible combinations which cannot all be considered in one experiment. During the first year we have concentrated our work on the stage 'time of shooting until flowering', during the second year - after consideration of the results from the first year - on the stage 'flowering until green maturity', and this year we are investigating the last development stage, 'green maturity until maturity'.

The experiment involves 208 'Mitscherlich' pots standing on so-called 'tapes of plates', each with a weighing balance and an automatic irrigation system. The water supply desired is regulated by the total weight of the pot, of the soil, of the plant and of the water that corresponds to the desired degree of supply.

The weight of the pots is checked twice a day, the difference between the estimated weight and real weight is then corrected by water supply from the automatic irrigation system. The weight of the plant will vary during the growth period and this is monitored by periodical harvests and considered accordingly.

The soil being used in the experiment at Völkenrode is a loamy sand soil with an effective water holding capacity of 10 mm/10 cm soil depth; the cultivar used is 'Quinta' which is widely grown commercially.

RESULTS

The results obtained are described in relative values, using the treatment '50% effective water holding capacity' throughout the period from shooting to maturity as control = 100.

This treatment has been chosen as standard because irrigation in Germany is only done when the soil water content has decreased to about 40% of the effective water holding capacity. The value of 50% offers sufficient water supply for all cultivars.

During the period 'time of shooting until flowering' the requirement for water supply of rape is not too high. Treatment with a water supply of 30% of effective water holding capacity until flowering do not influence negatively the yield of seed per plant or the compnents of yield, provided the water supply is sufficient between flowering and green maturity.

The yield parameter 'number of pods per plant' (Figure 2) is influenced considerably by water supply. The plants which received little water at the beginning of their growth period produced a high number of pods, whereas the plants with a higher rate of water supply did not produce a higher number of pods (treatments 4 - 8).

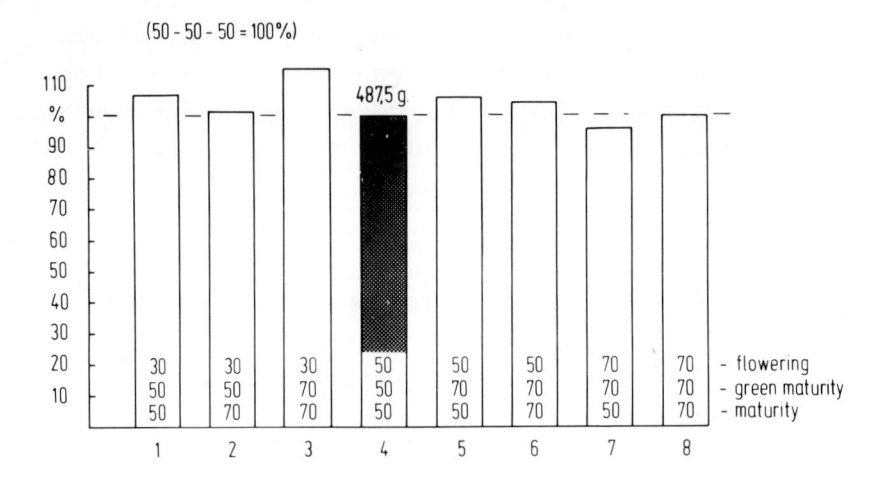

Fig.2. Number of pods per plant dependent on different water supply 1979

The number of seeds per pod was also positively influenced by a decreased water supply until flowering (treatments 2 and 3).

An increase of water supply to 50% and 70% of the effective water holding capacity leads to a small decrease and increase respectively in the number of seeds per pod (see treatments 3, 6 and 8). The increase in treatment 8 has been achieved with a considerable increase of water supply.

(50 - 50 - 50 = 100%)

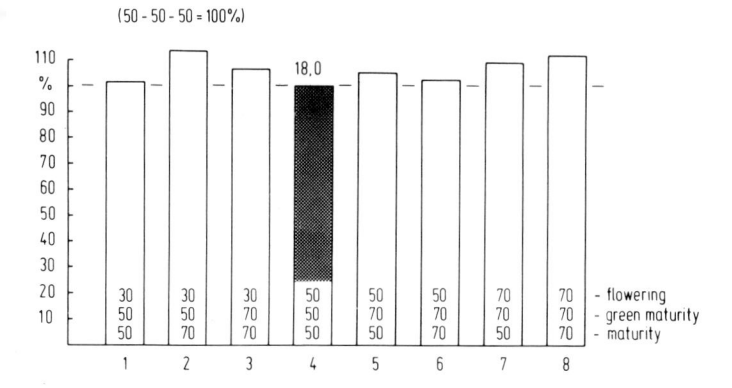

```
                    18,0
110 ┬───┬───────┬───┬───────┬───┬───┬───┬───┬───┬
 %  │   │       │   │███████│   │   │   │   │   │
 90 │   │       │   │███████│   │   │   │   │   │
 80 │   │       │   │███████│   │   │   │   │   │
 70 │   │       │   │███████│   │   │   │   │   │
 60 │   │       │   │███████│   │   │   │   │   │
 50 │   │       │   │███████│   │   │   │   │   │
 40 │   │       │   │███████│   │   │   │   │   │
 30 │   │       │   │███████│   │   │   │   │   │
 20 │ 30│   30  │ 30│   50  │ 50│ 50│ 70│ 70│  - flowering
 10 │ 50│   50  │ 50│   50  │ 70│ 70│ 70│ 70│  - green maturity
    │ 50│   70  │ 70│   50  │ 50│ 70│ 50│ 70│  - maturity
    └───┴───────┴───┴───────┴───┴───┴───┴───┘
      1    2      3     4     5   6   7   8
```

Fig.3. Number of seeds per pod dependent on different water supply 1979

The lowest rate of water supply until flowering also promotes seed weight per pod (Figure 4, treatments 2 and 3). Increased water supply during this period has no positive influence (treatments 4 - 8).

The thousand kernel weight, however, is only slightly influenced by water supply (Figure 5). Nevertheless, the lowest rate of water supply until flowering again has no negative effect on the thousand kernel weight when water supply is increased after flowering.

These yield structure parameters together give the yield of the single plant (seed weight per plant, Figure 6). The figure shows quite clearly that rape does not achieve highest yields under a high water supply during the period from shooting to maturity (treatments 5 - 8); a lower water supply until flowering leads to higher yields.

As we can see from the Figures, a good water supply is decisive for the time after flowering. Accordingly, the water supply during this development stage in the second year was only varied between 50% and 70% of the effective water holding capacity. In the first year - the results of which are shown in Figure 7 - the spectrum of variations was wider, to provide basic results that allow a reduction in the number of treatments for the experiments in the second year.

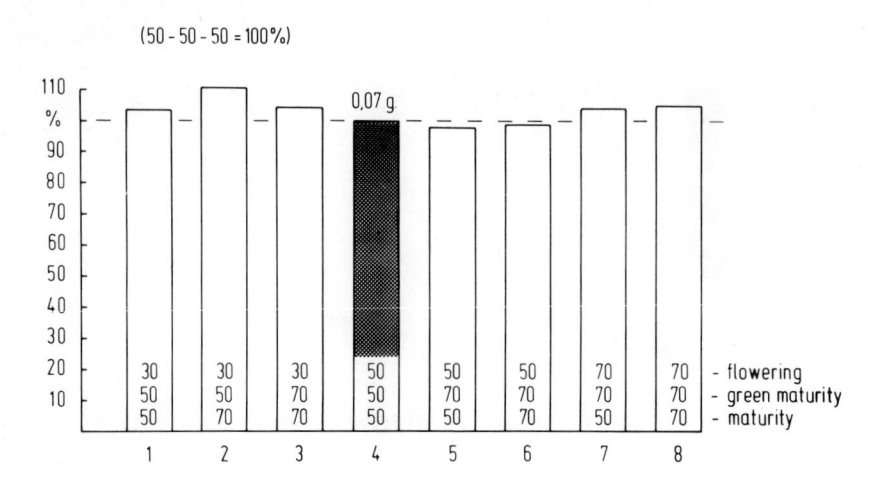

Fig.4. Seed weight per pod dependent on different water supply 1979

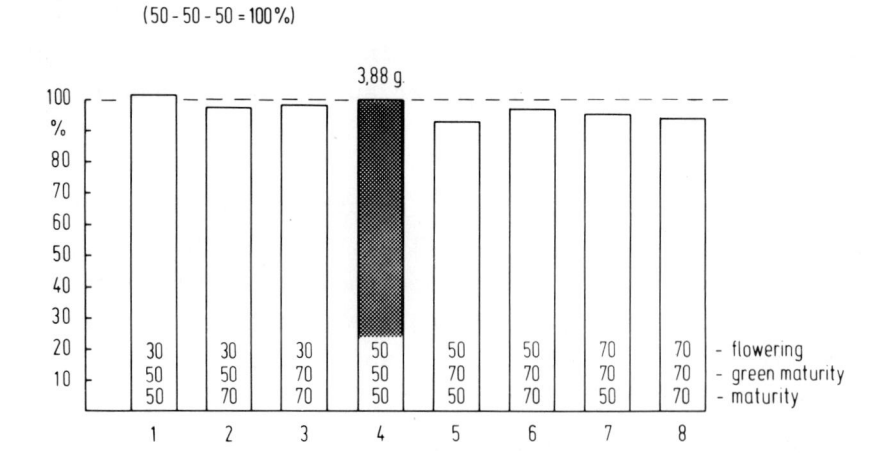

Fig.5. Thousand kernel weight dependent on different water supply 1979

(50 - 50 - 50 = 100%)

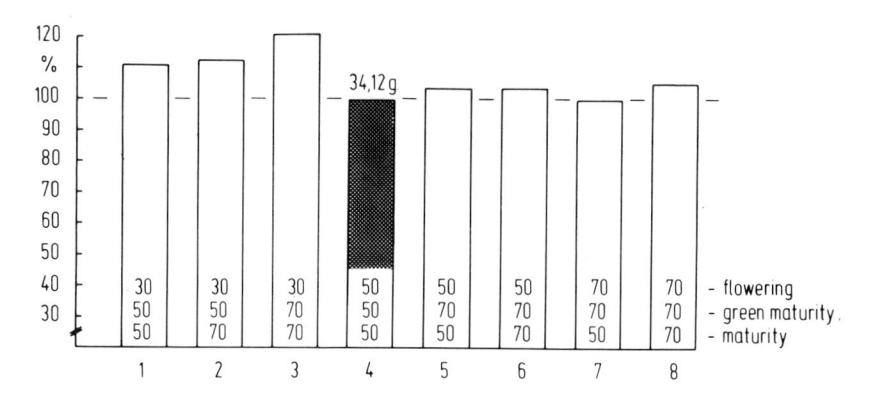

Fig.6. Seed weight per plant dependent on different water supply 1979

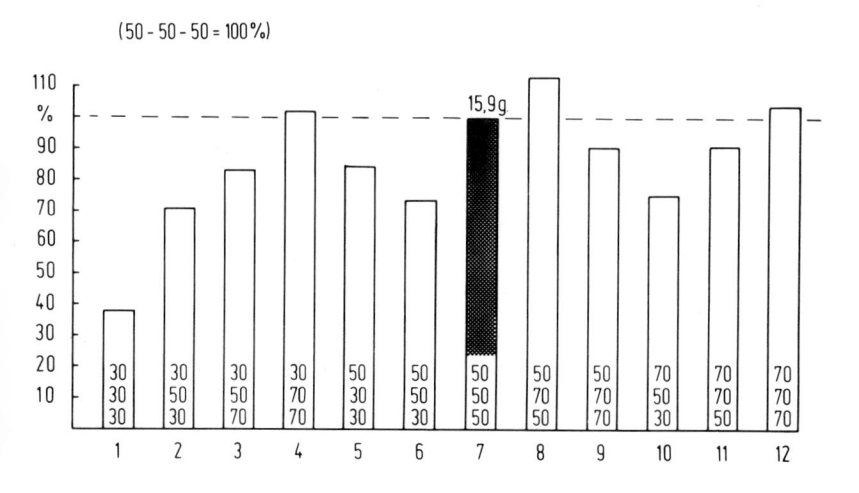

(50 - 50 - 50 = 100%)

Fig.7. Seed weight per plant dependent on different water supply 1978

The Figure shows that little water supply between shooting and maturity causes small yields (treatments 1,2,5,6) but increased water supply during this period did not improve productivity significantly in treatments 9 - 12; the exception was treatment 8 (50 - 70 - 50). Therefore this treatment was tested once more in the second year; the result was contrary (cf. Figure 6, treatment 5) to that of the first year, so in 1980 the treatment is being tested once more, but results are not yet available.

The present experiment in 1980 is concentrated especially on the question of optimum water supply after green maturity. The results are not yet available, but previous results suggest the following conclusions:

- Until flowering time the water requirement is relatively small, so that an irrigation before flowering time has no significant effect.

- The water requirement increases considerably after flowering - the period between flowering and green maturity seems to be the critical growing phase - so that, at least for light soils, supplementary irrigation is essential. During this development stage the water content of soils should be kept between 50% and 70% of effective water holding capacity; if it decreases below 50% a reduction of yield is likely.

Further results are necessary to make statements on the water requirement after green maturity. However, we expect that the water requirement will then decline so that irrigation from green maturity until harvest is not necessary.

Finally, I would like to point out that pot experiments, especially those concerned with 'plant-water relations of crop plants' , can only be a part of the total investigations. The difficulties of applying such results to field conditions are well known. Therefore rape is also being cultivated in the field, in trials parallel to the pot experiments and concentrating on questions of irrigation and of nitrogen fertilisation

In these experiments the drip irrigation system enables us to control exactly the supply of water and nitrogen in the form of dissolved urea. This exact control is necessary in order to find out, in connection with other investigations, the real water and nitrogen requirement for each growth stage of the plants.

DISCUSSION

D.G. Morgan (UK)

How long a time is it between flowering and green maturity?

A. Bramm (FRG)

It takes from 3 to 4 weeks.

D.G. Morgan

It coincides with that sensitive phase when pod numbers and seed numbers per pod are being fixed. Could you observe, within the plants, the distribution of any effects that you were getting? Was it on the terminal inflorescence or the axillaries? Could you observe any effects of that kind?

A. Bramm

No, I cannot remember any results about that.

N.J. Mendham (UK)

Can you speculate as to why, in your experiments, relatively low amounts of water application appear to be beneficial up until flowering? Was it an artifact of your experiments that too much water early on results in the soil being too wet? Is there any reason to suppose that a crop less well-supplied with water will yield better than a well-watered crop?

A. Bramm

I think this is because our experiments were done with winter rape. From the beginning of the growing season until flowering there is enough soil moisture in our climatic conditions here from the winter rainfall. It does not appear to be necessary to irrigate the rape; it gets enough water from the soil. From our pot experiments we can find a parallel result. In the development stage the rape does not need a great amount of water. A low level of soil water content is enough.

N.J. Mendham

This would seem to go against the results we get from the field where water stress in May, as the crop is expanding its leaves before it flowers, can reduce the leaf area to the point where the crop behaves in a very poor fashion. There is insufficient leaf area to make a good yield later on. I wondered if it had anything to do with your pot experiments. As leaves are expanding, as with nitrogen supply, good water supply is essential to get the crop growing properly.

M. Dambroth   (FRG)

I would like to give an additional answer. First of all we must define when water stress starts. With only 30% of field capacity we do not have water stress. Good results in seed production are achieved with low (but enough) water at the beginning of growth; we can control the amount of green material so the competition in the plant, between the leaf and the green material, and flowering and pod setting, is not so high as with a high amount of water in the young stages of plant growth. We have the same situation with potatoes and also with cereals: low water content in the early stages of plant growth is always better for seed production.

D.G. Morgan

May I make a plea here. I think that an interpretation of an effect is never easy. However, I think it helps a great deal if one can observe certain things in sequence. Can you say if there were any obvious differences in leaf area? It is also relevant to know which branch is being affected. It makes it easier to interpret these end effects if one has a picture of what is happening in passing.

A. Bramm

We did some measurements with the leaf area but I am afraid I do not have the results available.

M. Dambroth

Perhaps Dr. Bramm could let you have an additional figure showing these data; he had 4 or 5 harvesting times during the vegetation period and the growth curves would give an answer to your question.

E.S. Bunting *(UK)*

If there are no further questions we will end there. Thank you very much indeed, Dr. Bramm.

SHORT COMMUNICATION

G.J. de Jong
RYP, Smedinghuis,
Lelystad, The Netherlands.

Ladies and gentlemen, first of all I would like to explain why I am here. I am a government official in the Ministry of Traffic and Waterworks in Holland, working on the project of the big Polders in the centre of our country.

Fundamentally, we produce usable soil out of mud. In doing so, it has been proven that agriculture is a very good production system, producing soil which is ready to use at any destination. Our production capacity is about 4 000 ha/year. It takes about 5 years to produce this new soil, so our farm has an area of about 20 000 ha each year. On this farm we grow about 6 000 - 7 000 ha of rapeseed, and the total area of rapeseed in Holland is about 10 000 ha. Therefore, two thirds of the total area of rapeseed is grown on our farm.

Our farm is different in many aspects from normal farms; it calls for different management and organisation and different research. I am responsible for the research - additional to the national research - related to the requirements of this farm, and this involves rapeseed research.

Yesterday, Mr. Thomas mentioned the way we should work with models. We try to run this farm efficiently and in a modern way, which means that we cannot do without models. However, contrary to the experience of Mr. Thomas, some of my colleagues think that reality is merely an inferior replica of a model. But it is very attractive to compete with such colleagues.

I will not enlarge upon experimental results. I will content myself by indicating in which fields of research we are concerned. First of all: variety trials. It is of great importance for the best net result of farming to incorporate the latest varieties as soon as possible, and get rid of obsolete varieties as quickly as possible. This is why I am very eager to get any information on new prospective varieties.

A very important aspect of large scale farming is cost reduction. This means that it is necessary to get as much return as possible out of the cost of labour and machinery. However, this can be stressed to the extent that economising on labour and machinery directly influences yields. A specified activity needs to be done at an optimum time; to economise with labour and machinery, the activity must be spread over a certain period. In other words, you have to start too early and finish too late. For instance, windrowing rapeseed is an activity on our farm which may take about 12 days to spread it economically and wisely. We have to start about 9 days before the best date for windrowing and finish about 3 days after the best date. As you can imagine, we are very interested in rapeseed varieties with different maturity dates. For instance, if we only had Jet Neuf then, this year, we would be compelled to start windrowing about 5 days before the best date. But, in combination with an early-ripening variety, Rafal, we could start windrowing 2 days before the best moment for Rafal and, subsequently, end about 2 days after the best date for Jet Neuf. This is merely an indication of how we try to make the best use of resources. It is the same with soil tillage. Sometimes we plough as late as Christmas. It is the same with fertilisation, drilling and the harvesting of rapeseed and cereals.

We get as much information as possible from our research experiments so that we can determine the effects of inputs on variability of yield.

MAIN FACTORS LIMITING SUNFLOWER YIELD IN DRY AREAS

R. Blanchet[1], J-R. Marty[1], A. Merrien[2], J. Puech[1]

[1] INRA Toulouse, Station d'Agronomie, BP. no 12,
31320 Castanet Tolosan

[2] CETIOM, Service Etudes et Recherche, 174, avenue Victor Hugo,
75116 Paris, France
mis à la disposition de la Station d'agronomie INRA Toulouse

ABSTRACT

The sunflower crop is at present expanding in southern areas of France. It has several advantages, including the relatively simple cultivation requirements, favourable timing of field operations and modest production costs; but certain factors limit yield. Apart from diseases and problems associated with soil structure, the main limiting factor is water supply.

The sunflower crop has a high water requirement, especially in the critical period around flowering. Although the response to water is not so good as that of maize, and at times accompanied by some wastage, an improvement in water supply, particularly during the critical period, guarantees a higher yield and makes the gross return from the sunflower crop equivalent to that from maize.

The development of drought-tolerant varieties would provide the best assurance of satisfactory yields.

INTRODUCTION

After a long stagnation, in area as well as yield, the sunflower crop is now expanding in southern France. For the farmer, it now provides an additional annual spring-sown crop, for consideration along with maize, sugar beet perhaps, and even sorghum and soyabeans in the warmest areas.

Studies of agronomic characteristics of sunflower show both advantages and disadvantages relative to the possible alternative crops, and also reveal the main factors limiting yield in southern France. Among these the most important is the water requirements of the plant. Sunflower is susceptible to water shortage, especially during its critical developmental period. Nethertheless, when adequate supplies are made available, the sunflower crop is capable of making good use of them, although it has a tendency to be wasteful.

In Southern France, tolerance to drought is one of the factors influencing varietal choice in sunflower and, naturally, this is also an important breeding objective.

We shall attempt to clarify the various points at issue in the light of our experience with the crop, and the results from a number of experiments will be given to illustrate the main points.

GENERAL AGRONOMIC CHARACTERISTICS OF SUNFLOWER

Main advantages

In addition to the quality of the oil and protein produced, the main advantages of the sunflower crop in commercial farming systems are:
1. little specialised machinery is required for crop production (for sowing, harvest etc.); farms producing cereals and maize, for example, possess the necessary equipment.

2. The timing of production procedures (seed bed preparation, sowing, chemical weed control, harvest) are of interest on farms involved in intensive cereal production, because the sunflower operations come at relatively slack periods: being later than comparable operations for small-grain cereal crops and somewhat earlier than those for maize.

3. Sunflower is harvested quite early (September), giving ample time for preparation of the land for wheat, the normal following crop; it is not as good as soyabeans as a preceding crop, but the soil structure after sunflower is generally satisfactory and it is a suitable precursor for wheat.

4. Production costs are not high: only moderate rates of N and P fertilisers are required, and though K requirements are high, a considerable amount of it is returned to the soil. The cost of seed, however, is quite high and is a significant factor in production costs.

5. Finally, sunflower often gives a better response than maize to soils of average quality (in depth, water reserves), as we shall be considering later.

Weak points of sunflower, and main factors limiting yield

Resistance to mildew has been incorporated in the most recent varieties, but among the remaining problems are:

1. the high requirements for water; although sunflower can give quite appreciable yields under dry conditions, it has a very high transpiration rate which may lead to a waste of water whenever this is available, and a rather complex response to irrigation. Although there is a less clearly defined critical period of drought susceptibility in sunflower than in maize, during the periods covering flowering and grain-filling the crop is quite sensitive to drought. Later we shall discuss in more detail these water relationships.

2. Next comes the problem of crop establishment; poor emergence caused by lack of seedling vigour, leading too often to crops being ploughed-in, and restricted root development caused by deficiences in soil structure (Rollier and Manichon, 1980). These structural problems adversely affect seedling development, insufficient roots are produced to colonise the soil satisfactorily, plants remain stunted and are very susceptible to summer drought. The development of the tap-root is quite often stopped by such accidents, and the resultant decline in yield is generally more severe than in the other main spring-sown crops (soya, maize, sorghum).

3. In third place, the susceptibility to fungal attack should be mentioned. As resistance to mildew has been incorporated into the new varieties, *Botrytis cinerea* is the most frequent cause of damage with *Sclerotinia sclerotiorum* needing to be closely watched. These fungal attacks are worst in humid seasons. One of the advantages of our region, however, is that *Verticillium dahliae* and *Macrophomina phaseoli* are only weakly parasitic on sunflowers.

4. Finally, it should be mentioned that seed quality is often defective in some hybrids, leading to low plant populations and irregular development. This problem is made more serious by the fact that the sunflower plant, with neither tillers nor branches, is less able to make compensatory growth than many other crop species, although some adjustments in number and weight of grains/capitulum do occur.

Points of less importance include bird damage, which can be severe on small plots, and shedding of grain, either naturally or as losses during harvest (but often this consists of empty achenes, a consequence of seed abortion after fertilisation or failure of pollination etc.).

A better partitioning of assimilates between vegetative and reproductive organs remains a hope for the future. At present, the ratio of grain/total shoot dry matter is rather low in sunflower (0.3 - 0.4), and although the leaves have a high photosynthetic potential the total biomass is not very large (about 10 t/ha dry matter, little more than half of that of maize). In part, at least, this is explained by the high content of oil, but the transfer of assimilates and their remobilisation in the achenes is not very active. A different plant structure might perhaps be considered, with a reduction in stem length, and the development of several capitula/plant with the same flowering time.

This tentative ranking of limiting factors applies to existing varieties; for our area the suggested order of importance is:

    high water requirements
    sensitivity to soil structure
    susceptibility to disease
    other factors (bird damage, shattering, harvest loss etc.)

This ranking might well differ in other areas where, for example, *Sclerotinia* is rampant.

WATER RELATIONS AS RELATED TO WATER SUPPLY - COMPARISONS BETWEEN SUNFLOWER AND OTHER ANNUAL SPRING-SOWN CROPS

Review of water relations in sunflower

General trends

When water is readily available, the intake by sunflower is very high (see Figure 1). Leaf permeability is high and water tension in the plant stays low (5 - 10 bars).

Photosynthetic activity is generally high, but seems to plateau at ambient temperatures around 27°C while transpiration continues to increase rapidly at higher temperatures. Despite the moderate leaf area index (3 - 4 at full flower), the high

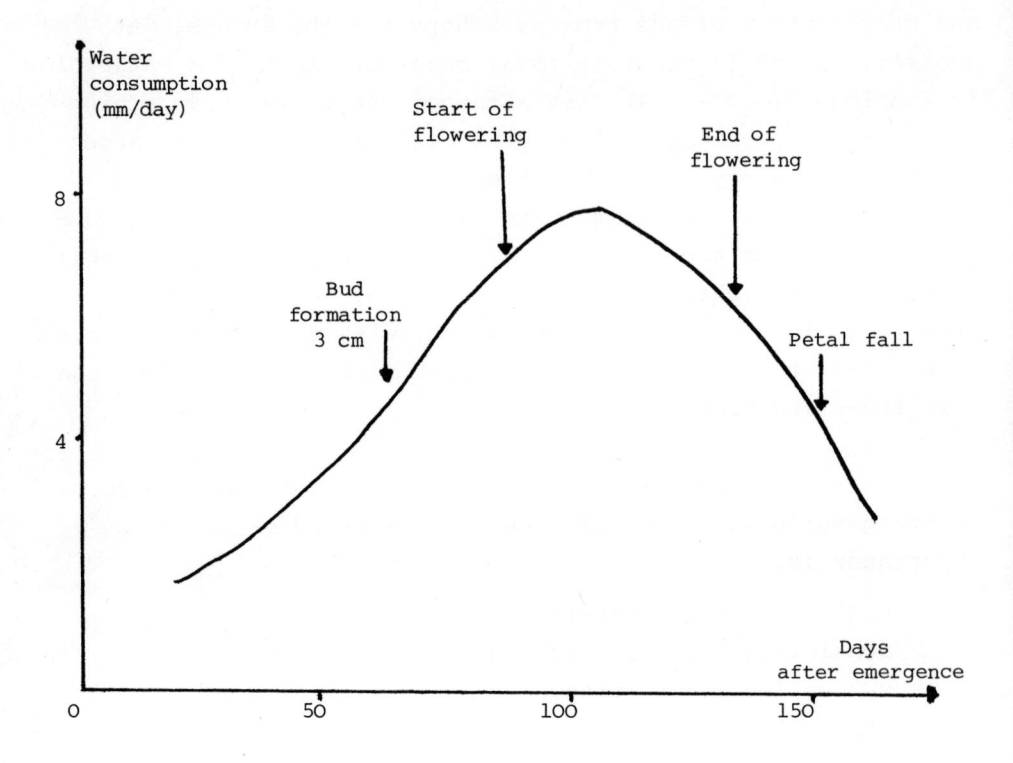

Fig. 1. Trend in water consumption in the sunflower crop (cv Relax)

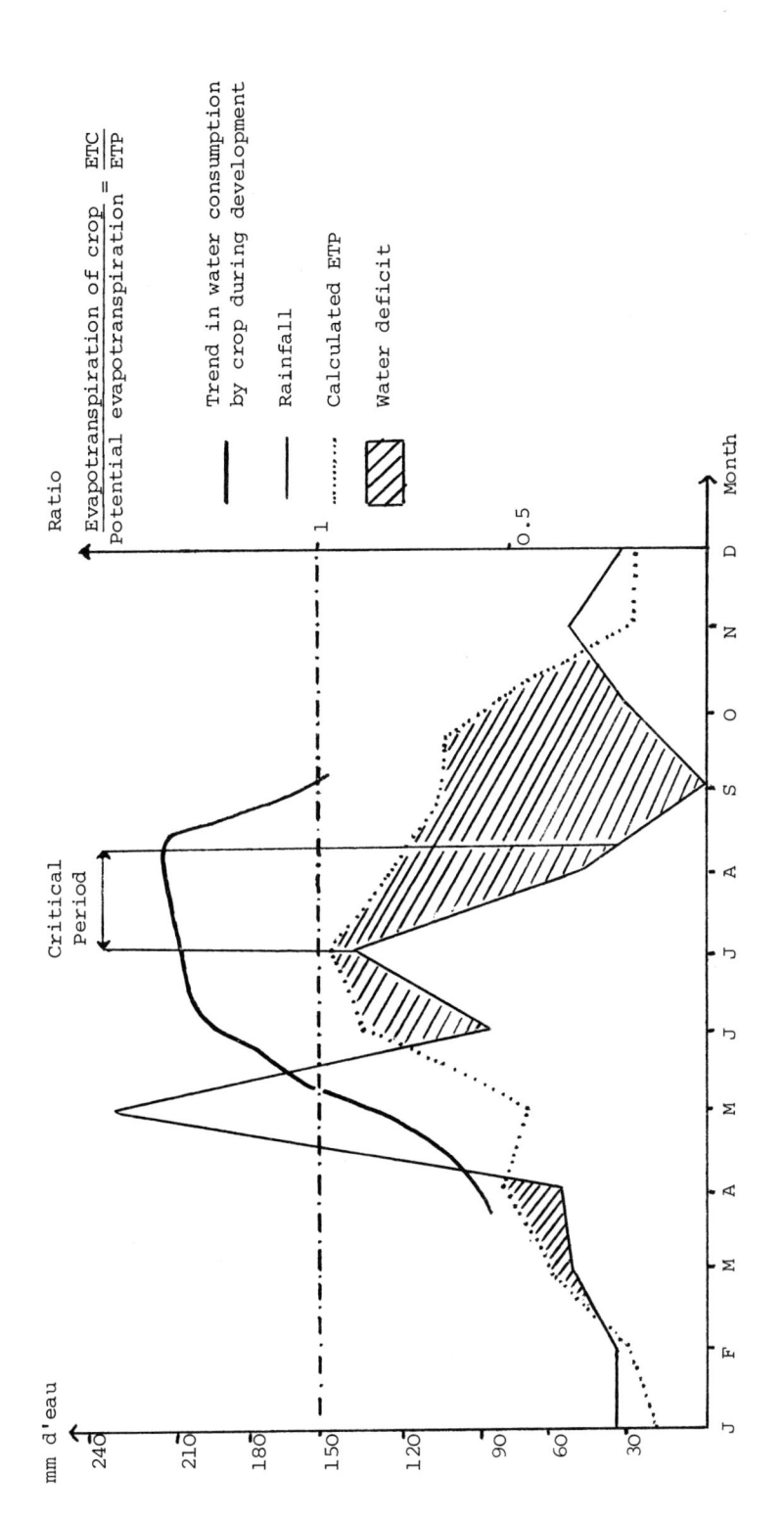

Fig. 2. Relation between the critical period for sunflower and rainfall in the Toulouse region (Climatologie - INRA Toulouse - 1977)

water uptake leads to a relatively low efficiency (2 - 3 kg/m^3) in water utilisation (dry matter produced/water consumed, or net photosynthesis/transpiration).

The same applies to water efficiency for grain production (dry matter grain produced/water consumed), which is less than 1 kg/m^3, less than half the comparable efficiency rating in maize.

Nevertheless, it should be emphasised that when the water supply is not unlimited the efficiency of utilisation is somewhat improved, both for total dry matter and grain production, becuase water stress affects photosynthesis less than it affects transpiration.

Critical period

The ratio between crop and potential evapotranspiration (ETC/ETP) is low up to the beginning of flowering, exceeds 1 at full flower and reaches its maximum at grain formation (Puech and Hernandez, 1973; Figure 2). Over this period leaf area index is at its maximum, fertilisation of the seed occurs and oil formation begins. It is during this pre- and post-flowering period of about 40 days that the sunflower is most sensitive to water shortage (Robelin, 1967). In the south-west of France, this critical period coincides with the seasonal drought; it is the time when irrigation is most fully justified.

Comparitive response to water of sunflower and some other major crop species

There is an essentially linear response between total dry matter production and the extent to which water requirements are met (Robelin, 1958); i.e. $MST = a \frac{ETR}{ETRM} \pm b$, where

MST = total crop dry matter production,
ETR = true crop evapotranspiration, and
ETRM = maximum true crop evapotranspiration.

Using this response of total dry matter production and water availability, one can derive another linear relationship:

$$\frac{MST \cdot ETR}{MST \cdot ETRM} = a' \frac{ETR}{ETRM} + b'$$

and this can be verified for maize, sunflower, soya and sorghum. Similarly, considering grain production (PG) only, the following relation has been verified:

$$\frac{PG \cdot ETR}{PG \cdot ETRM} = a'' \frac{ETR}{ETRM} + b''$$

The response to water of these four crop species differs (Maity et al., 1975; see Figure 3). Maize gives a very strong yield response to water; sunflower responds less strongly than maize, but the response to a given amount of water is increased if it is applied mainly around flowering time (the critical period) rather than evenly distributed throughout vegetative development.

Soya has a similar response as sunflower to irrigation during flowering: it is most sensitive to water shortage in the period from flowering to grain-fill.

Sorghum has a relatively weak response to water, in the conditions of southern France; its most sensitive period is during fertilisation of the inflorescences of the main stem and tillers (Langlet, 1973). The yield response of the four species to water supply is summarised in Figure 4. In this context, 'water supply' is taken to mean the total resources available during the vegetative cycle, i.e.

availability = utilisable soil reserves + (rainfall - drainage) + subsequent irrigation

The data presented in Figure 4 are the average yields obtained, over a number of years, in experiments with water uptake controlled with the aid of neutron probes. In contrast with maize, which received irrigation treatments essentially satisfying ETRM (maximum true evapotranspiration) subsequent irrigations for other crops were limited to the critical stage of development. The main conclusions are:

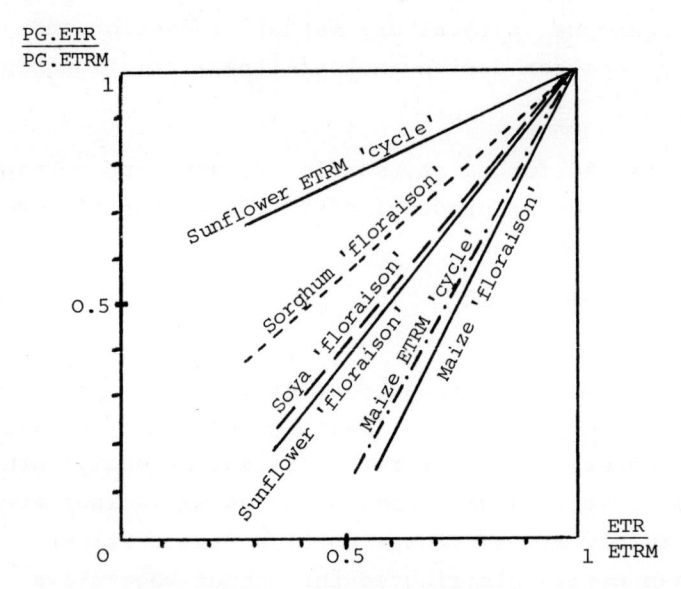

Fig. 3. Patterns of response to water at different levels of requirements

ETR = Real evapotranspiration   PG = Grain yield
ETRM = Maximum real evapotranspiration

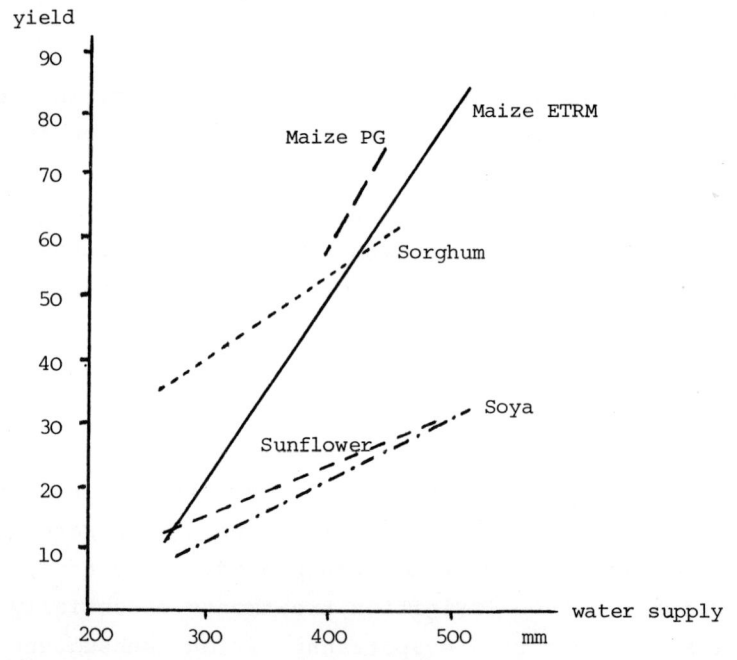

Fig. 4. Yield of dry grain (qu/ha) as a function of water supply (water consumed, mm)

- The strong response of maize to water
- The higher yield potential of maize (> 80 qu/ha of dry grain) and sorghum (60 qu/ha with 450 mm water) compared with sunflower and soya (30 - 35 qu/ha with 500 mm water).

However, the commercial value of these products differ considerably, and it is necessary to compare the gross income and gross margins in these crops. As a first approximation, we have assumed that, under the same environmental (soil - climate) conditions, production costs would not differ very much for the different crop species (with additional costs for seed or herbicides largely compensated for by lower fertiliser requirements; Marty and Cabelguenne, 1980). Also an examination of comparative gross returns is an interesting indication of the extent to which the economic potential of the conditions can be realised by the different species at present levels of production and prices.

The gross return is defined here as the product of grain yield (of commercial grade) and the selling price prevailing in 1979-80. As Figure 5 shows sunflower and soya produce the highest return when the total water supply is around 400 mm. In drier conditions (300 - 250 mm) sorghum and sunflower seem to be best: both have the ability to exploit such dry conditions, even though yields are rather low.

It is also possible to define the amount of water required by the crops to give an equivalent gross return (Cabelguenne et al., 1980). For example, with a relative price for sunflower : maize of 2.7 : 1 and for a water supply of 450 mm, there would be an equivalent return for the sunflower yielding 25 qu/ha and maize yielding 60 qu/ha. When there is particular interest in response to irrigation, then in certain situations, sunflower seems to be equally capable of utilising limited quantities of water economically (Cabelguenne et al., 1980).

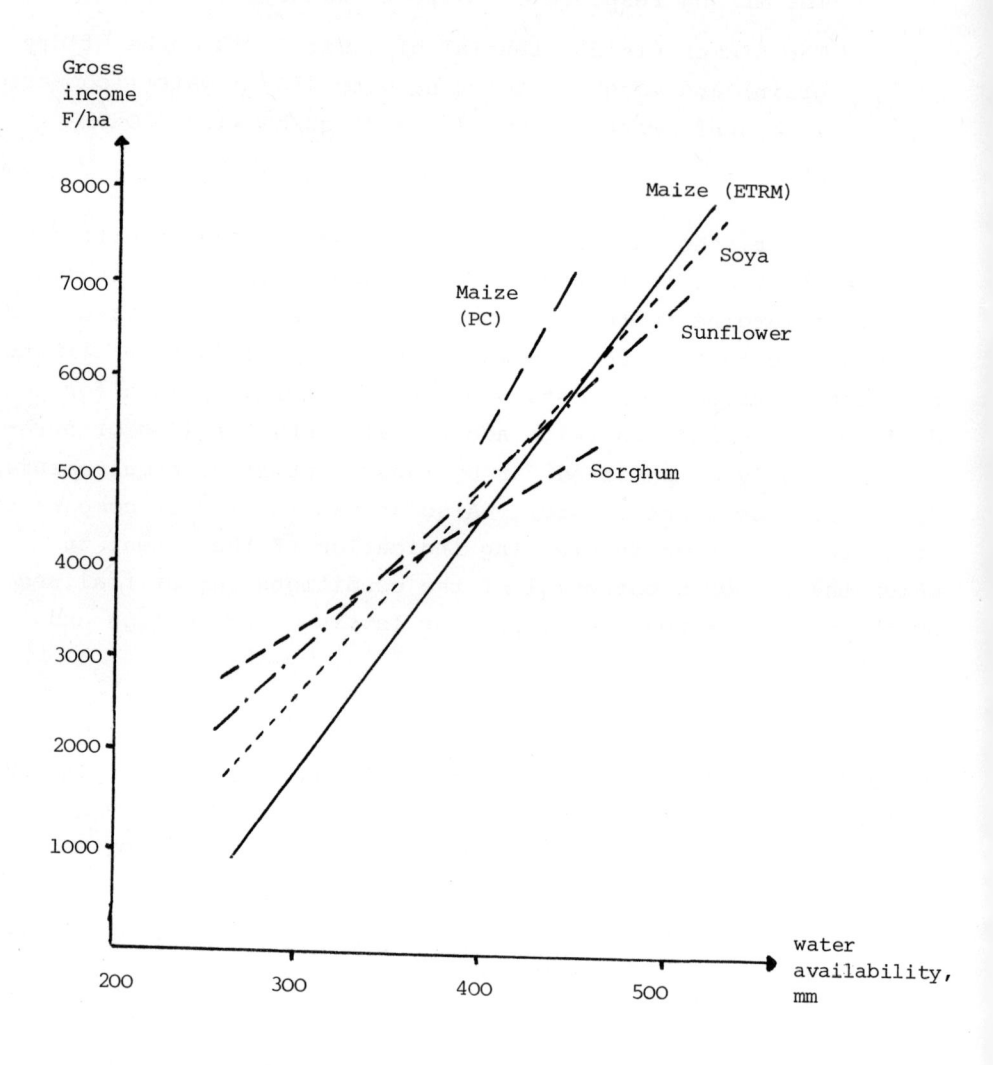

Fig. 5. Relation between gross income produced (PB) and water availability

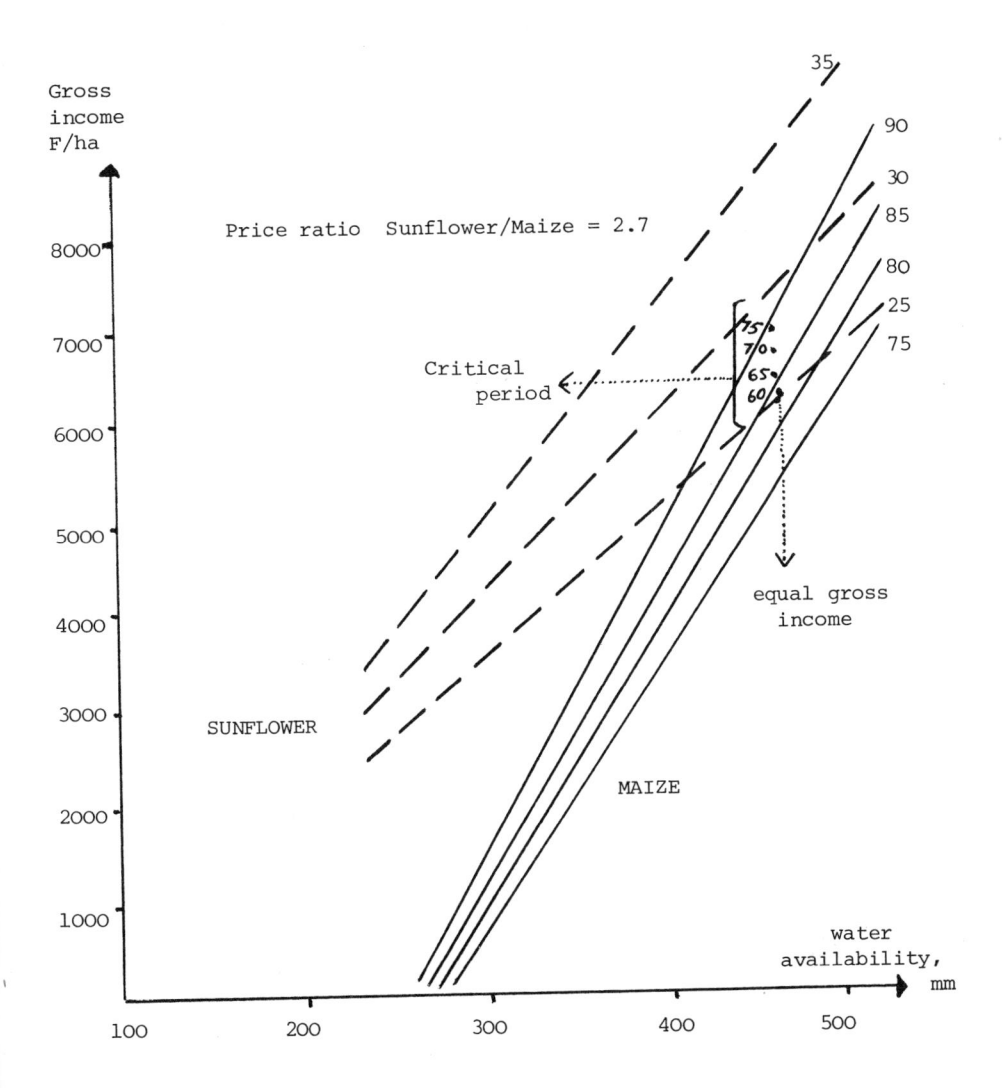

Fig. 6. Trend in gross income as a function of water availability.

The farmer should bear these facts in mind to make the best use of situations with different soil - moisture conditions.

Tolerance to drought

The main effect of drought is to reduce plant height and leaf area. A reduction in the length of the roots also occurs (Pujol et al., 1972).

More generally, water stress at any time disturbs the trend in dry matter accumulation; the effect is most marked in the period from 40 to 80 days after emergence (Rollier and Gachon, 1972). A general rule is that any water stress during the critical period will depress grain yield.

Physiological processes are also adversely affected by water stress; photosynthetic rate is reduced, transpiration limited etc. So far as grain production is concerned, Table 1 shows that the main effect depends on the variety, and may be an increase in sterile area of the capitulum (cv. Wielkopolski an increase in the proportion of empty achenes (Peredovik, Stadium, Relax), or a reduction in average grain weight (Wielkopolski, Stadium).

Main factors determining drought tolerance

Different reactions between varieties have been observed in trials subjected to considerable water stress during ripening, but not too severe to prevent reasonable grain yields being obtained (Blanchet et al., 1980). The results, given in Table 2, show that there was a reduction in plant dry weight in Peredovik, the negative balance being caused by respiration exceeding assimilation; the reduction in dry weight occurred in all plant components (stem, leaf and capitulum) but sufficient accumulated reserves were translocated to the grain to give an appreciable yield.

With Wielkopolski and Stadium, total dry weight remained essentially constant; the apparent transfer of assimilates from vegetative organs to the capitulum and grain gave final grain

TABLE 1

EFFECT OF STRONG WATER STRESS APPLIED AFTER FLOWERING ON VARIOUS CHARACTERISTICS AFFECTING GRAIN PRODUCTION

Water stress	Low	Heavy after flowering	Heavy after flowering	Heavy after flowering	Low	Heavy after flowering
Characteristics / Variety	Yield, q/ha	Yield, q/ha	% of surface sterile	% empty achenes	Weight 1000 grains, g	Weight 1000 grains, g
Wielkopolski	49	30	17	25	65	45
Peredovick	-	33	8	38	57	51
Stadium	59	45	3	36	113	81
Relax	46	43	11	33	58	53
Mirasol	-	44	6	7	37	34

TABLE 2

CHANGES IN DRY WEIGHT OF DIFFERENT PARTS OF THE SHOOT FOLLOWING HEAVY WATER STRESS APPLIED ATER FLOWERING, AND YIELDS OBTAINED

| | Dry weight 27/7, g/pl | Changes in dry weight between 27/7 and harvest (g/plant) |||| Main Characteristic | Yield qu/ha |
		Total crop	Stem and petioles	Leaf	Capitulum and grain		
Perodovick	122	-20.6	-12.8	-6.8	- 1.0	Loss in DW	33
Wielkopolski	110	+ 2.0	- 2.6	-2.4	+ 7.0	DW stable	30
Stadium	165	+ 0.6	- 5.4	-4.8	+10.8	Very vigorous variety	45
Relax	132	+22.5	- 1.4	+0.6	+21.5	Gain in DW	43
Mirasol	118	+16.4	-11.0	+6.0	+21.4	Gain in DW	44

yields which, although reduced (Table 1), were still substantial - especially in the very vigorous variety, Stadium.

Relax and Mirasol were characterised by continued leaf activity, a gain in total crop dry weight, and only a slight decline in final grain yield. This good response seems to be attributable to a strong root system in Relax, and to a collection of characteristics in Mirasol that favour both assimilation and seed production (Table 1).

It is clear from this study that there are many possible methods by which varieties can produce a reasonable yield despite the effects of water stress.

The main determining factors appear to be: resistance of leaves to senescence, a strong root system, percentage fertilisation and seed setting, the ability to transfer assimilates to the grain; all these characteristics should be incorporated into varieties being developed for drought tolerance.

CONCLUSIONS

In order to continue its expansion in dry areas such as the south of France, sunflower must be competitive with the major spring-sown crops, maize, soya, sorghum.

A good preceding crop for wheat, sunflower still has a number of limiting factors which must be eliminated in the course of varietal improvement. Among these, reference must be made to water requirements, and susceptibility to disease and to physical constraints in the soil.

The choice of soil/climate conditions suitable for sunflower production, with or without irrigation, and the varieties adapted to these conditions, are important agronomic problems. For irrigated production, it is especially necessary to eliminate wastage of water, by rates of applications in accordance with requirements for the stage of crop development and with soil water reserves.

The future of the crop will depend on agronomic management practices, progress in genetic improvement, and also the trends in prices of oilseed crops relative to cereals, as is illustrated in Figure 6 (Cabelguenne et al., 1980). Sunflower may find a suitable niche in these relatively dry areas, in production systems based on autumn-sown cereals and oilseed rape, and eventually other oilseed and forage crops; the improvement of drought tolerance in sunflower will be an important factor in determining the extent of the success achieved.

REFERENCES

Blanchet, R. et al., 1980. Some factors determining the tolerance to severe drought of different varieties of sunflower after flowering. 9th International Conference on Sunflowers. Malaga, June 1980.

Cabelguenne, M. et al., 1980. Avantages résultant du choix de la culture du mais ou du tournesol dans diverses conditions de disponibilité en eau; influence du progrés agronomique et des rapports de prix. C-R. Acad. Agric. de France, 7 p. (in press).

Langlet, A., 1973. Effets de la sécheresse sur la croissance et la production du sorgho grain. Ann. Agron., 24 (3), 307-338.

Marty, J-R. and Cabelguenne, M., 1980. Modèle de raisonnement du choix entre cultures d'été selon les disponibilités en eau, le niveau de rendements, les rapports de prix et les coûts de production. C.E.E. Séminaire Toulouse, 7-9 May, 1980. Méthodologie d'étude des systèmes de culture. (in press, C.E.C. - Commission of the European Communities, Brussels).

Marty, J-R., Puech, J., Maertens, C. and Blanchet, R., 1975. Etude expérimentale de la réponse de quelques grandes cultures à l'irrigation. C-R. Acad. Agric. de France, 61 (10), 560-575.

Robelin, M. and Collier, D., 1958. Evapotranspiration et rendements culturaux. C-R. Acad. Sciences, 257, 1774-1776.

Robelin, M., 1967. Action et arrière action de la sécheresse sur la croissance et la production du tournesol. Ann. Agron., 18 (6), 579-599.

Rollier, M. and Gachon, L., 1972. The nutritional needs of sunflower. 5th International Conference on Sunflowers. 25-29 July, 1972, Clermont-Ferrand, p. 63-76.

Rollier, M. and Manichon, H., 1980. Results in press. 9th International Conference on Sunflowers. Malaga, Spain, June 1980.

DISCUSSION

J.M.F. Martinez *(Spain)*

As regards the comparisons between cultivars under your conditions, what is the difference in flowering time? This could explain some differences.

M.A. Merrien *(France)*

The first variety to flower is Perodovick. The others are later and flower at much the same time.

D.G. Morgan *(UK)*

Could you offer an explanation as to why sunflower performs better than maize under moisture stress?

M.A. Merrien

Perhaps it is due to the root system. In my paper I said that sunflower is able to use water to a depth of 2 m. I do not know the depth of the root system for maize but I think it is not so great.

J. Böhm *(FRG)*

Do you know how much water is required for producing 1 kg DM in sunflowers?

M.A. Merrien

The efficiency of the sunflowers is about 2 - 3 kg/m^3. Therefore, with 1 m^3 you could make about 2 - 3 kg DM.

E.S. Bunting *(UK)*

This is much less than maize. In comparative tests, the efficiency of water utilisation by sunflower is usually about two-thirds that of maize.

D.G. Morgan

Were the differences in yield associated more with the number of fruits or with the unit weight of the fruit?

M.A. Merrien

I do not remember the number of achenes per plant or per unit of area. However, if you refer to my Table 1, you will see that for Relax the weight of 1 000 grains (achenes) was 58 in low stress and 53 in high stress. But I do not remember the number.

J.M.F. Martinez

Do varieties which have more seeds per head have less weight per 1 000 seeds?

M.A. Merrien

For Stadium, an open-pollinated variety, the weight of 1 000 grains is over 100 g; for these varieties with very big grains the negative correlation with seed numbers can be found.

E.S. Bunting

The critical factor seems to be what Dr. Merrien calls 'the sterile part'. That is to say that in some genotypes the middle of the capitulum does not in fact set seed. It seems clear that Mirasol is much less affected in this respect than most of the other varieties. I would think it is this response to drought that is significant.

T.M. Thomas *(Ireland)*

Can you tell us what percentage of the dry matter is present as husk in the achenes?

M.A. Merrien

That often depends upon the climatic conditions when the crop is harvested.

E.S. Bunting

Are all these varieties of high oil content and with a low husk content?

M.A. Merrien

I do not have the data on oil content with me, but for most varieties - and especially for Stadium - the oil content is over 50%.

E.S. Bunting

I think it would be reasonable to suggest that you are talking of a husk content of about 20 - 25% with all these varieties. These would all be varieties with a relatively low husk content.

If there are no further questions the discussion will end here. Thank you very much indeed, Dr. Merrien.

YIELD STABILITY IN SOYA BEAN IN NORTH-EASTERN ITALY

L. Toniolo and G. Mosca
Institute of Agronomy, University of Padua,
Via Gradenigo, 35100 Padua, Italy.

INTRODUCTION

For various reasons, the cultivation of soya beans may be of future interest in different parts of Italy, and especially in the North-Eastern regions.

This species has never been cultivated in the past in Italy; there are no indigenous types and all the material has to be introduced from abroad.

To improve our knowledge of the behaviour of soya bean, the Institute of Agronomy at the University of Padua has started a series of experiments to evaluate the adaptation of foreign commercial varieties in our region.

This paper presents some conclusions on grain yield and its stability.

MATERIAL AND METHODS

Ten soya bean cultivars belonging to the maturity groups 00, 0, I, II and III, were included. These varieties were chosen from 107 cultivars, previously tested in trials involving several sowing dates in different years.

The cultivars were grown at the Experimental Farm of the Agricultural Department (Legnaro, province of Padua) on clay soils; of neutral or sub-basic pH reaction, rich in limestone, poor in organic matter, with medium quantities of N and K_2O, and medium to high levels of available P_2O_5.

Temperature, rainfall and water depth are reported in Table 1 and Figure 1.

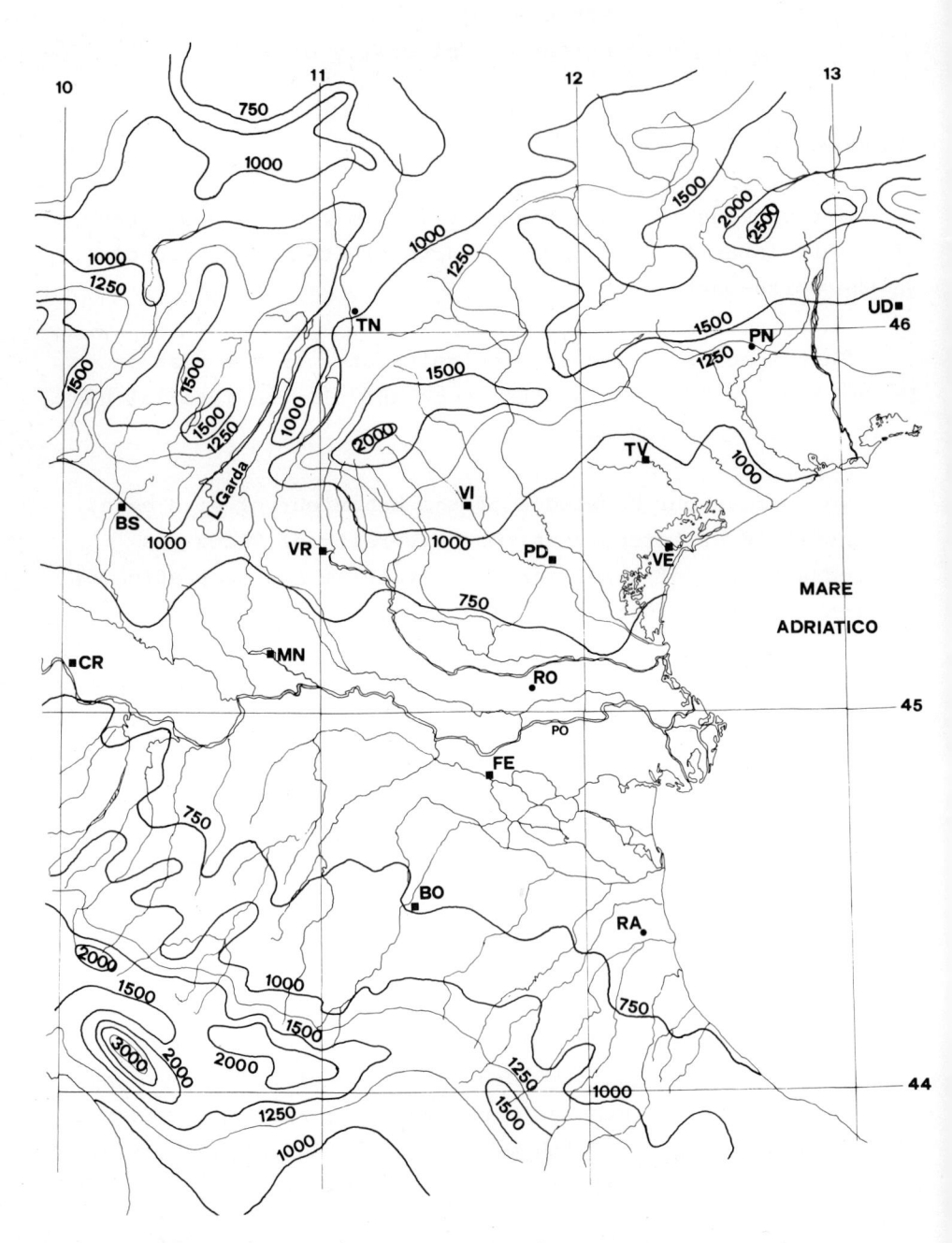

Fig. 1. North-Eastern Italy map of annual rainfall (thirty year average, mm/year). PD = Padua; VE = Venice.

TABLE 1

METEOROLOGICAL DATA AND WATER TABLE DEPTH OBSERVED AT LEGNARO (PADUA - ITALY).

Month	Average temperature (°C) (1)	Rainfall (mm) (1) (2) 1^a decade	2^ad.	3^ad.	Total	Average water table depth (cm) (3)
April	10.7	21.9	20.4	24.5	66.8	127
May	15.4	25.9	17.1	27.5	70.5	132
June	19.5	30.6	28.4	14.4	73.4	149
July	21.5	37.6	23.2	24.8	85.6	156
August	20.7	20.4	32.6	39.9	92.9	163
September	17.1	23.0	29.3	25.1	77.4	156
October	11.7	20.5	21.4	19.8	61.7	148

(1) - Mean values for 1964-1979.
(2) - The annual rainfall in the last 16 years was 863 mm.
(3) - Mean data for 1970-1979.

The planting dates varied from April to July (see Figure 2) over a five year period from 1974 to 1978.

A randomised block design with three replications was used for all the trials. Plots were of four rows, with rows 50 cm apart. Several characteristics of the two centre rows were measured; in this paper only grain yield will be considered.

In the statistical analysis of yield data, differences among cultivar means were tested at 0.05 probability level with Duncan's new multiple range test. The method of Finlay and Wilkinson (1963) was used to analyse the 'genotype x environment' interaction; the yield of each cultivar, for each environment (years and planting dates), was regressed on the mean yield of all cultivars at that environment, to determine whether differences in stability exist among cultivars.

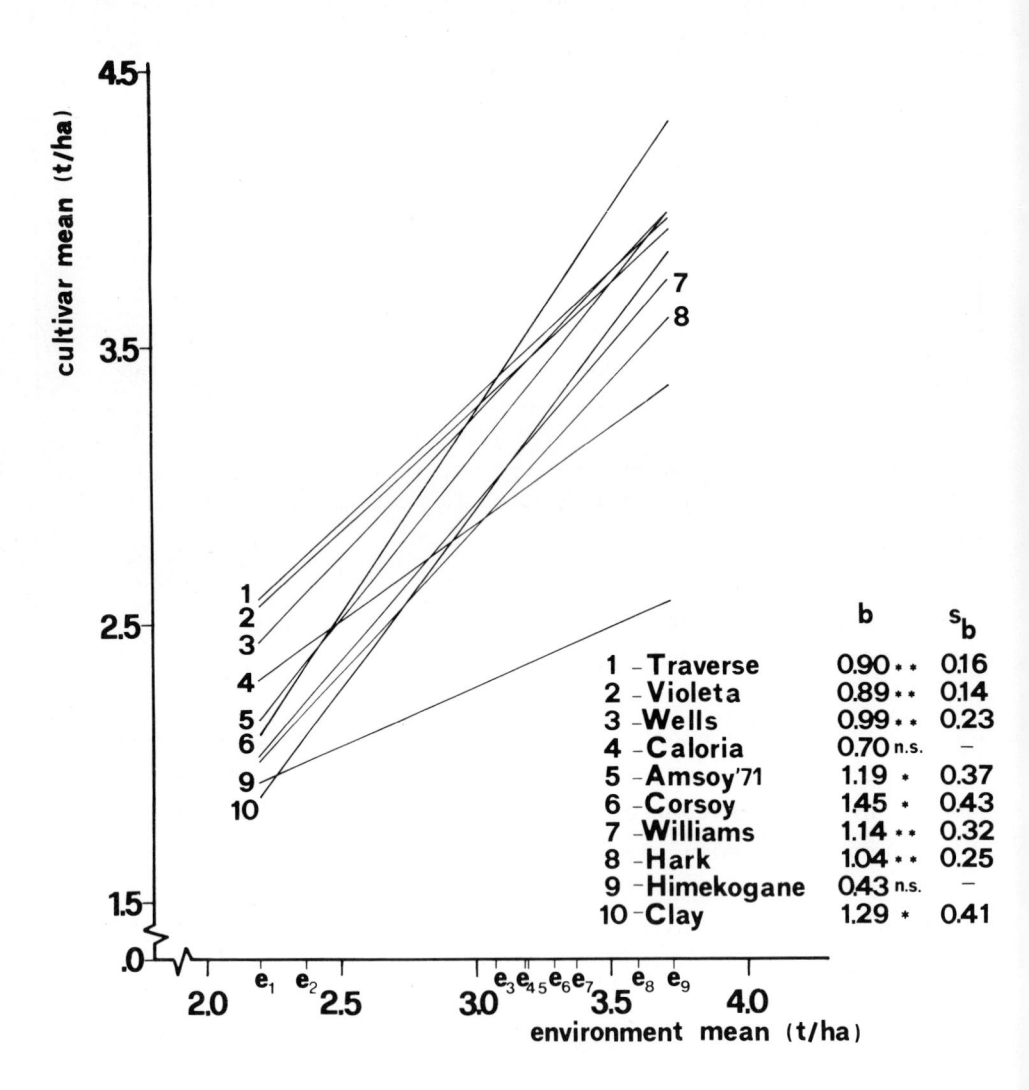

Fig. 2. Regression lines of ten cultivar means on year-sowing time means.
e_1 = 5/7/74; e_2 = 15/6/76; e_3 = 27/4/77; e_4 = 24/4/78;
e_5 = 14/6/77; e_6 = 23/6/75; e_7 = 14/6/74; e_8 = 10/5/76;
e_9 = 15/5/75.
**: significant at 0.01P; *: significance at 0.05P; n.s.: not significant.

EXPERIMENTAL RESULTS

The average grain yield obtained (Table 2) was 3.11 t/ha at 13% of moisture. The cultivars with the highest production were, in order, 'Corsoy', 'Traverse','Violeta', 'Wells' and 'Amsoy 71'.

TABLE 2

SEED YIELD (t/ha AT 13% OF MOISTURE) OF 10 SOYA BEAN CULTIVARS

Cultivars	Maturity group	Country of origin	Grain yield (1) (t/ha)
Corsoy	II	USA	3.42 a
Traverse	O	USA	3.41 ab
Violeta	II+	Romania	3.38 ab
Wells	II	USA	3.35 ab
Amsoy '71	II	USA	3.24 ab
Williams	III	USA	3.05 ab
Clay	O	USA	3.05 ab
Hark	I	USA	2.96 ab
Caloria	OO	West Germany	2.94 b
Himekogane	OO	Japan	2.32 c
Mean			3.11

(1) Cultivar means followed by the same letter do not differ at the 0.05 probability level.

The mean squares of the ANOVA of the yields of 10 varieties in 9 different environments are given in Table 3. The mean square for environments is very much higher than that for varieties, showing that environments have a stronger effect on yield than varieties. This greater effect of environments was much more marked in summer sowings than in spring sowings.

The interaction of genotype with environment (G x E) was high is spring sowings, indicating that varieties can better express their different yield potential when sown in spring. The significance of the G x E interaction is illustrated in

Figure 2, which shows the regression lines of the yield of each cultivar plotted against the average production of all the cultivars in the various environments.

TABLE 3

VARIANCE IN YIELD ATTRIBUTABLE TO ENVIRONMENT, GENOTYPE AND 'GENOTYPE x ENVIRONMENT' INTERACTION IN SOYA BEAN TRIALS, 1974-1978.

Source of variation	Trials where environment included different sowing times	Spring sowings	Summer sowings
Environment	794.1	304.1	638.4
Genotype	307.0	285.1	119.5
G x E	65.1	86.9	50.6

The analysis of adaptation enables us to make several observations: eight cultivars had a stability index significantly greater than zero; the high stability index confirms that the yield of this species is very variable.

The coefficients of seven cultivars did not differ significantly from 1.0 - indicating that these varieties produced a uniform increase in yield as the productive capacity of the environment increased. The exception was Corsoy, which gave an above average response to improvements in the environment.

To discover the environmental factors which had influenced both grain yield and stability index, the results from spring and summer sowings were considered separately (given that the sowing date is obviously a stability factor). As Figure 3 illustrates, cultivars showed a change in behaviour in response to changes in sowing date. Corsoy and the late maturing variety, Williams, demonstrate a tendency to greater yield stability in spring sowings, whereas the early varieties from the OO and O groups (Caloria and Clay) were less stable than in summer sowings. In the summer sowings, the greater difficulty created by the environment reduces the average productivity

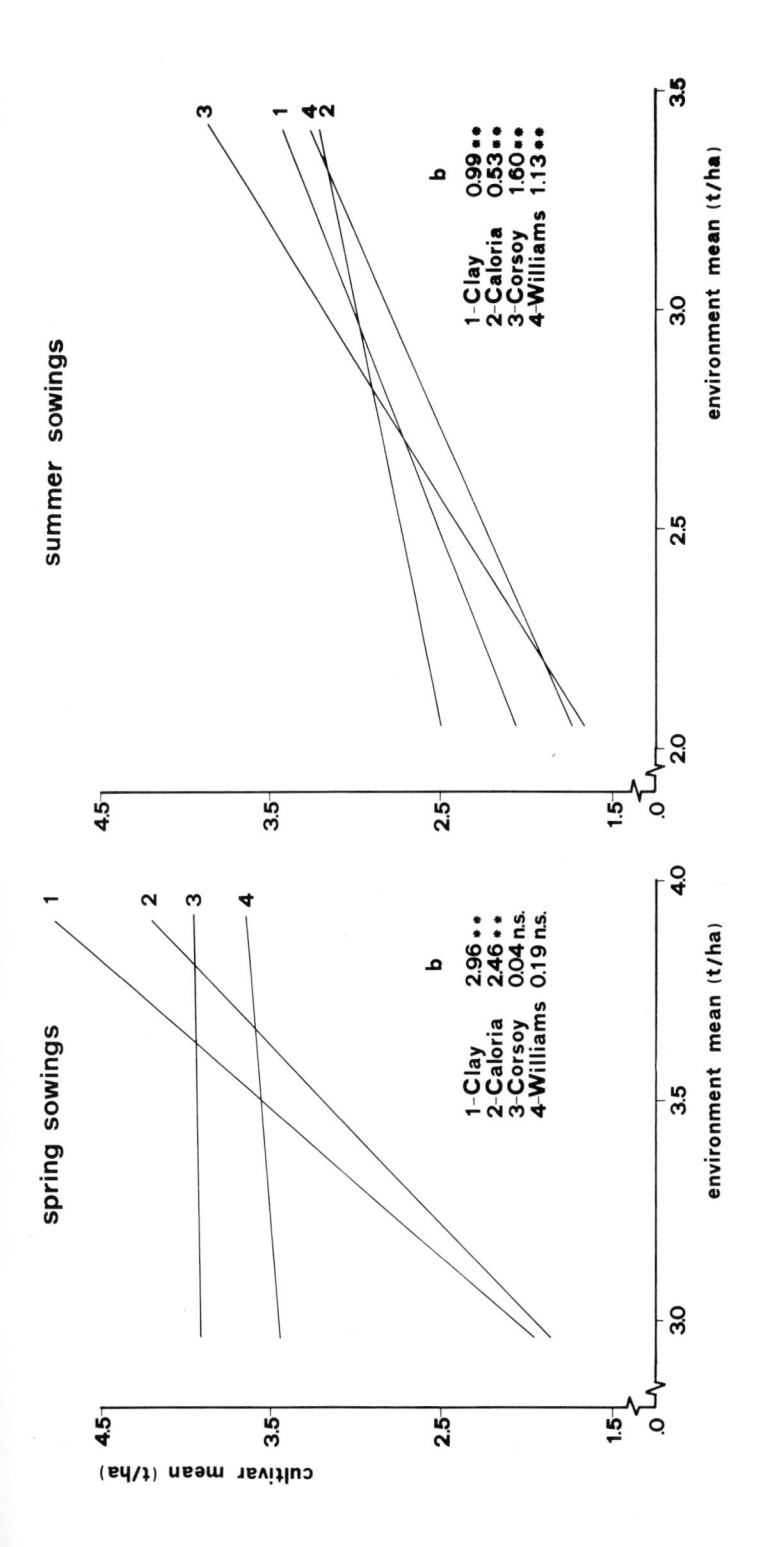

Fig. 3. Regression lines of mean yields for 'Caloria', 'Clay', 'Corsoy' and 'Williams' on mean yields of each sowing-time.
**: significance at 0.01P; *: significance at 0.05P; n.s.: not significant.

of the varieties. The late cultivars lose the stability which was demonstrated during the spring sowings (Figure 3, e.g. Corsoy and Williams), while the early ripening cultivars show increased stability.

The data from the two different sowing times help us to study the relationship which exists between grain yield and rainfall (see Figure 4). In the spring (April - May) sowings, significant positive correlations were found for the I and II maturity groups (r = 0.978* and r = 0.504*, respectively) between yield and rainfall for the period from sowing until the beginning of flowering only. For the OO group, however, a significant correlation with yield was obtained for rainfall over the entire cycle of vegetation (r = 0.750*).

This behaviour can be explained by considering the depth of the water table during the cycle of vegetation (see Table 1). During the June - July period, when the plant begins to flower, the depth of the water table is such that a part of the root system can reach it, as was shown in soya bean trials on the same farm by Giardini and Giovanardi (1979). This situation may be true so far as the later varieties are concerned (groups I and II) because they have a large root system which is already complete at this period of growth. It is not true, however, for the earlier varieties, which have a less extended root system and so are unable to take advantage of the above situation.

In the summer sowings, the early varieties (i.e. O group) have the same behaviour as in the spring sowings (r = 0.870**) while for the late varieties (i.e. II group) the situation changes because of excessive rainfall (r = -0.543*) and lowering of temperature before they are ripe.

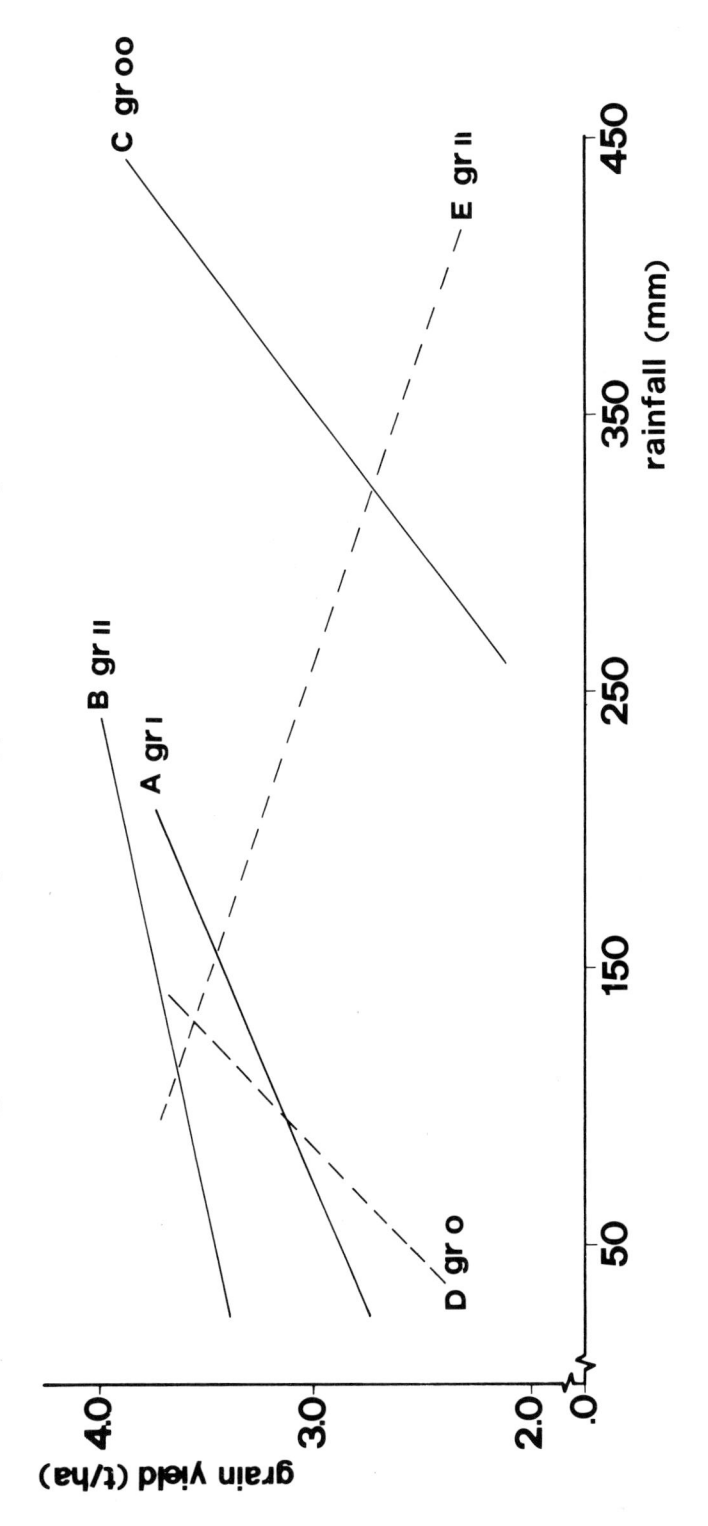

Fig. 4. Effect on yield of rainfall measured from sowing date to beginning of flowering (A: group I y = 26.074 + 0.054x, r = 0.978*; B: group II y = 33.288 + 0.028x, r = 0.504*; D: group O y = 19.536 + 0.122x, r = 0.870**) and from sowing date to physiological maturity (C: group OO y = -3.871 + 0.096x, r = 0.750*; E: group II y = 41.193 - 0.0043x, r = -0.543*) on grain yield of soya bean cultivars.

CONCLUSIONS

Generally there is a very marked influence of environment on soya bean grain yield.

If the results are considered according to a sub-division of sowing times, it is noted that the later cultivars which are sown in the spring are stable, while in summer sowings they prove to be less stable.

Among the various environmental factors, rainfall and, above all, the height of the water table, influence grain yield. The water bearing soil stratum may be reached by the late varieties, especially if they are sown during the spring. The earlier cultivars, however, can only take advantage of water obtained from precipitation.

REFERENCES

Finlay, K.W. and Wilkinson, G.N., 1963. The analysis of adaptation in a plant-breeding programme. Australian Journal Agric. Research, 14, 742-754.

Giardini, L. and Giovanardi, R., 1979. Evapotranspiration and productive response of soybean in dependence on soil mixture. Rivista Agronomia, 1, 91-106.

DISCUSSION

N.J. Mendham *(UK)*

Did you say that the elevation at Padua is 750 m?

L. Toniolo *(Italy)*

No, it is 24 m above sea level.

E.S. Bunting *(UK)* (Chair)

Dr. Mendham is thinking of the rainfall diagram (Figure 1).

L. Toniolo

Yes, that was rainfall: between 750 and 1 000 mm.

N.J. Mendham

At your latitude, which group on the American system, would you expect to be best adapted - group one, or thereabouts?

L. Toniolo

Yes, I think it would be group one, and the earlier group two varieties. When we grow soya beans in rotation, I think it is best to grow winter barley, to harvest it at the beginning of June, and to make soyabeans the second crop. This is done in Sicily: there is a crop during the winter and soya beans are grown afterwards - two crops per year.

N.J. Mendham

Group one should be the most stable then, in your latitude. Corsoy is group two, is it?

L. Toniolo

Yes, group two.

N.J. Mendham

And yet it was the most unstable, with the highest yield?

L. Toniolo

Yes, but if you grow it in the spring it is more stable.

E.S. Bunting

But what would be the case if you were going to grow the crop in a double-cropping system?

L. Toniolo

Another possibility is to sow soyabeans after wheat, in July. In this case the early variety from Giessen can be best, but production can be low from such late sowings.

N.J. Mendham

Is the early, double 0 group, less stable from spring sowing? Is that because it flowers prematurely in some years?

L. Toniolo

It is not less stable - but it gives lower yields.

J. Böhm (FRG)

Do you have problems with flower-drop?

L. Toniolo

If you had estimated the production of seeds on the basis of the flowers produced three days ago when I left Padua, it would have been enormous, but afterwards most of the flowers drop off. This is something we must study, and it is our intention to make deeper physiology studies. But our first step is to give information to the farmers, to give them ideas about the varieties and production methods. There are many problems to be solved - including some technical problems. For instance, we have reached the conclusion that 50 cm between rows could be best, but there are difficulties with farm machinery because the drills are designed for 75 cm for corn. Still, there can be 30 quintals per hectare from a second crop in the north, and this is also of interest.

C. Calet (*France*)

Professor Toniolo stressed the effect of soya beans in the rotation to improve the soil structure and also to spare fertiliser. Other legume seeds have the same effect. In Italy, when a farmer has to choose, what is the most preferable for him? Should he choose peas, horse beans or soya beans?

L. Toniolo

It depends on which part of Italy it is. In the north we have not chosen peas for winter crops because it is too cold. But there are some new winter varieties of peas and we hope to try these in the north. For several years we have been trying *Vicia faba* minor. For four years we had good results with French, German and Dutch varieties but in the last year they all failed to survive the low temperatures. We need to find a leguminous crop - not for forage but for protein, for example, *Vicia faba* minor or soya beans. But I cannot say if one is better than the other as regards nitrogen fixation. Perhaps *Vicia faba* minor leaves more nitrogen than soya bean; there is some data from Blanchet in France about this but we do not have any.

E.S. Bunting

Thank you very much indeed Professor Toniolo. That closes the session.

SESSION IV

ASPECTS OF ANIMAL NUTRITION

Chairman : A. Rérat

NUTRITIVE VALUE OF PROTEIN FEEDINGSTUFFS
FROM OILSEED CROPS

E. Schulz and U. Petersen

Institut für Tierernährung der Bundesforschungsanstalt
für Landwirtschaft (FAL), Braunschweig-Völkenrode,
Bundesallee 50, D-3300 Braunschweig,
Federal Republic of Germany

1. INTRODUCTION

For many years there has been wide-ranging discussion of the claim that protein supply may not be adequate to meet the expanding world needs for human- and animal nutrition.

In Europe, especially in the EEC-member countries with a very intensive livestock production industry, there is a wide gap between demand and production of concentrated protein feeding stuffs. A comparison of total consumption and EEC production of the most important protein feedingstuffs is given in Table 1. The data are taken from Eurostat edition 1980. The figures for 1970/71 and 1976/77 were chosen to give an indication of development during recent years.

Total consumption in 1970 was about 12.8 million metric tons, and in 1976 it was about 16.6 million metric tons. This is an increase of about 3.8 million metric tons in six years, an average increase of round about 650 000 metric tons/year. On the production side there was a small decrease in the EEC, from 734 000 metric tons in 1970/1 to 689 000 metric tons in 1976/77. If we calculate the proportion of the EEC production in relation to the total amount, then we have a very serious decrease in self-sufficiency from 5.7% in 1970/71 to 4.1% in 1976/77.

The development in the consumption of different crops is quite variable. The most important source is soya bean with an increase in proportion from 55% in 1970/71 to 65% in 1976/77.

TABLE 1
BALANCE OF CONCENTRATED PROTEIN FEEDING STUFFS OF PLANT ORIGIN IN THE EEC-9[1])

Origin	Total amount (1000 t) 1970/71	1976/77	Change[2])	EEC-Production (1000 t) 1970/71	1976/77	Change[2])
1. Soya bean	7100	10743	+3643	-	2	+2
2. Groundnut	943	1312	+369	-	-	-
3. Copra	694	1024	+330	-	-	-
4. Rapeseed	686	943	+257	191	328	+137
5. Palmkernel	332	436	+104	-	-	-
6. Sunflower	589	547	-42	14	56	+42
7. Linseed	847	581	-266	5	10	+5
8. Cottonseed	859	582	-277	-	-	-
9. Pulses	784	469	-315	524	293	-231
Total	12834	16637	+3803	734	689	-45

[1]) Eurostat 1980

A positive development is also indicated for groundnut, copra, rapeseed and palmkernel, whereas sunflower, linseed, cottonseed and pulses were declining in importance. The reduced proportion of pulses is mainly due to a lower production in the EEC, and not to increased consumption. On the other hand we find a very impressive increase in the production of rapeseed, sunflower and linseed in the EEC.

Corresponding to the intention of this meeting the aim of this paper is to present data about the general nutritional value of the most promising EEC-producible protein feedingstuffs. In our opinion, these are rapeseed, sunflower and - hopefully for the future - soya bean.

2. CRUDE NUTRIENT COMPOSITION

An overview of the crude nutrient content of the different protein feedstuffs is given in Table 2 (DLG, 1972; Nehring et al., 1970; NN, 1976/78; Petersen and Schulz, 1980).

The highest crude protein content is found in soya bean meal with more than 50% of the dry matter, followed by rapeseed meal with about 40%. Sunflower products show quite a variation depending on the degree of dehulling; dehulled seeds contain well above 40% crude protein; partly dehulled seed may contain only about 30% crude protein. This is the same range as for pulses; for example, crude protein content of sweet lupins is well above 40% and in field beans it is only about 30%.

Ether extract is between 1% and 2% in the oilmeals depending on the degree of extraction; the figure for sweet lupins is three to four times higher.

The crude fibre content is relatively low in soya beans and field beans (6 - 8%) and relatively high in rapeseed and sweet lupins (14 - 17%). As expected there is a wide range indicated for sunflower depending on the degree of dehulling.

TABLE 2
COMPOSITION OF DIFFERENT PROTEIN FEEDSTUFFS (DRY MATTER)

Feedstuff	Crude Protein %	Ether Extract %	Crude Fibre %	NFE %	Ash %	Gross Energy KJ/g
Soya bean oilmeal[1]	51.9	1.4	6.2	33.6	6.9	20.3
Rapeseed oilmeal						
Convent.[1]	41.0	2.7	14.0	34.0	8.3	19.5
Erglu[3]	40.4	1.8	15.2	34.6	8.0	19.4
Sunflower oilmeal						
Dehulled[2]	44.7	1.0	18.1	28.3	7.9	19.9
Partly dehulled[1]	32.1	1.8	26.6	32.2	7.3	20.1
Sweet lupin[3]	43.2	5.6	16.7	29.5	5.0	19.7
Field bean[3]	29.5	1.6	8.1	57.2	3.6	19.1

[1] Ann. Report Inst. Angew. Botanik., Univ. Hamburg, 1973-76
[2] DLG-Futterwerttab., 1972
[3] Petersen and Schulz, 1980

The crude fibre content may possibly be a limitation for the utilisation of a crop in intensive animal production.

The content of NFE in field beans is near to 60%. This is essentially due to the low protein and fibre content. The other feedstuffs contain about 28 - 35% NFE.

The gross energy content is very similar for all feeds. The range is between 19.1 kJ/g and 20.3 kJ/g.

3. DIGESTIBILITY AND NUTRIENT CONTENT

For practical feeding purposes it is necessary to have additional information about the digestibility of the crude nutrients. In this review paper we will restrict our interest to the digestibility of the organic matter and of crude protein for ruminants, pigs and poultry. Table 3 summarises data from different feedstuff tables and from our laboratory (DLG, 1972; Nehring et al., 1970; Schulz and Petersen, 1978; Vogt, 1980). On the basis of such data it is possible to make a first decision about the probable use of a feedingstuff.

Generally speaking, the apparent digestibility of organic matter decreases in the order ruminant>pig>poultry. In addition a comparable range of the digestibility of the different feedstuffs within animals is indicated, the order being soya bean, pulses (sweet lupin, field bean), rapeseed, and sunflower. The difference between the figures for the three animal groups is mainly a reflection of their different potential to digest and utilise the fibre fraction. This is especially clear if we compare the data for dehulled and partly dehulled sunflower oilmeal.

There are some doubts about the reliability of some figures in Table 3. These data (soya bean for poultry, dehulled sunflower for pigs) have been put in brackets.

TABLE 3

APPARENT DIGESTIBILITY OF ORGANIC MATTER AND CRUDE PROTEIN OF DIFFERENT PROTEIN FEEDSTUFFS

	Organ. matter			Crude protein		
	Ruminant[1] %	Pig %	Poultry[2] %	Ruminant[1] %	Pig %	Poultry[1] %
Soya bean oilmeal	91	87	(65)	93	87	81
Rapeseed oilmeal						
Convent	77	69	43	85	79	72
Erglu	-	70	-	-	78	-
Sunflower oilmeal						
Dehulled	70	(80)[1]	66	86	89[1]	85
Partly dehulled	72	46[1]	-	88	82[1]	-
Sweet lupin	88	84	67	90	88	92
Field bean	86	80	71	83	79	81

[1] DLG-Futterwerttab., 1972 and Jahrbuch F. Geflugelw., 1978/80
[2] Futtermitteltabellenwerk, Nehring et al., 1970

The apparent digestibility of crude protein is in the range from 80% to 90%, with a decrease in the order ruminant> pig>poultry. The digestibility of crude protein is generally a little higher than for organic matter. Special attention should be paid to the remarkably good figures for sunflower protein.

The content of digestible protein and energy is presented in Table 4 (DLG, 1972; Schulz and Petersen, 1978; Vogt, 1980). The figures are calculated on the basis of data from the previous Tables. The content of digestible crude protein ranges between 240 g and 420 g for ruminants, 230 g and 480 g for pigs, and 240g and 420 g for poultry. The gradation of feedingstuff is mainly dependent on the crude protein content of the material, the influence of differences in digestibility are of minor importance.

In contrast, the differentiation between the energy values of the feedingstuffs is predominantly the result of differences in digestibility, because the gross energy contents were very similar, as was shown in Table 2. The energy values for the different animal species are not directly comparable. The definitions are net energy for ruminants (SE = starch equivalents = 9.862 kJ/g), a kind of digestible energy for pigs, and metabolisable energy for poultry. The energy value of rapeseed and sunflower is significantly lower than for soya bean and pulses. This is mainly an effect of the high crude fibre content.

4. PROTEIN QUALITY

The estimation of the content and digestibility of the crude nutrients is only a first step in the evaluation of the nutritional quality of a feedingstuff. For a more detailed evaluation it is necessary to have additional information about the content and bio-availability of essential nutrients. For a protein feedingstuff the content and availability of the amino acids is of utmost importance for monogastric animals like pigs and poultry, whereas for ruminants aspects of protein degradability in the rumen should be discussed.

TABLE 4

CONTENT OF DIGESTIBLE PROTEIN AND ENERGY VALUE (DRY MATTER)

	Dig. protein			Energy		
	Ruminant g/kg	Pig g/kg	Poultry g/kg	Ruminant[1] SE/kg	Pig TDN/kg	Poultry[2] MJ/kg
Soya bean oilmeal	480	450	420	800	815	11.1
Rapeseed oilmeal (Conv.)	350	320	290	670	665	7.0
" " (Erglu)	-	310	-	-	660	-
Sunflower - dehulled	380	400	380	620	700[1]	9.4
" - partly dehulled	280	260	-	575	430[1]	-
Sweet lupin	380	380	390	840	860	9.6
Field bean	240	230	240	790	790	11.8

[1] DLG-Futterwerttabelle, 1972
[2] Jahrbuch Geflugelw., 1980

4.1. Amino acid content

The amino acid composition of the protein feedingstuffs discussed is shown in Table 5 (AEC, 1972; Angelova et al., 1976; Degussa, 1973/77; DLG, 1976; Ergül and Schiller, 1971; Ergül and Schulz, 1980; Petersen and Schulz, 1980). In addition, the proposed composition of feed protein for broiler and pigs is presented in order to indicate the potential of the different protein sources to fulfil their needs. Only the most important amino acids for meat production with pig and poultry are chosen. All data are given as g amino acid/16 g N.

The lysine content is very low, and insufficient, in sunflower protein; in rapeseed and sweet lupins it is well within the range proposed for the reference animals, whereas soya bean and especially field beans show a remarkable surplus and can therefore compensate for lysine deficits in other feed proteins of a ration. The figure for the German double-low rapeseed variety 'Erglu' from our laboratory is surprisingly high. This is based on the few data so far available and therefore it is difficult to make a conclusive statement with regard to its reliability.

The content of the sulphur containing amino acids (methionine and cystine) is low in the legumes (soya bean, field bean and sweet lupin), and does not meet the proposed content for feed protein for broiler and pigs. Fortunately there is some surplus of sulphur amino acids in cereals, and also supplementation with synthetic dl-methionine is cheap. Moreover, a suitable combination of methionine-poor legumes with methionine-rich rapeseed or sunflower might be used advantageously.

A special problem in broiler nutrition is their high need for glycine. In all three legumes (soya bean, field bean, sweet lupin) an insufficient content is obvious, but there is a good chance for compensation by other feedstuffs, as - for example - rapeseed and sunflower.

TABLE 5

AMINO ACID COMPOSITION OF DIFFERENT PROTEIN FEEDSTUFFS AND PROPOSED COMPOSITION OF FEED PROTEIN FOR BROILER AND PIGS (g/16 g N)

	Lys	Met	Cys	Gly	Try[3]	Arg	His	Ile	Leu	Thr	Val
Soya bean oilmeal[1]	6.0	1.5	1.6	4.2	(1.5)	7.2	2.4	4.6	7.5	3.8	4.9
Rapeseed oilmeal											
Convent[2]	5.5	2.0	2.7	6.0	(1.3)	6.0	2.9	3.9	6.8	4.2	5.1
Erglu[1]	6.1	2.0	2.3	5.4	-	6.8	2.7	4.2	7.4	4.6	5.4
Sunflower oilmeal[1]	3.4	2.3	1.8	5.9	(1.2)	8.3	2.3	4.2	6.6	3.6	5.0
Field beans[1]	6.7	0.9	1.5	4.4	(0.9)	10.4	2.6	4.3	7.8	3.6	4.8
Sweet lupin[1]	5.0	0.9	1.9	4.0	(1.0)	11.2	2.6	3.8	7.8	3.2	3.5
Broiler	5.3	3.8		5.0	(1.0)	5.5	2.0	3.8	7.0	3.5	4.2
Pig (20 - 40 kg LW)	5.0	3.5		-	(0.8)	1.2	1.3	3.5	4.2	2.9	3.1

[1] Inst. Tierernahr., Fal., 1977-1980
[2] Degussa, 1973-1977
[3] AEC-Techn. Dokument Nr. 111, 1972

The figures for tryptophan are presented in brackets. The reliability of averages from the literature is questionable because different analytical methods are used by different authors. Standardisation of methods would be helpful.

For the other important essential amino acids cited in Table 5, shortages are unlikely. This statement also holds true if we look at the protein quality of cereals, which are the most important constituents of rations for poultry and livestock in Western European countries. This is shown for example in Table 6 (AEC, 1972; Angelova et al., 1976; Degussa, 1973/77). No deficits are indicated for isoleucine and leucine; the threonine content of wheat and possibly of barley might be limiting. Against that a clear deficit is demonstrated for lysine and glycine in all three cereal proteins, as well as a surplus of methionine and cystine as discussed in connection with the previous Table.

TABLE 6

AMINO ACID COMPOSITION OF CEREAL PROTEIN AND PROPOSED COMPOSITION OF FEED PROTEIN FOR BROILER AND PIGS (g/16 g N)

	Lys	Met + Cys	Try[2]	Arg	His	Gly	Ile	Leu	Thr	Val
Cereals:										
Barley[1]	3.6	3.7	(1.2)	5.1	2.3	4.0	3.6	7.0	3.3	5.1
Wheat[1]	2.7	4.0	(1.1)	4.5	2.3	3.9	3.5	6.7	2.8	4.4
Corn	2.8	4.5	(1.0)	4.5	2.7	3.8	3.4	13.1	3.6	4.7
Feed protein										
Broiler	5.3	3.8	(1.0)	5.5	2.0	5.0	3.8	7.0	3.5	4.2
Pigs (20-40 kg lw)	5.0	3.5	(0.8)	1.2	1.3	–	3.5	4.2	2.9	3.1

1) Degussa, 1973/77 and DLG-Futterwerttabelle, 1976
2) AEC-Techn., Document Nr. 111, 1972

4.2. Biological value

In practical feeding, protein feedstuffs are mainly used as supplements in order to increase the protein level and to balance the amino acid supply of a ration. As mentioned earlier in the Western European countries the origin of the basal protein is mostly from cereals and by-products. In order to demonstrate the potential supplementary effect of soya bean, rapeseed or sunflower protein, experiments were made with a combination of 60% cereal protein and 40% protein from the respective protein feedstuff. Some results from work with rats in our laboratory are presented in Table 7.

TABLE 7

SUPPLEMENTARY EFFECT OF DIFFERENT PROTEIN FEEDSTUFFS WITH BARLEY OR HIGH-PROTEIN CORN

Protein Feedstuff	Barley Corn	DN %	BV %	NPU %	PER
Soya bean oilmeal	$-$	82	79	71	3.4
"	$+ B^{1)}$	75	79	66	3.4
Rapeseed oilmeal (Erglu)	$-$	75	91	76	3.9
"	$+ B^{1)}$	78	74	64	2.9
Sunflower oilmeal	$-$	75	63	53	2.3
"	$+ C^{1)}$	82	54	50	1.7
	Barley	73	66	55	2.1
	Corn	81	54	48	1.6

1) Combination of 60% cereal protein plus 40% oilseed meal protein

The data clearly show a pronounced difference in biological protein quality between soya bean, rapeseed and sunflower. The best results were for rapeseed protein (76% NPU), followed by soya bean protein (71% NPU) and then sunflower protein, for which a NPU of only 53% was estimated. This means that sunflower protein is not better than barley protein (NPU = 55%) or corn (maize) protein (NPU = 48%) as a sole protein source.

The combination of barley protein with soya bean protein resulted in an increase of the protein quality if we take barley as a basis. Presumably this is predominantly a result of the surplus of lysine in soya protein previously mentioned. Consequently the results of the combination of rapeseed protein with barley should have been in the same order. Actually the effect was slightly smaller. It is difficult to find an explanation for this.

Sunflower was only tested as a supplement to corn (maize) protein. The results clearly show no improvement of protein quality (BV, NPU, PER) of the sunflower-corn combination as compared to corn protein alone. This is not surprising because, as was shown before, both feedstuffs are deficient in lysine and therefore a combination is of limited practical significance. But if we take an improved high-lysine corn variety instead of common corn then an increase in the protein quality characteristics should be expected. This effect is demonstrated in the upper part of Table 8 (Angelova et al., 1976). The BV, NPU and PER values of the combination with high-lysine corn are remarkably higher than with common corn. But for both combinations lysine is still the first limiting amino acid, as is demonstrated by the effects of lysine supplementation in the lower part of Table 8.

TABLE 8

EFFECT OF LYSINE SUPPLEMENTATION ON PROTEIN QUALITY OF SUNFLOWER-CORN MIXTURES[1] (RAT)

Feedstuff	DN %	BV %	NPU %	PER
Sunflower oilmeal	75	63	53	2.3
" + corn (2.6 g Lys/16 g N)	82	54	50	1.7
" + corn (4.2 g Lys/16 g N)	81	62	56	2.4
" + corn (2.6 g Lys/16 g N) + Lys.	81	73	66	2.7
" + corn (4.2 g Lys/16 g N) + Lys.	82	82	75	3.3

1) Combination of 60% corn protein plus 40% sunflower protein

As a background to these results we would like to say that breeders should not only concentrate their activities on improvi the protein quality of high concentrated protein feedingstuffs. The improvement of cereal protein should also be taken into consideration. Presumably this will require much time, but in the long run it could be important for the economic independence of the Community.

4.3. Protein quality for ruminants

For ruminants, generally speaking, the amino acid content and balance of feed protein is of limited importance. The nutrition of the high yielding dairy cow might be an exception because there is some indication of an insufficient methionine supply due to the limited microbial protein syntheses in the fore-stomachs.

There is another more general aspect in ruminant feeding with respect to rumen fermentation. Normally large amounts of feed protein are degraded in the rumen and the ammonium is re-utilised by the microbes. However, especially with high producing cows, the amount of degraded protein is often higher than the potential of the microbes for re-utilisation, thus resulting in increased N-losses which, in turn, means a lower feed-N utilisation. Table 9 gives an example of the differences in the degradability of oilmeal proteins (INRA, 1978). The highest degradation was measured for sunflower and groundnut protein (70%); the lowest figure is given for coconut protein (35%); soya bean and rapeseed being somewhere in the middle of that range (55%). There is work in progress to reduce the degra dability of feed protein by means of chemical processes; for instance, treatment with formaldehyde has been successful.

In concluding this part of the paper, it should be mention that things become more complicated the more we go into details. For instance, the balance of amino acids in monogastric nutri- tion seems to be a very promising field for future research.

TABLE 9

DEGRADATION OF OILSEED PROTEIN IN THE RUMEN *(in vitro)*

Oilmeal	Degradation % [1]
Sunflower	70
Groundnut	70
Soya bean	55
Rapeseed	55
Coconut	35

1) Alimentation des ruminants, INRA, 1978

5. ANTINUTRITIONAL FACTORS

With respect to practical feeding, antinutritional factors may limit the tolerable inclusion rate of feedstuffs. Some of these factors only affect digestibility and absorption of nutrients resulting in low contents of utilisable nutrients; others may also have pharmacological or even toxicological effects. Such compounds are especially well-known in rapeseed and soya bean. Examples are the trypsine inhibitor or haemagglutinins in soya; both factors are inactivated by proper toasting of the feedstuff (Liener, 1973). A special problem with rapeseed is the variable content of glucosinolates (Brak, 1978; Liener, 1973; MacGregor, 1978; McKinnon and Bowland, 1979). This is a collective name for a number of different chemical compounds. For special information about glucosinolates we would like to refer to the papers of Gland and Sørensen at this meeting.

Glucosinolates may have an effect on feed consumption and on the function of some internal organs like the thyroid gland, the liver and kidney. In our opinion, the reduction of such antinutritional constituents is a matter for further breeding activities, because it is not easy and not fully satisfactory or safe to overcome this sort of problem with processing techniques in the feedstuff industry.

Another serious problem which depresses the nutritional value of a feedstuff is the tannin content. Tannin is the collective name for a group of phenolic compounds. The result of high tannin content in a feedstuff is a depression of feed intake and/or a lowering of protein utilisation. Relatively high tannin contents are reported for horse beans (Jahreis and Gruhn, 1980; Liebert and Gebhardt, 1980) and for rapeseed (Fenwick and Hoggan, 1976).

In sunflower, antinutritional chemical compounds are of little significance. The content of chlorogenic acid or caffeic acid may be mentioned, but their concentration is low and negative effects on protein utilisation are unknown (Canella et al., 1976; Felice et al., 1976).

The technique of oil extraction may possibly have an effect on the protein quality. For example, it is well known that the availability of lysine might be impaired by factors in the management of steam, pressure and temperature.

6. RECOMMENDED INCLUSION RATES

6.1. Soya bean oilmeal

Provided that the processing has been adequate (trypsine inhibitor destroyed, extraction medium evaporated, availability of amino acids not impaired) no restrictions for the inclusion of soya bean in livestock and poultry feeding are necessary. In our laboratory we very often use soya bean oilmeal as a standard protein feedingstuff in feeding trials.

6.2. Sunflower oilmeal

The inclusion rate for sunflower oilmeal in rations for monogastric animals is 5% to 15%. Limitations are due to the content of the nearly indigestible hulls, which results in lower energy concentration and to the low lysine content (Afifi, 1972; Bourdon and Baudet, 1976; Stojanov, 1975).

In ruminant feeds, sunflower oilmeal can make up to 25% of the concentrate, although aspects of protein solubility may gain importance in high yielding cows.

6.3. Rapeseed oilmeal

The inclusion of rapeseed oilmeal in animal feeds is still limited by the glucosinolate content. Dependent on age and production purposes, 4% to 8% is recommended (Bourdon and Baudet, 1979; Bowland, 1976; Vogt, 1974). In feeds for breeding pigs only small proportions of rapeseed oilmeal should be included because of the effects of glucosinolates on the metabolism of some internal organs. More specified details about rapeseed feeding will be given by the other speakers in this session, and we will only refer to an example from our own experience with oilmeal from the double-low variety 'Erglu'.

Table 10 summarises our results (Petersen and Schulz, 1980). The design was a step-wise isonitrogenous exchange of soya bean oilmeal against rapeseed oilmeal, starting with a soya-barley diet and ending with a rapeseed-barley diet; the feeding regime was *ad libitum*. Feed intake was increased with increasing rapeseed content of the diet; however, the energy consumption was the same for all groups. Daily gain was not affected by the rapeseed content of the diet. Accordingly there is an increase in the feed conversion index on a feed basis and no effect at all on a TDN basis.

These results are very promising and if commerce is able to differentiate between conventional and new varieties different recommendations for the inclusion of rapeseed meal in animal feeds might be suitable.

In our experiments with growing-finishing pigs an inclusion of up to 25% 'Erglu' rapeseed oilmeal in the diet has had no adverse effect on the performance of the animals.

TABLE 10

FEEDING OF 'ERGLU' RAPESEED OILMEAL TO GROWING-FINISHING PIGS

Rapeseed	%	-	4.1	8.2	12.3	16.4	20.5	24.6	Regr.
Feed intake	Kg/d	2.51	2.54	2.56	2.56	2.57	2.64	2.63	L
	TDN/d	1780	1790	1790	1780	1770	1810	1780	-
Gain	Kg/d	0.84	0.84	0.84	0.82	0.85	0.85	0.82	-
Feed conversion	Feed/gain	3.00	3.04	3.04	3.12	3.02	3.13	3.20	L
Index	TDN/gain	2.14	2.15	2.13	2.17	2.08	2.14	2.17	-

REFERENCES

AEC, 1972. Techn. Dokument. Nr. 111, Commentry, France.

Afifi, M.A., 1972. Arch. Geflügelk., 36, 129-133.

Angelova, L., Schulz, E. and Oslage, H.J., 1976. Landbauforschung Völkenrode, 26, 23-39.

Bourdon, D. and Baudet, J.J., 1979. In: Journées Rech. Porc. INRA, Paris, France, 283-290.

Bowland, J.P., 1976. In: Feed Energy Sources for Livestock (Eds. Swan, H. and Lewis, D.) Butterworth, 129-192.

Brak, B., 1978. Diss. Univ. Kiel

Canella, M., Gastriotta, G. Mignini, V. and Sodoni, G., 1976. Riv. Ital. Sostanze Grasse, 53, 156-160.

Degussa, 1973-77. Aminosäuren-Zusammensetzung v. Futtermitteln, Hanau

DLG, 1972. Futterwerttabelle, DLG-Verlag, Frankfurt

DLG, 1976. Aminosäurengehalte in Futtermitteln, DLG-Verlag, Frankfurt

Ergül, M. and Schiller, K., 1971. Landbauforschung Völkenrode, 21, 113-116.

Ergül, M. and Schulz, E., 1980. Landbauforschung Völkenrode (in press)

Felice, L.J., King, W.P. and Kissinger, P.T., 1976. J. Agric. Food. Chem., 24, 380-382.

Fenwick, R.G. and Hoggan, S.A., 1976. Brit. Poultry Sci., 17, 59-62.

Gland, A., 1980. In: Proceed. Product. a. Utilizat. of Protein in Oilseed Crops, Braunschweig, 8.-10.7.

INRA, 1978. Alimentation des Ruminants, Versailles

Jahreis, G. and Gruhn, K., 1980. Arch. Tierern., 30, 381-390.

Liebert, F. and Gebhardt, G., 1980. Arch. Tierern., 30, 363-372.

Liener, J.E., 1973. In: Proteins in Human Nutrition, (Eds. Porter, J.W.G. and Rolls, B.A.), Academic Press, London - New York, 48L-500.

McGregor, D.J., 1978. Proceed. 5. Int. Rapeseed Conference, Malmö, 12.-16.6., 2, 64-67.

McKinnon, P.J. and Bowland, J.P., 1979. Can. J. Anim. Sci., 59, 589-596.

Nehring, K., Beyer, M. and Hoffman, B., 1970. Futtermitteltabellenwerk, VEB Deutscher Landw.-Verlag. Berlin.

NN, 1976-78. Jahresbericht 1973-76, Inst. f. Angew. Botanik, Hamburg

NN, 1980. Eurostat, Stat. Amt-EG, Brüssel.

Petersen, U. and Schulz, E., 1978. Landw. Forschung, 31, 269-289.

Petersen, U. and Schulz, E., 1980. In: *Vicia faba*, Feeding value, processing and viruses (Ed. Bond, D.A.), Publ. Martinus Nijhoff, 45-65.

Schulz, E. and Petersen, U., 1978. Landw. Forschung, 31, 218-232.

Sørensen, H., 1980. In: Proceed. Product. a. Utilizat. of Protein in Oilseed Crops, Braunschweig, 8.-10.7.

Stojanov, V., 1975. Arch. Tierern., 25. 707-716.

Vogt, H., 1974. In: Proceed. 4. Int. Rapskongress, Giessen, 4.-8.7., 453-462.

Vogt, H., 1980. Jahrbuch Geflügelwirtschaft, Verlag Eugen Ulmer.

USE OF OIL MEALS FOR DETERMINING PROTEIN REQUIREMENTS
OR AS SUPPLEMENTS FOR RAT AND PIG DIETS

A.A. Rérat

Laboratoire de Physiologie de la Nutrition,
INRA - CNRZ, Jouy-en-Josas 78350, France.

Studies on this topic made several years ago at the Pig Research Station of INRA (France) can be divided into two main groups according to whether the oil meals were considered (1) as a purified source of protein used to determine the protein and amino acid requirements of various categories of animals (growing pigs, breeding sows) or (2) as a supplement of cereals. The value of these oil meals used as supplements was estimated as well as the changes caused by different technological treatments.

USE OF VARIOUS OIL MEALS FOR DETERMINING PROTEIN REQUIREMENTS IN THE RAT AND THE PIG

Variations in the spontaneous energy intake according to the protein supply

It has been established that the level of feed intake is controlled by the energy requirement of the animal (Cowgill, 1928; Adolph, 1947; Mayer, 1955), and that it varies in particular according to the energy level of the diet (Hill and Dansky, 1954). However, other factors may be involved, such as the level and nature of the dietary crude protein (Osborne and Mendel, 1919).

Consequently, it may be asked whether the animal adjusts its level of feed intake according to its energy or its protein requirement. The application of the so-called 'separate feeding' method may contribute to answering this question. This method consists of a separate and simultaneous administration of two diets, one supplying all the proteins and the other the whole energy (Abraham et al., 1961). The proteins used in this experiment were supplied by fish, peanut and gluten (rat) or soyabean (pig).

Variations in the spontaneous energy intake according to the nature and the amount of crude protein were studied in 400 rats and subsequently in 200 pigs. In both cases the energy was offered *ad libitum*. Two cases were considered in relation to the protein supply:

- when the protein was offered *ad libitum* as was the energy, the animals were subjected to the 'free choice' method. In those conditions the rats and the pigs did not adjust their protein intake to their protein requirement very well (Henry and Rérat, 1963a; Rérat, 1970; Rérat, 1972). In the two species, the spontaneous level of protein intake depends on the nature of the latter and on the presence of more or less large amounts of carbohydrates in the protein diet. Thus the taste of this diet seems to have a marked effect on its level of intake.

- when the protein supply is not large enough to ensure a maximum growth, the spontaneous energy intake increases with the protein supply; for a given amount of protein it is all the higher as the quality of the protein source is better (Abraham et al., 1961). An increase in the energy intake is related to an improvement of growth performance, i.e. whatever the level and the nature of protein, energy intake raised to a given power is proportional to weight, and it is directly proportional to the amount of nitrogen retained in the tissues (Henry and Rérat, 1963b; Rérat and Henry, 1963a; Rérat and Henry, 1963b; Rérat et al., 1963). In addition, use of protein free diets with various lipid and crude fibre concentrations allows the specification that the adjustment involves the digestible fraction of the energy ingested (Henry and Rérat, 1966). From the results of these studies it may be concluded that in the growing rat, the feed intake is controlled by the requirement for energy which depends on the level of growth rate and the degree of nitrogen retention permitted by the quality and quantity of protein ingested by the animal.

TABLE 1

INFLUENCE OF THE LEVEL OF PROTEIN SUPPLY ON THE SPONTANEOUS ENERGY INTAKE AND THE GROWTH PERFORMANCES OF PIGS SUBJECTED TO SEPARATE FEEDING BETWEEN 27 kg and 90 kg (HENRY AND RERAT, 1968)

Sex		Females			Castrated males (2)	
Amount crude protein offered g/d (1)	152	229	305	152	229
Number of animals	7	8	5	7	4
Mean gain/d, g	451	586	667	481	628
Crude protein intake, g/d	141	197	255	142	198
Dry matter intake, kg/d	1.51	1.75	1.95	1.58	1.88
Crude protein % dry matter (DM)	9.4	11.3	13.1	8.9	10.5
Feed conversion ratio (DM)	3.34	2.99	2.89	3.29	2.99
Body composition (% net weight)					
Ham + loin	50.3	49.5	50.7	46.0	49.1
Backfat + flarefat	19.6	20.1	18.4	24.1	20.8
Backfat mean thickness (Loin + backfat)/2, mm	29.1	29.4	29.0	37.0	31.1

(1) Soyabean oil meal
(2) The refusals being too large, the highest level was not taken into consideration

Similarly, in the pig the spontaneous energy intake varies proportionally to the growth rate allowed by the protein supply as long as the requirement for protein is not fully satisfied. However, the adjustment of the spontaneous energy intake is much better in females than in castrated males i.e. at the same protein level, their intake and adiposity are lower and this adiposity remains constant whatever the protein supply, within the previously determined limits (Henry and Rérat, 1964; Henry and Rérat, 1968).

However, the general rules for the adjustment of the energy intake established by the technique of separate feeding cannot be applied to the classical mode of feeding without some modifications. This was shown by comparing results obtained in separate and in mixed feeding trials using the same amount of protein intake. The influence on the amount of energy intake of the proportion and the amount of crude protein could thus be dissociated. When the supply does not completely meet the requirement the spontaneous energy intake of rats subjected to separate feeding is lower than that of their congeners subjected to the corresponding mixed feeding. The reduced adiposity of the former rats leads to a slight growth rate decrease as compared to the latter, but their nitrogen retention is not affected. On the other hand, when the protein requirement is fully satisfied, the adjustment is the same in separate and mixed feeding and the body composition is similar for the two groups of animals (Henry and Rérat, 1965). The same conclusions can be drawn with pigs, but only with females (Rérat and Henry, 1964; Henry and Rérat, 1968). The original aspects of this work can be summarised as follows: when the rat and the female pig are subjected to a hyponitrogenous mixed feeding, they tend to increase their level of feed intake so as to obtain a better satisfaction of their requirement for protein. In other words, protein regulation modulates energy regulation which plays the main part when the diet contains a sufficient amount of amino acids.

Protein requirement during growth

The purpose of these studies, using semi-synthetic diets with a constant energy level, was to determine the 'conversion rate' of various protein feeds as compared with a reference protein of very good quality. Norwegian fish meal was used as a reference protein because of its constancy of composition and mode of preparation. Besides, its biological value is only a little changed by the technological treatments. It is very digestible and the availability of its amino acids is apparently high (Rérat and Jacquot, 1956; Rérat and Lougnon, 1963).

A series of experiments performed on 400 rats fed various protein feeds (Jacquot and Rérat, 1966) showed that for a given protein, the growth rate and nitrogen retention increase with the protein level up to a maximum which varies according to the nature of the protein. The crude protein level leading to maximum growth is correspondingly lower and the growth rate correspondingly higher as the protein source is of better quality. Thus, the maximum daily mean gains and corresponding crude protein levels are:

for fish meal	5.0 g and 16%
for sunflower oil meal	4.7 g and 20%
for peanut oil meal	4.4 g and 24%
for wheat gluten	4.0 g and 44%

Accordingly, a protein of poor quality does not give the same performance as a protein of good quality even when its concentration in the diet is increased. Furthermore, elevation of the dietary crude protein level to the optimum, which varies according to the protein, leads to an increase in the percentage of protein in the tissues.

Requirements for essential amino acids

The overall requirement for protein defined in previous studies does not give any information about the requirement for essential amino acids. These studies had therefore to be complemented by other investigations aimed at determining a protein of 'ideal' composition. This can be done by using either synthetic diets in whch the protein supply is represented by a mixture of synthetic amino acids in given proportions or semi-synthetic diets containing a protein of known amino acid availability and whose imbalances are compensated for by addition of various synthetic amino acids. We chose the latter method as it is less expensive.

TABLE 2

COMPARISON BETWEEN MIXED AND SEPARATE FEEDING (THE SAME AMOUNT OF PROTEIN INTAKE) IN GROWING-FINISHING PIGS (HENRY AND RERAT, 1968). INITIAL LIVE WEIGHT : 28.8; FINAL LIVE WEIGHT : 92.0 kg

Sex	Females		Castrated males	
Feeding	Mixed(x)	Separate	Mixed(x)	Separate
Mean gain/d, g	515	416	523	507
Crude protein intake, g/d	158	153	161	157
Dry matter intake, kg/d	1.63	1.42	1.67	1.64
Crude protein, % DM	9.7	10.8	9.7	9.5
Feed conversion ratio (DM)	3.22	3.48	3.56	3.28
Body composition (% net weight)				
Ham + loin	47.9	50.3	46.7	45.6
Backfat + flarefat	23.2	19.6	23.5	22.5
Backfat mean thickness $\frac{(Loin + backfat)}{2}$, mm	33.5	28.4	36.2	35.4

(x) semi-purified diet including 8% fish protein

In order to obtain results concerning the different amino acids more rapidly, we made these studies by means of various proteins in which the nature and the importance of the limiting factor was different. Each of these protein feeds was introduced in various proportions into the semi-synthetic diets and was admixed with the presumed first limiting amino acid at various levels and with or without addition of the second limiting factor. The satisfaction of the requirement was estimated according to various criteria (growth rate, nitrogen retention) (Rérat et al., 1956; Rérat and Jaquot, 1956). This original method was developed in the rat and used to determine the requirement for lysine and sulphur amino acids; it was thereafter applied to the pig for the determination of the requirements for sulphur amino acids, lysine, threonine and isoleucine.

Determination of the requirement of lysine for growth

These experiments were performed by means of sunflower oil meal whose first limiting factor is lysine. The exclusive use of this protein source allows us to estimate the requirement for this amino acid. However, the availability of its amino acid has not been as well established as in fish meal and should therefore be further examined.

In the rat (Rérat and Henry, 1963c), there is a linear relation between the amount (Y) of lysine and the level (X) of proteins in the diet; in the case of lysine, this relation is the following:

$$Y = 8.15 - 0.018\ X$$

applicable for protein levels which do not allow maximum growth to be obtained even with a lysine supplementation. The lysine requirement for maximum growth in the white rat (Wistar strain) is 0.8% of a diet including 4 300 Kcal/kg.

In the pig (Pion et al., 1966; Henry et al., 1971), factorial experiments (3 sunflower protein levels: 12, 16 and 20%; 3 supplementary lysine levels: 0, 1.5 and 3% of the proteins) using semi-synthetic diets made it possible to calculate the requirements of females and castrated males between 20 and 50 kg liveweight as, respectively, 2.9 and 2.5 g/1 000 Kcal digestible energy. Here again, an excess of the added amino acids reduces the maximum utilisation of the diet. This excess, especially when combined with low protein levels, leads to a decrease in the feed intake which may be used to reduce the adiposity of the carcases.

TABLE 3

LYSINE SUPPLEMENTATION OF A SUNFLOWER DIET IN THE GROWING PIG (HENRY ET AL., 1971). BETWEEN 20 AND 90 kg LIVEWEIGHT, 12 ANIMALS PER GROUP (6 CASTRATED MALES, 6 FEMALES)

Crude protein	12			16			20		
Lysine supplement % crude protein	0	1.5	3.0	0	1.5	3.0	0	1.5	3.0
Daily mean gain (g/d)	588	638	623	603	693	632	632	667	670
Feed intake (kg/d)	2.05	1.99	1.93	2.03	2.14	1.97	2.05	2.10	2.09
Feed conversion ratio	3.53	3.14	3.10	3.38	3.10	3.14	3.27	3.19	3.13
Dressing percentage, %	74.2	74.4	74.9	73.4	73.7	74.2	74.2	73.8	73.8
Lean cuts, %	46.8	50.3	50.3	50.3	50.2	51.4	50.1	50.4	50.6
Fat cuts, %	24.0	19.9	20.0	20.0	20.2	18.3	19.9	19.5	19.5
Backfat mean thickness, mm	36.5	31.1	31.9	28.3	29.1	26.5	29.6	29.2	29.9

TABLE 4

INFLUENCE OF A LYSINE DEFICIENCY ON THE REPRODUCTIVE PERFORMANCE OF PRIMIPAROUS SOWS (DUEE AND RERAT, 1973)

Physiological status % Lysine in the diet	LOT 1 V 0.44	LOT 2 G 0.23	LOT 3 G 0.44	LOT 4 G 0.64	Statistical (1) significance s̄x
Gestation total gain (kg)	30.7 a	39.7 ab	44.8 bc	49.8 c	3.5 (22.3) **
Gestation net gain (kg)	30.7	25.6	28.2	33.7	3.0 (27.1) NS
Weight loss during lactation (kg)	—	-1.3	5.5	16.3	5.3 (192.4) NS
Number of piglets born alive/litter	—	8.4	8.5	7.5	0.6 (18.3) NS
Average weight of piglets at birth (kg)	—	0.98 a	1.25 b	1.24 b	0.07 (15.54) *
Total weight of the litter at birth (kg)	—	8.11	10.55	9.42	0.75 (21.23) NS
Number of piglets weaned/litter (at 35 days)	—	7.5	7.5	7.3	1.0 (33.4) NS
Average weight of piglets at weaning (kg)	—	5.73 a	7.25 b	8.10 b	0.35 (12.20) **
Total weight of the litter at weaning (kg)	—	44.43	53.30	58.40	7.10 (33.4) NS

(1) s̄x : SEM (variation coefficient). Figures followed by the same letter do not differ at the threshold of $P < 0.05$

NS : non significant difference
* : difference significant at the threshold $P < 0.05$
** : difference significant at the threshold $P < 0.01$
V : non pregnant sows
G : pregnant sows

Determination of the requirement for lysine in the breeding sow

Studies on amino acid requirements during the reproductive cycle first involved lysine, as this amino acid, together with threonine, is the only strictly essential amino acid, and as the basal diet in pigs is composed of lysine deficient cereals. The same experimental principle was used as that applied to studies on growth (semi-synthetic diets). The protein source was oil meal supplemented with various amounts of lysine. In addition to the reproductive performance of the animals (number of ova produced, number and weight of piglets born and weaned) the variations in the aminoacidaemia patterns were also studied to show the nutritional status and especially the nature of the limiting factor of the diet. A method previously developed for estimating the threonine requirement during growth (Pion et al., 1971) was used.

In our experiment, lysine deficiency during pregnancy led to a lowering of the mean weight of the piglets at birth (without, however, affecting their number) as well as to a reduction in subsequent milk production, probably because of a smaller mobilisation of the body reserves during lactation. This results in a decrease in the growth of the suckled piglets (Duée and Rérat, 1974; Duée and Rérat, 1975).

These investigations are being pursued to determine the requirements for other amino acids during the different phases of life of the breeding sow (prepuberty growth, pregnancy, lactation).

Relationships between protein requirement and energy supply

Because of the large variations in the spontaneous feed intake depending on the nature and the level of the energy supply, and also because of the feed restrictions applied to animals to reduce their adiposity, a systematic study was made of the possible variations in the requirement for protein and amino acids resulting from these factors. The example given concerns the

influence of the dietary protein level on the necessary supply
of protein, and in particular on the lysine requirement (Rérat
et al., 1970).

Instead of studying the variations in the overall protein
requirement this factorial experiment only involved the requirement for lysine (3 energy levels: 3 500, 3 200 and 2 900 Kcal/kg;
4 lysine levels: 0,1,2 and 3% of the proteins; 2 categories of
animals: females and castrated males) using sunflower meal (poor
in lysine) as a protein source. In those conditions, the optimum level of lysine in the diet varied in the same way as the
energy level of this diet. In practice, therefore, it appeared
to be necessary to express the amino acid requirements relative
to the energy value of the diet.

Chronology of the dietary supply of amino acids

It is generally admitted that the best possible metabolic
use of amino acids depends on two factors: a proper balance
between these amino acids and their simultaneous supply at the
sites of protein synthesis. Thus, it has been stated that in
the case of a dietary protein deficiency, the addition of amino
acids - in the form of proteins or in a synthetic form - is only
fully efficient if it is made simultaneously. So far, only a
few experiments have been made on that topic in the pig, which
has a digestive tract anatomically and physiologically different
from that of the species in which these facts have been demonstrated (rat: Geiger, 1947; dog: Elman, 1939). Thus, the speed of
transit and absorption of the proteins studied may be different
in the pig. It seemed of interest to study two aspects of this
problem:

- Comparison of the absorption rate of amino acids offered
  in the free form or as a protein complex

- Influence of the length of time necessary to re-balance
  the diet on its biological efficiency (growth rate feed
  efficiency). Since the physiological processes, especially the digestive ones (enzyme, pancreatic and hepatic activities, blood flow rate) are subjected to 24 h cycles,

TABLE 5

PERFORMANCES OF GROWING PIGS AS AFFECTED BY ALTERNATE PROTEIN * AND ENERGY * FEEDING (RERAT AND BOURDON, 1972)

	Mean weight	Mean age
Beginning of the experiment	18.9 kg	70 days
End of the experiment	89.1 kg	187 days

Group	1	2	3	4
Protein levels (1)	16 16	23 9	9 23	37 9 9 9
Feeding time (2)	M E	M E	M E	M E M E
Growing period (20-60kg)				
Daily mean gain (g)	540	527	527	483
Feed conversion ratio (kg fresh matter)	2.85	2.99	2.95	3.26
Finishing period (60-90kg)				
Daily mean gain (g)	778	750	804	804
Feed conversion ratio (kg fresh matter)	3.60	3.87	3.53	3.72
Growing-finishing period (20-90kg)				
Daily mean gain (g)	620	601	613	573
Feed conversion ratio (kg fresh matter)	3.14	3.28	3.18	3.45
Body composition				
Yield %	71.22	71.15	70.77	71.06
Lean cuts %	52.88	52.60	52.29	52.06
Fat cuts %	17.27	17.64	17.25	18.33
Backfat thickness $\frac{Loin + backfat}{2}$, mm	23.7	24.2	24.6	27.1

* Meals either composed of barley (9% proteins) or of various mixtures of barley and soya bean (16, 23 and 37% proteins)

(1) These figures indicate the protein level of the meals used. Feeding chronology is 24 h (groups 1, 2, and 3) and 48 h (group 4).

(2) Feeding time : M : morning; E : evening.

TABLE 6

DELAYED LYSINE SUPPLEMENTATION, 1975. GROWING PERIOD : 20 - 60 kg.

Number of animals/group : 12									Mean weight/kg			Mean age/d		
		Beginning of experiment							19.5			73		
		End of experiment							60.1			140		
Growth-feed intake														
Group		1			2			3			4			
Feeding time		M	E		M	E		M	E		M	E		
Diet sequence		C	C		S	NS		NS	S		L	NS	NS	NS
Daily mean gain, g/d	CM	605			614			630			557			
	F	714 A			610 BC			672 AB			569 C			
	M	660 A			612 AB			651 A			563 B			
Average feed intake	CM	1.77			1.66			1.79			1.66			
kg/d	F	1.86			1.73			1.82			1.70			
	M	1.81 a			1.69 b			1.81 a			1.68 b			
Feed conversion ratio	CM	2.92			2.71			2.85			2.98			
	F	2.61 C			2.84 AB			2.78 B			3.00 A			
	M	2.76 A			2.77 A			2.82 A			2.99 B			

M : morning   E : evening
C : control   NS : non-supplemented (0.4% lysine)   S : supplemented (0.8% lysine)
L : supplemented (1.6% lysine)
CM : castrated males   F : females   M: mean

the interaction between this length of time and the moment of the main protein meal during the day (morning or evening) should be taken into account.

The answer to these questions varies according to the type of supplementation used. When the supplement is a complex protein (addition of soyabean to a barley diet) the supplementation fully plays its part provided that the interval between the intake of the two protein sources does not exceed 10 h. If the interval until the administration of the supplement extends to 48 h, then a depression of the growth rate and the feed conversion ratio is observed (Rérat and Bourdon, 1972). When the supplementation is based on a single amino acid (addition of lysine to a semi-synthetic diet containing sesame oil meal) the effect of the supplementation decreases when the latter is delayed by at least 10 h, the reduction of performances increasing with the interval between the intake of the meals and that of their supplement. It has to be emphasised that the moment when the meal containing the supplement is offered seems to be important, being less efficient in the morning than in the evening (Rérat and Bourdon, 1975). This may be related either to the rhythm of the hepatic enzyme activities studied by one of the teams of the Nutrition Research Centre (Peret et al., 1978) or to the different intervals between meals from the morning to the evening (10 h) or from the evening to the morning (14 h).

RESEARCH ON THE NUTRITIVE VALUE OF OIL MEALS USED AS SUPPLEMENTS AND ITS VARIATIONS ACCORDING TO THE TECHNOLOGICAL TREATMENTS

Research in this field should be considered in a wider context involving examination of the nutritive value of various feeds relative to their chemical composition. Special attention has been paid to various feed resources of potential interest in pig production, especially in terms of import savings, with the aim of reducing the incorporation of soya bean into the diet to the benefit of plants cultivated in France.

Supplementation value of peanut and sunflower oil meals admixed with lysine

In the absence of an accurate knowledge of amino acid requirements the rules of a rational pig feeding established between 1940 and 1950 preconized the use of complex diets, the protein supply being based on a mixture of plant and animal proteins. In the best conditions, this method corresponded to an empirical application of the principle of supplementation, i.e. a reciprocal improvement of the value of the proteins. Accordingly, the efficiency of the mixture exceeded the arithmetical mean of the efficiency of each component. The presence of proteins of animal origin in animal diets was justified by their good amino acid balance, improving the efficiency of plant proteins which are not so well balanced. However, the yield of animal protein from plant protein being very low, it did not seem to be advisable to have another cycle of transformations before reaching the definitive consumer, i.e. man, except in the case of by-products which cannot be used in human nutrition.

On account of the very good amino acid balance of some plant proteins (soya bean), a logical next step is their suppression in pig feeding in order to use them in human feeding. The future trends of pig feeding will be the utilisation of cereals suitably supplemented with synthetic amino acids, or industrial by-products re-balanced in the same way, or new sources of cereals rich in some amino acids (Opaque maize 2, hiproly barley) This explains the chronology in our work on amino acid supplementation in pig feeding: first using complex diets, thereafter simplified plant diets and then diets based exclusively on cereals.

The simplified plant diets included a cereal (barley) and various oil meals (peanut, sunflower) admixed with synthetic amino acids (lysine, methionine). A feeding formula was developed (barley + sunflower + lysine) and the results obtained in practice were as good as those observed with protein mixtures including animal proteins. The results recorded with peanuts, however, were poorer because of other amino acid deficiencies in this protein feed (Rérat and Lougnon, 1965).

TABLE 7

COMPARISON BETWEEN BARLEY DIETS SUPPLEMENTED WITH FISH MEAL OR WITH PEANUT OIL MEAL + LYSINE (RERAT AND LOUGNON, 1965)

Diet	Experiment A Barley fish	Barley peanut	Barley peanut	Experiment B Barley + peanut (0.2% lysine)	Barley + peanut (0.4% lysine)
Daily mean gain (g)	573	487	494	591	566
Daily feed intake (kg)	1.99	1.96	1.89	2.11	1.94
Feed conversion ratio (kg)	3.49	4.03	3.85	3.57	3.44
Backfat mean thickness (mm)	28.7	30.3	33.0	32.1	29.1

Influence of industrial treatments on the composition and feeding value of oil meals (Rérat et al., 1956; Jacquot et al., 1958, 1959; Abraham et al., 1971; Lougnon and Rérat, 1971)

A certain number of protein rich feeds must undergo technological treatments before they can be incorporated into human or animal diets. They may, for instance, be concentrated by separation from the rest of the product (oil meals from seeds whose oils are extracted by pressure or by solvents). Most of these treatments require a more or less long heating of the product, and it has been known since the beginning of this century (Maillard, 1912) that such treatment may lead to the formation of irreversible chemical bonds between the amino radicals of some amino acids (lysine for instance) and the ketone or aldehyde groups of some sugars (lactose, for instance). Bonds may also be formed inside the proteins so that they become partly unusable from a biological point of view (Ford, 1975).

The purpose of the studies made in this field was on the one hand to determine the most suitable treatments in terms of influence on the nutritive value, and on the other hand to find out which amino acids were liable to become unavailable and the magnitude of the deficits.

Several series of studies were performed in the rat:

- according to a request of the FAO (peanut: influence of intensity of the heat treatment)

- according to a request of the ITERG* (various oil meals: influence of the nature of the extraction solvents and of the heat treatments)

- within the INRA projects (heat treatments of soya bean)

These studies show that:

- the nature of the solvent used to extract the lipids affects the nutritive value of the oil meals. Lipid extraction by means of acetone gives oil meals of a higher quality than those obtained after classical delipidation with petrol; this is the case for peanut, for linseed, and for rapeseed. Use of trichlorethylene is absolutely noxious.

- oil meal proteins are susceptible to excessive heat treatments. This can be observed both in rapeseed and peanut. In the latter case, a more or less marked destruction of lysine, thiamine and pantothenic acid takes place, leading to a lowering of the nutritive value of the oil meal. As regards soya bean, a very detailed study shows that beyond $120°C$, the treatment has a deleterious effect on the nutritive value of the oil meal. This effect increases with the temperature and length of treatment. Overheating is characterised by an unavailability or a large destruction of lysine. This then becomes the limiting amino acid whereas when the oil meal has been subjected to moderate heating the limiting factor is methionine.

* Technical Institute for Lipid Research

TABLE 8

HEAT TREATMENTS OF EXPERIMENTAL OIL MEALS

Oil meal	Temperature (°C)	Heating duration (mn)	Humidity (%)
41	110	20	20
42	120	20	20
43	130	20	20
44	140	20	20
51	100	10	20
52	110	20	20
53	110	40	20
54	120	20	20
55	130	10	20
56	130	40	20

TABLE 9

EXPERIMENT IA : MEAN RESULTS (7 ANIMALS/GROUP). INITIAL LIVEWEIGHT : 48.6 g
DURATION : 28 DAYS

Parameter	Methionine supplement (g/kg)	Oil meals 41	42	43	44	$S\bar{x}$ (3)
Daily mean gain (g/d)	0	2.41	2.48	1.91	0.90	
	1.5	3.28	3.84	2.47	0.97	0.21
	3	2.45	3.56	2.65	0.82	
Dry matter intake (g)	0	292	293	284	207	
	1.5	336	349	294	206	14.9
	3	269	326	308	205	
Feed conversion ratio (IC) (1)	0	4.36	4.24	5.39	8.73	
	1.5	3.86	3.26	4.33	7.91	0.46
	3	4.35	3.30	4.19	9.36	
Protein efficiency ratio (CEP) (2)	0	2.03	2.07	1.66	1.07	
	1.5	2.28	2.60	2.00	1.17	0.11
	3	2.09	2.68	2.12	0.98	

(1) $IC = \dfrac{\text{Dry matter intake (g)}}{\text{Weight gain (g)}}$ (3) $S\bar{x}$ = SEM

(2) $CEP\ (PER) = \dfrac{\text{Weight gain (g)}}{\text{Crude protein intake N} \times 6.25\ (g)}$

TABLE 10

EXPERIMENT IB : MEAN RESULTS (7 ANIMALS/GROUP). INITIAL WEIGHT : 70.5 g
DURATION : 28 DAYS

Groups	DM	DL	DT	QM	QL	QT	$\bar{S_x}$
Oil meal number	42	42	42	44	44	44	
DL-methionine supplementation (2g/kg)	+	+	+	+	+	+	
L-lysine supplementation (2g/kg)		+	+		+	+	
L-threonine supplementation (1g/kg)			+			+	
Daily mean gain (g)	3.78	3.19	4.24	0.73	2.05	2.90	0.29
DM intake (g)	367.4	321.4	360.3	229.1	284.8	323.6	27.1
Feed conversion ratio	3.36	3.69	3.08	12.27	5.26	4.00	**
Protein efficiency ratio	2.40	2.21	2.75	0.73	1.64	2.14	**

** $\bar{S_x}$: not indicated because of the heterogeneity of the error according to the oil meals studied.

TABLE 11

EXPERIMENT II : MEAN RESULTS (8 ANIMALS/TREATMENT). INITIAL WEIGHT : 83.3 g
DURATION : 28 DAYS

Parameter	DL-methionine supplement (g/kg)	\multicolumn{6}{c}{Oil meal number}	$\bar{S_x}$
		51	52	53	54	55	56	
Daily mean gain (g)	0	3.19	3.09	3.00	2.67	2.23	1.54	0.19
	2	4.90	4.39	4.65	4.87	3.78	1.91	
Dry matter intake (g)	0	357	359	353	342	327	306	10.8
	2	395	374	400	407	391	316	
Feed conversion ratio	0	4.06	4.20	4.31	4.64	5.50	7.26	0.23
	2	2.90	3.05	3.10	3.02	3.69	6.01	
Protein efficiency ratio	0	2.10	2.05	1.99	1.88	1.65	1.21	0.08
	2	2.96	2.84	2.84	2.85	2.27	1.45	

PRESENT STUDIES ON THE DIGESTION OF PROTEINS AND CARBOHYDRATES
(Rérat et al., 1978; Rérat et al., 1978, 1979)

The aim of these studies is to determine quantitatively the kinetics of nutrients in the body during digestion. They are based on a methodology used to estimate the nutrient enrichment of the intestinal efferent blood (analysis of the differences in the porto-arterial concentration), the calculation of the amount of nutrients per time unit being based on the measurement of the blood flow rate in the portal vein.

This methodology has been used to study the digestion of various sugars (maize starch, sucrose, lactose, glucose) and various feeds (fish meal, barley, wheat, maize) in the pig. Further investigations will be made using this technique to examine the availability of the amino acids in milk powder and proteinaceous plants.

REFERENCES

Abraham, J., Calet, C., Rérat, A. and Jacquot, R., 1961. Solidarité des besoins énergétique et protéique de croissance: le phénomène d'ajustement entre protéines et calories. CR Acad. Sci., 253, 2768-2770.

Abraham, J., Adrian, J., Calet, C., Geneviève Charlet-Léry, Delort-Laval, J., Guillaume, J., Gutton, M., Lougnon, J. and Rérat, A., 1971. Traitement thermique et qualité des protéines du soja. VIII - Sensibilité de trois espèces animales (Rat, Porc, Poulet) et de divers tests biochimiques à l'intensité de la cuisson du tourteau. Ann. Zootech., 20, 75-87.

Adolph, E.F., 1947. Urges to eat and drink in rats. Am. J. Physiol., 151, 110-125.

Cowgill, G.R., 1928. The energy factor in relation to food intake: experiments on the dog. Am. J. Physiol., 85, 45-64.

Duée, P.H. and Rérat, A., 1974. Etude du besoin en lysine de la truie gestante nullipare. Journées Rech. Porcine en France, INRA. ITP ed., Paris, 49-56.

Duée, P.H. and Rérat, A., 1975. Etude du besoin en lysine de la truie gestante nullipare. Ann. Zootech., 24 (3), 447-464.

Elman, R., 1939. Time factor in retention of nitrogen after intravenous injection of a mixture of amino acids. Proc. Soc. Exptl. Biol. Med. 40, 484-487.

Ford, J.E., 1975. Some effects of overheating of protein on its digestion and absorption. In: Matthews, D.M. and Payne, J.W. (eds.). Peptide transport in protein nutrition, 183-203, North-Holland Publ. Company, Amsterdam.

Geiger, E., 1947. Experiments with delayed supplementation of incomplete amino acid mixture. J. Nutr. 34, 97-111.

Henry, Y. and Rérat, A., 1963a. Etude de l'ingestion spontanée d'éléments énergétiques et de protéines chez le rat en croissance par la méthode du libre choix. Ann. Biol. anim. Bioch. Biophys., 3(2), 103-117.

Henry, Y. and Rérat, A., 1963b. Influence de la qualité et de la quantité des matières azotées ingérées sur la consommation spontanée d'énergie chez le rat en croissance. Ann. Biol. anim. Bioch. Biophys., 2, 267-276.

Henry, Y. and Rérat, A., 1964. Untersuchung über die Bestimmenden Faktoren der Nahrungsaufnahme von Ratte und Schwein mit Hilfe getrennte Mahlzeiten. Landwirtschaftliche Forschung, 19, (special number).

Henry, Y. and Rérat, A., 1965. Ingestion spontanée d'éléments énergétiques en régimes mixtes et séparés chez le rat en croissance. Ann. Biol. anim. Bioch. Biophys., 5, 283-292.

Henry, Y. and Rérat, A., 1966. Evolution de l'ingestion spontanée de principes énergétiques en fonction de la vitesse de croissance et de la protéinogénèse chez le rat blanc. Cahier AEC, 6, 237-262.

Henry, Y. and Rérat, A., 1968. Voluntary intake of energy by the growing pig given separately the energy and protein components. Proc. 2nd. Wld. Conf. Anim. Prod. College Park, Maryland, session 13 paper 89, 460-461.

Henry, Y., Rérat, A. and Tomassone, R., 1971. Etude du besoin en lysine du porc en croissance finition - Application de l'analyse multi-dimensionnelle. Ann. Zootech., 20, 521-550.

Hill, F.W. and Dansky, L.M., 1954. Studies on the energy requirements of chickens. I-The effect of dietary energy level on growth and feed consumption. Poult. Sci., 33, 112-119.

Jacquot, R., Adrian, J. and Rérat, A., 1958. Influence de la nature des solvants d'extraction sur la valeur alimentaire des tourteaux. Revue Francaise des Corps Gras, V, 3-8.

Jacquot, R., Abraham, J., Adrian, J., Bourdel, G., Charconnet, F., Raulin, J. and Rérat, A., 1959. Normes de qualité et préparation des farines alimentaires d'arachide basée sur la valeur nutritive. Publications UNICEF, FAO, Ac. Sci. New York and Revue Francaise des Corps Gras.

Jacquot, R. and Rérat, A., 1966. La valeur biologique des protéines. Congrés Mondial de Nutrition animale, Madrid, vol. 1, 327-407.

Lougnon, J. and Rérat, A., 1971. Traitement thermique et qualité des protéines du soja. IV. - Effet de la cuisson des tourteaux et de leur supplémentation par des amino acides sur la croissance et la rétention azotée chez le Rat. Ann. Zootech., 20, 41-52.

Maillard, L.C., 1912. Action des acides aminés sur les sucres: formation des melanoïdines par voie méthodique. CR Accd. Sci., Paris, série D, 154, 66.

Mayer, J., 1955. Regulation of energy intake and the bodyweight: the glucostatic theory and the lipostatic hypothesis. Ann. N.Y. Acad. Sci., 63, 15-43.

Osborne, T.B. and Mendel, L.B., 1919. The nutritive value of the wheat kernel and its milling products. J. Biol. Chem., 37, 557-601.

Peret, J., Chanez, M. and Bois-Joyeux, B., 1978. Schedule of protein ingestion and circadian variations of glycogen phosphorylase, glycogen synthetase and phosphoenol pyruvate carboxykinase in rat liver. J. Nutr., 108, 265-272.

Pion, R., Fauconneau, G. and Rérat, A., 1966. Etude de la digestion des protéines chez le porc par la mesure de l'aminoacidémie porte. Cahiers AEC number 6, 327-339.

Pion, R., Prugnaud, J., Henry, Y. and Rérat, A., 1971. Influence de la teneur en thréonine du régime sur l'aminoacidémie libre du porc en croissance. Xe Congrès Int. Zootechnie, Versailles - Thème I porcs.

Rérat, A., 1970. La qualité des carcasses de porc et ses variations en fonction de l'alimentation durant la croissance. Rec. Med. Vet., CXLVI, Vigot Frères Ed., 1243-1295.

Rérat, A., 1972. Recent advances in Swine feeding. 2nd World Congr. Anim. Feeding, Madrid, IV, 39-155.

Rérat, A., Bouffault, J. and Jacquot, R., 1956. Double limitation de l'efficacité protidique par carence ou par excès du même acide aminé indispensable démontrée par l'aptitude de la DL-lysine à supplémenter le gluten de maïs ou de blé. CR Acad. Sci., 243, 192-194.

Rérat, A. and Bourdon, D., 1972. Supplémentation retardée de régimes à base d'orge chez le porc en croissance finition. Journées Rech. Porcine en France, INRA, ITP ed. 215-224.

Rérat, A. and Bourdon, D., 1975. Supplémentation retardée à l'aide de la lysine industrielle d'un régime déficient en cet acide aminé. Journées Rech. porcine en France. INRA, ITP ed., Paris, 27-37.

Rérat, A., Corring, T. and Laplace, J.P., 1978. Quelques aspects des recherches en physiologie digestive chez le porc: applications possibles. Journées Rech. Porcine en France 10, 95-119, INRA, ITP ed., Paris.

Rérat, A. and Henry, Y., 1963a. Variations de la consommation spontanée d'énergie en fonction de la nature et de la quantité des matières azotées ingérées chez le rat en croissance. Ann. Biol. anim. Bioch. Biophys., 3(3) (number HSI), 81-95.

Rérat, A. and Henry, Y., 1963b. Etude de l'ajustement de la consommation spontanée d'énergie en fonction de l'apport azoté chez le rat en croissance. Ann. Biol. anim. Bioch. Biophys., 3, 263-298.

Rérat, A. and Henry, Y., 1963c. Besoin en lysine du rat en croissance: principe d'une méthodologie et résultats expérimentaux. CR Acad. Sci., 257, 3045-3048.

Rérat, A. and Henry, Y., 1964. Consommation spontanée d'énergie en alimentation mixte et séparée chez le Porc en croissance. Ann. Biol. anim. Bioch. Biophys., 4, 441-444.

Rérat, A., Henry, Y. and Jacquot, R., 1963. Relation entre la consommation spontanée d'énergie et la rétention azotée chez le rat en croissance. CR Acad. Sci., 256, 787-789.

Rérat, A. and Jacquot, R., 1956. Comparaison entre les effets supplétifs de la DL-Lysine et la farine de poisson vis-à-vis des protides de céréales. Conf. Interafricaine sur la Nutrition, 3rd session, Luanda, 1956.

Rérat, A., Jacquot, R. and Adrian, J., 1956. Influence de la nature des solvants utilisés pour l'extraction de l'huile sur la valeur alimentaire du tourteau d'arachide. Etude expérimentale. L'industrie des Corps Gras, 7-8, 25-29.

Rérat, A., Lerner, J., Henry, Y. and Bourdon, D., 1970. Etude du besoin en lysine du porc en croissance en relation avec le taux énergétique du régime. Journées Rech. Porcine en France. INRA, ITP ed., 79-84.

Rérat, A. and Lougnon, J., 1963. Influence de la conduite du séchage à la flamme sur l'efficacité des protéines de farine de hareng. Ann. Biol. anim. Bioch. Biophys., 3, (HS1), 71-74.

Rérat, A. and Lougnon, J., 1965. Supplémentation par la L-lysine d'un régime simplifié chez le Porc en croissance. Ann. Zootech., 14, 247-260.

Rérat, A., Vaissade, P. and Vaugelage, P., 1978. Kinetics of absorption of some carbohydrates in the pig. XIe Congrès Intern. Nutrition, Rio de Janeiro, 27 August - 1 September, 1978.

Rérat, A., Vaissade, P. and Vaugelage, P., 1979. Absorption kinetics of amino acids and reducing sugars during digestion of barley or wheat in pig: preliminary data. Ann. Biol. anim. Bioch. Biophys., 19, 739-747.

DISCUSSION

C. Calet *(France)*

I appreciated Dr. Petersen's talk and I have three comments. The first is about energy value. I am older than Dr. Petersen and when I speak of metabolisable energy I use the term kilo-calorie and not joules. The energy value of feedstuffs is very important. I think it is the deciding factor when choosing a feedstuff. I agree with you when you say that the energetic value of field beans is higher than soyabean oil meal for poultry. But I think that peas are better. We have found that for soyabean oil meal with a value of 2 300 kilo-calorie per kg, there is 2 600 kilo-calorie for peas. This is why rapeseed oil meal and peas is a very good combination. Using the computer, when we increase the amount of rapeseed in the diet (for pigs, for instance) immediately we observe an increase of pea content in the diet. They are very complementary.

My second comment is on rapeseed. I am very glad to see that you can obtain very good results with rapeseed for pigs. When you put it in the diet up to a level of 8% there was no difference with the control, according to your last figures. This is very important for practical purposes. Generally, rapeseed is said to be very bad for the appetite, not only for pigs but for poultry and ruminants too. However, in this experiment there was no bad effect on intake. Can you explain this?

U. Petersen *(FRG)*

I will try to give an explanation. First of all, the feeding regime was *ad libitum*, so if there was any negative effect in the diet there should have been a decrease in feed consumption. The results showed the opposite; with an increased proportion of rapeseed in the diet there was a significant linear increase in feed consumption, on a feedstuff basis; on an energy basis there was almost no effect. The explanation could be that the glucosinolate content was very low in these feedstuffs and, in addition, the feed was given as a dry meal.

W. Eeckhout *(Belgium)*

The protein digestibility of rapeseed meal seems to be very low, at least at the faecal level. Do you have any information about protein digestibility or perhaps protein availability of rapeseed meal at the terminal ileal level - with a reasonably low crude fibre content in the diet?

U. Petersen

I do not think that the protein digestibility figures were so low in our Table. The figure for rapeseed protein digestibility in pigs was about 8%. This was about the same figure as for field beans, but I agree it is lower than for soyabeans. However, with respect to ileal digestibility, I have no information at the moment.

A. Rérat *(France)*

There was some data from Bayley and his team in Canada about 8 years ago concerning amino acid digestibility at the ileum. However, this was very early work on this topic and the technique may not have been very accurate. I do not remember the exact results; perhaps it is just as well!

G. Röbbelen *(FRG)*

Being a plant breeder I am impressed by all these different values which have been shown. I have learned that the two major questions concern energy and protein, and availability in particular. However, if you ask the plant breeder what to do the answer is generally directed towards glucosinolates and lysine. If you compare many of your figures you will find that this is not the total answer. There are two other factors. In the native seeds, for example in sweet lupin, your first Table showed some figures which were rather high and which were not in accord with the very low methionine content. The second point is a question: to what extent does the processing method contribute to all these changes? Have your rapeseed meals been processed by large-scale industrial methods or by laboratory extractions? Surely this could make a difference. I can see real problems for the plant breeder because the amino acid story is not the protein story. Many of the unexplained differences may come from tertiary structure of the protein, for example, or for other reasons. In the long term we should go into feedstuffs in more detail and see whether the plant breeder can help in some way or another.

U. Petersen

That is a very complex comment. It is very difficult to supply answers to the various aspects of it. You should direct your attention to the energy content of feedingstuffs. If we make a computer diet, then there is a very strong limitation for energy content; supplementation of energy, by fat for instance, is more expensive, at this time, than supplementation with methionine or lysine. This is one aspect and I hope it gives at least one answer.

As for availability problems and amino acid composition, it is very difficult to differentiate between these aspects. The result of a biological experiment depends on both. I have no idea what to propose in this respect.

A. Rérat

It is a very interesting question: we do not always agree in animal nutrition - on availability, for example. The level of consumption is very important; some of the factors present in oil meals are anti-consumption. There are also anti-nutritive factors which affect the use of the protein. Afterwards, there is the protein balance: the balance between amino acids. In rapeseed the balance is not bad; it is better than sunflower. In sunflower there is very little lysine. The other aspect of the balance of amino acids is the fact that they are digestible. We can discuss availability, but availability is a time factor. Digestibility concerns whether or not the nutrient is absorbed. With the pig, the time between meals and the length of the gut is such that it is difficult to ascertain availability. I agree with Dr. Eeckhout, digestibility in the ileum is important.

The first priority for animal nutrition is to destroy all the anti-nutritive factors by technological treatments. If technological treatments are easier, we must do it; if it is easier to do it by breeding - you must do it! At this time I do not think the balance of amino acids in the rapeseed is so very important.

U. Petersen

I agree with you. However, I think it would be difficult to breed for proper balance of amino acids because it depends on which animal the rapeseed protein is for. If it is for poultry you must decide whether it is for a broiler or a laying hen; with pigs you must decide whether it is for piglets or full-grown pigs or for sows. It is the job of the nutritionist to make proper use of the different feedingstuffs by combining the different sources to meet specific requirements. I agree with Dr. Rérat, the feedstuff must be consumed by the animal and, in the long term, it is important to have a good energy content in feedingstuffs for our western European livestock industry.

G. Röbbelen

I like this more than the search for the anti-nutritive factor. We are always hunting after the negative evil; we must not have this factor 'X' which we were looking for in the oil story, or we must not have glucosinolates.

A. Rérat

Perhaps we can return to this after Dr. Eggum's paper.

SHORT COMMUNICATION

NUTRITIONAL VALUE OF RAPESEED MEAL

C. Calet

INRA, 149 rue de Grenelle, 75341 Paris Cedex 07, France.

Many compounds contained in rapeseed meal are said to explain its poor nutritional value, causing a reduction in intake and/or in conversion rates. Often the substances contained in the central part of the seed are questioned. Glucosinolates have long been the most studied of these substances, but more recently a lot of data have been reported on the involvement of sinapine.

The poor nutritional effect of rape is often mentioned, but it is not always observed. In the years 1965 to 1969, I did not notice any set-back in growth, or any reduction in feed conversion, with chickens given diets containing 12 to 18% of normal rapeseed meal, and poorer performances were obtained with Bronowski rapeseed meal, which contains a very low level of glucosinolates.

INRA scientists have just completed experiments on pigs (1) and on doe rabbits (2) which are revealing. They compared meals processed from seeds low in erucic acid content (cv. Jet Neuf) and from a Canadian variety low in glucosinolates (cv. Regent). In addition, meals were prepared from intact or dehulled seeds. It was shown that dehulling had a better effect on pig growth than a reduction in glucosinolate content. Meals from intact rapeseed had a bad effect on rabbit reproduction, reducing the average litter size by 0.5 units, while meal from dehulled rapeseed increased litter size by one unit.

It seems, therefore, that the most detrimental compounds in rapeseed are located in the seed coat, which is easily removable, and that the products in the kernel which it is considered necessary to eliminate are not the most dangerous for monogastric animals.

REFERENCES

1. Bourdon, D., Perez, J.M. and Baudet, J.J., 1981. Utilisation de nouveaux types de tourteaux colza par le porc en croissance-finition. Influence des gluconsinolates et du depelliculage. Journées de la Recherche Porcine en France, 4/5 February 1981 (in press)
2. Lebas, F. Personal communication. Laboratoire de Recherches sur L'Elevage du Lapin. INRA - Toulouse 31320 Castanet-Tolosan.

NUTRITIONAL PROBLEMS RELATED TO DOUBLE LOW RAPESEED IN ANIMAL NUTRITION

Bjorn O. Eggum
National Institute of Animal Science
Rolighedsvej 25, DK-1958, Copenhagen V,
Denmark.

INTRODUCTION

The increasing national and global need for protein has stimulated research workers to search for new protein sources as well as to improve the protein quality of the more traditionally grown crops. Rapeseed has attracted much attention as a potential protein source in countries with moderate to cool climatic conditions.

However, plants of the Cruciferae family, including rapeseed, contain glucosinolates which exert antinutritional effects. Because of the nature and number of these compounds and their autolysis products, the nutritional problems are very complex. In addition to the problems due to glucosinolates, rapeseed also has a high content of crude fibre and tannin, both of which exert a negative influence on protein as well as energy utilisation (Eggum and Christensen, 1975).

In the 'double-low' rapeseed varieties, the glucosinolate content is significantly reduced but these compounds still cause nutritional problems. On the other hand, the erucic acid content in these varieties is so low that it is no longer considered to be a problem in animal feeding.

The main deleterious factors when rapeseed is used in animal nutrition are crude fibre, glucosinolates and tannins. Despite these difficulties, rapeseed production attracts much interest as a staple crop. The main reason for this is the excellent protein quality.

Chemical composition and protein quality of rapeseed meal compared to soyabean meal

It appears from Table 1 that 'double-low' Erglu rapeseed meal contains less protein and readily-hydrolysable carbohydrates than soyabean meal, which indicates a lower energy density. Furthermore, the high crude fibre level (13.0%) in the rapeseed and the high tannin content (3.6%) predict a much lower digestible energy value than in soyabean meal.

TABLE 1

PROXIMATE ANALYSIS (% DRY MATTER) OF ERGLU RAPESEED MEAL COMPARED TO SOYABEAN MEAL

	Erglu rapeseed meal	Soyabean meal
Crude protein	43.7	50.7
Crude fat	3.6	3.3
Crude fibre	13.0	6.6
Ash	7.9	6.5
Nitrogen-free extractives	31.8	32.9
Readily-hydrolysable carbohydrates	10.3	15.5
Tannin	3.6	0.4

The proteins in rapeseed meal, however, have a high level of lysine, methione+cystine, threonine and tryptophan, and this is very favourable for animal production. In rat balance trials (Eggum, 1973) a biological value (BV) of 89.1% was obtained for rapeseed meal, compared to 67.8% for soyabean meal. Due to the high content of crude fibre and tannin, true protein digestibility (TD) in rapeseed meal is only 83.3% compared to 93.4% in soyabean meal. However, because of the very high BV in rapeseed meal, net protein utilisation (NPU) is much higher (74.2%) than in soyabean meal (63.3%). This counteracts the lower protein content in rapeseed meal so the same amount of utilisable protein (UP) is obtained from Erglu rapeseed meal as from soyabean meal.

TABLE 2

PROTEIN QUALITY OF ERGLU RAPESEED MEAL COMPARED TO SOYABEAN MEAL

Protein source	Erglu rapeseed meal (g/16gN)	Soyabean meal (g/16gN)
Threonine	4.28	3.61
Valine	5.17	5.07
Isoleucine	4.07	4.64
Leucine	7.18	7.49
Lysine	5.88	5.99
Histidine	2.71	3.38
Arginine	6.54	7.19
Methionine	2.13	1.60
Cystine	2.39	1.57
Tryptophan	1.47	1.18
	(%)	(%)
Protein (CP = N x 6.25)	43.7	50.7
True protiin digestibility	83.3	93.4
Biological value	89.1	67.8
Net protein utilisation (NPU)	74.2	63.3
Utilisable protein (CP x NPU/100)	32.4	32.1

Experiments with poultry

As mentioned earlier, the tannin content in rapeseed meal is relatively high (2 - 4%), even in the double-low varieties. Yápar and Clandinin (1972) studied the effects of the tannins in rapeseed meal (RSM) on the nutritive value of this feedstuff for chicks. In the first experiment, tannins were extracted from RSM and the metabolisable energy (ME) content of the resulting meal was compared with that of normal RSM. The extracted tannins were added to a diet containing soyabean meal (SBM) and the ME content of this diet was compared with that of the same diet without added tannins. Results showed that removing the tannins from RSM significantly increased ME value of the chicks from 1.171 to 1.844 kcal/kg. This is in agreement with the work of Eggum and Christensen (1975) in a study on rats, when a highly significant negative correlation between protein digestibility and the dietary tannin content was found.

Gaardbo Thomsen (1979) used increasing amounts of double-low rapeseed meal (Erglu), or of meal from a normal winter rapeseed variety in comparative diets. Some of the results are given in Table 3. Energy (ME) and protein were balanced in all diets.

TABLE 3

THE EFFECTS OF INCREASING AMOUNTS OF DOUBLE-LOW OR NORMAL (WINTER) RAPESEED MEALS IN DIETS FOR BROILER CHICKS.

Rapeseed meal (Erglu) %	0.0	4.0	8.0	16.0	0.0	0.0
Rapeseed meal (Winter type) %	0.0	0.0	0.0	0.0	4.0	16.0
Body wt. ratio	100	98	98	90	98	91
kg feed/chicken	2.14	2.09	2.12	1.95	2.11	1.96
kg feed/kg chicken	1.82	1.81	1.84	1.84	1.81	1.82
kcal ME/kg chicken	5489	5314	5439	5404	5401	5374
Thyroid (mg/100g chicken)	7.6	7.3	10.7	10.1	8.7	11.6

The two varieties of rapeseed meal resulted in identical reductions in daily gain with increasing dietary concentrations. In both experiments the reduction in daily gain was due to a reduction in feed consumption as feed utilisation was not affected. It appears from Table 3 that the thyroids were significantly heavier in diets highest in rapeseed meal. It should be stressed that this was also the case with diets made up of the Erglu variety.

In work with laying hens and broilers, Leeson et al. (1978) showed that laying hens consuming 10% dietary raw whole rapeseed from Tower produced significantly ($P<0.05$) more eggs than did birds fed on an isocaloric isonitrogenous corn-soyabean meal control diet. Egg production of birds fed either the control diet, autoclaved whole rapeseed or autoclaved ground rapeseed, was not significantly ($P>0.05$) different. Birds fed rapeseed produced smaller eggs (P<0.001), while those birds offered autoclaved ground rapeseed produced eggs with significantly (P<0.05) superior shell quality compared to eggs from the control birds. Broiler chickens consuming raw rapeseed (10 - 20% in the diets) had larger thyroids than did control birds, while autoclaving, and to a lesser extent dry-heating, corrected this anomaly. Body weight gain to 4 weeks of age was not influenced by dietary treatment, although birds eating 20% dry-heated rapeseed showed a significantly (P<0.05) inferior feed intake : body weight gain in comparison to that calculated for control birds. It is concluded that whole Tower rapeseed can be well utilised by both broiler chickens and laying hens at inclusion levels of 10 - 20% and 10% respectively, and that no advantages accrue from grinding the seed. For broiler chickens, there is an indication that rapeseed should be heat-treated.

Fenwick et al. (1979) demonstrated that treatment of rapeseed meal with calcium hydroxide suspension decreased the sinapine content by up to 90%. Smaller decreases were obtained by autolysis, steaming, and treatment with ammonia. When this treated meal was fed to susceptible ('tainting') hens the concentration of trimethylamine in the eggs was decreased to much less than that required to cause taint.

Perosis seems to occur when rapeseed meal is fed to fast growing birds. Thus March and MacMillan (1980) showed that incidence of perosis was higher with rapeseed meal diets than with soyabean diets. Addition of choline to the soyabean meal diet affected neither growth nor incidence of perosis, but supplementary choline increased growth and decreased the incidence of perosis in birds fed the rapeseed meal diets. In a second experiment to study availability of choline from rapeseed meal, chicks were fed a diet containing 28% rapeseed meal and 2310 mg choline per kilogram. The freeze-dried contents of the lower intestine contained 4.92 mg choline per gram from the chicks fed rapeseed meal and 2.45 mg/g when soyabean meal was fed. The results of a third experiment showed that unabsorbed choline in the intestine was associated with an increase in bacterial production of trimethylamine. The caecal contents of chicks fed 28% of dietary rapeseed meal contained significantly more trimethylamine than those of chicks fed a control soyabean meal diet although both diets contained 2310 mg choline per kilogram.

Experiments with pigs

Singam and Lawrence (1979) carried out acceptability and nitrogen utilisation studies with pigs on diets containing barley and either extracted soyabean meal (SBM) or extracted rapeseed meal from one of the two low-glucosinolate varieties, Tower (TRSM) amd Erglu (ERSM). In acceptability studies, SBM, TRSM, and ERSM were substituted isometrically at 20% in the diets which were offered *ad libitum* for 8 weeks to pigs of 23 kg initial live weight. For the first 2 weeks, intake per unit metabolic weight ($W^{0.75}$) was greatest, but thereafter was least, for the SBM diet. In metabolism studies three diets were computed and fed so that 54.5% of the daily intake of 170 g of crude protein was derived from SBM, TRSM or ERSM. Apparent digestibility of crude protein was higher for the SBM diet compared with the TRSM ($P<0.05$) and ERSM ($P<0.05$) diets. Nitrogen retention decreased significantly from the SBM to the ERSM to the TRSM diets. The SBM fed pigs grew significantly faster and required significantly less crude protein per unit of gain than those fed TRSM or ERSM, between which there were no significant differences. This is in agreement with the work of Fernandez et al. (1980). These workers determined the content of Scandinavian feed units in Erglu rapeseed meal to be 94/kg dry matter whereas soyabean meal contains 131 (Anderson and Just, 1975).

TABLE 4

APPARENT DIGESTIBILITY COEFFICIENTS AND NITROGEN RETENTION DATA FROM DIETS FED IN A METABOLISM EXPERIMENT

Diet Apparent digestib. coef. (%)	SBM	TRSM	ERSM	LSD (P=0.05)
Dry matter	76.4	72.3	71.7	1.0***
Crude protein	75.3	71.9	73.0	2.8*
Gross energy	76.1	72.8	71.9	1.4***
DE content (MJ/kg DM)	14.91	14.44	14.47	0.57 NS
ME content (MJ/kg DM)[a]	13.74	13.32	13.05	0.27***
ME as percentage of DE	92.1	92.2	92.2	-
Nitrogen retention:				
% of intake	42.1	34.1	38.1	4.0***
g/day	11.3	9.2	9.9	0.3***
g/day/kg body weight	0.48	0.37	0.43	0.06***
Mean pig live weight (kg)	24.8	25.7	24.7	2.0 NS

[a]Corrected to zero nitrogen balance

An experiment was conducted by Grandhi et al. (1979) with 96 Yorkshire barrows and gilts, to study the effects of feeding corn-soyabean meal (SBM), corn-SBM-Tower RSM, and corn-SBM-Candle RSM diets, in the form of mash and steam-processed pellets, on average daily gain (ADG), efficiency of feed conversion (feed to gain ration = F/G), average daily feed intake, and carcass backfat thickness of pigs reared from 23 kg to 91 kg live weight. Tower RSM or Candle RSM at the 15% level in corn-SBM based diets did not result in any adverse effects on ADG, feed intake or backfat thickness, but did increase F/G (P<0.05). There were no differences in F/G between pigs fed Tower RSM and Candle RSM diets. Steam pelleting enhanced ADG (P < 0.01) and F/G (P < 0.05) across all three diets, but did not affect feed intake or backfat thickness. Barrows had higher (P < 0.01) ADG and backfat thickness than gilts, but did not differ in feed intake or F/G. Similar results were earlier reported by Kennelly et al. (1978).

Hansen et al. (1978) fed pigs with increasing amounts of double-low Erglu rapeseed meal (0, 5.5, 11.0 and 22%, at the expense of soya-bean meal), in the period 20 to 90 kg liveweight. The results show that rapeseed meal can be used as the only protein concentrate in the period from 20 - 90 kg liveweight without any negative effect on daily gain, feed utilisation or slaughter quality, provided the rapeseed meal is given the correct energy value. Furthermore, Hansen et al. (1979a) demonstrated that increasing amounts of Erglu rapeseed meal in the diets had no significant influence on bacon or meat quality. There were no consistent differences in the fatty acid composition and there was no erucic acid in the products. However, increasing amounts of dietary Erglu rapeseed meal caused an increase in liver weight, but much less than with normal rapeseed meal. Thyroxine determinations in the plasma indicated that the thyroids were somewhat restrained.

Macroscopic evaluation of heart, liver, lungs and kidneys showed no pathological or anatomical changes that could be related to rapeseed feeding, although the thyroids were significantly affected by dietary rapeseed meal (Hansen et al. 1979b). The results of Lewis et al. (1978) indicate that Tower RSM (containing 0.98 mg glucosinolates/g meal) may be used as partial or complete replacement for SBM in the diets of pregnant and lactating swine for at least two reproductive cycles with no apparent reduction in sow reproductive performance.

Experiments with ruminants

Papas et al. (1979) studied the effect of feeding rapeseed meals (RSM) containing low (Tower) or high (Target/Turret) levels of glucosinolates on thyroid status, iodine and glucosinolate contents of milk, and other parameters in dairy cows and young calves. RSM (Tower and Turret) fed to dairy cows at 25% of the grain mixture reduced iodine content of milk. Diets containing Tower and Turret RSM tended to reduce plasma thyroxine (T_4) in cows and to increase the size of thyroids in rats receiving their milk. In calves, diets containing Target and Tower RSM resulted in increased liver and thyroid weights, but only those containing Target tended to reduce plasma T_4 levels. Feed intake, weight gain, haemoglobin, blood cell volume and erythrocyte count were not affected by diets containing

Tower RSM, but Target RSM reduced all these parameters. In addition, diets containing Target caused more pronounced histological changes of the calves' thyroids than those containing Tower RSM. No measurable amounts of intact glucosinolates were detected in milk of cows fed RSM. Similarly, the glucosinolate aglucones, isothiocyanates or vinyl oxazlidinethione, were not transferred to milk although small amounts of unsaturated nitrile (1-cyano-2-hydroxy-3-butene) and inorganic thiocyanate were detected in milk. Rats receiving milk from cows fed Turret RSM developed larger thyroid than those receiving milk from control-fed cows. Supplemental iodine (61.0 µg/d) prevented the thyroid enlargement in the rat.

TABLE 5

EFFECTS OF DIETS CONTAINING TOWER AND TURRET RSM ON FEED INTAKE, MILK YIELD AND MILK COMPOSITION

Treatment	Feed intake				Milk composition		
	Hay	Silage	Grain mix.	Milk yield	Fat	Prot.	SNF[1]
	(kg dry matter/day)			(kg/day)	(%)	(%)	(%)
Soyabean meal	1.9	9.1	8.32	21.8	3.8	3.7	8.8
Tower RSM	1.8	8.7	8.60	23.5	3.9	3.8	8.7
Turret RSM	1.8	9.0	8.12	21.0	3.9	3.7	9.0
SE	0.05	0.10	1.08	2.17	0.18	0.09	0.13

[1] Solid non fat

In experiments with lactating cows, Højland Fredriksen and Andersen (1979) fed increasing amounts from 0 to 75% of rapeseed expeller meal in the protein concentrate. The experimental period lasted 10 weeks and all animals received the same basal diet. The main conclusion from the trial was that the cows consumed only slowly the mixtures containing 50 or 75% rapeseed expeller meal - while the concentrates containing 0 or 25% were consumed immediately after feeding. There were no significant differences in milk yield between the four groups. However, the taste panel registered a significant off-flavour in both milk and butter from the cows consuming more than 25% rapeseed expeller meal in the concentrates. Thus the authors recommend that rapeseed expeller meal should not be fed at rates exceeding 1.2 kg per cow/day.

The same authors also carried out digestibility trials with sheep and the results are given in Table 6.

TABLE 6

THE RELATIONSHIP (y=a+bx) BETWEEN DIGESTIBILITY COEFFICIENTS (y) AND THE PERCENTAGE CONTENT OF RAPESEED EXPELLER (x) IN CONCENTRATES, AND DIGESTIBILITY COEFFICIENTS IN RAPESEED EXPELLER

	a	b	Rapeseed expel.
Crude protein	92.5 ± 0.68	-0.129 ± 0.015	79.6 ± 1.28
Crude fat (Stoldt)	88.2 ± 2.71	-0.071 ± 0.057	81.1 ± 5.13
Crude fibre	67.9 ± 4.24	-0.563 ± 0.091	11.6 ± 8.02
NFE	94.9 ± 1.23	-0.157 ± 0.026	79.3 ± 2.33
Org. matter	92.1 ± 0.89	-0.206 ± 0.019	71.5 ± 1.68
Energy	91.3 ± 0.93	-0.201 ± 0.199	71.3 ± 0.75

It can be seen that there exists a negative relationship between the contribution of dietary rapeseed expeller, to the concentrate mixture, and the digestibility of the various concentrate fractions. This is especially true for crude fibre, as the crude fibre in rapeseed expeller is only about 12% digestible. In general, the digestibility of rapeseed expellers is quite low when measured in sheep.

Conclusions

It can be concluded that the protein quality (amino acid composition) of double-low rapeseed is excellent, with a biological value of 90% when tested on rats. However, the high contents of crude fibre and tannin reduce protein as well as energy digestibility. This is demonstrated in experiments with poultry, pigs, and ruminants.

In spite of the low glucosinolate content, double-low rapeseed meal still seems to cause nutritional problems. In particular, the thyroids and liver are significantly affected. In poultry, leg problems can occur when rapeseed meal is fed. Pigs are apparently less affected by the glucosinolates in double low rapeseed meal than either poultry or ruminants. When more than 1.2 kg rapeseed expellers are fed to lactating cows, the appetite is reduced and off-flavour is registered in both the milk and the butter.

REFERENCES

Andersen, P.E. and Just, A., 1975. Tabeller over fodermidlers sammensaetning m.m. Kvaeg - Svin. Det kgl. Danske Landhusholdningsselskab.

Eggum, B.O., 1973. A study of certain factors influencing protein utilisation in rats and pigs. (Thesis) 406. beretn. National Institute of Animal Science. Copenhagen. 173 pp.

Eggum, B.O. and Christensen, K.D., 1975. Influence of tannin on protein utilisation in feedstuffs with special reference to barley. pp. 135. In 'Breeding for Seed Protein Improvement using Nuclear Techniques'. International Atomic Energy Agency. Vienna.

Fenwick, G.R., Hobson-Frohock, A., Land, D.G. and Curtis, R.F., 1979. Rapeseed meal and egg taint; treatment of rapeseed meal to reduce tainting potential. Br. Poult. Sci. 20:323.

Fernández, J.A., Just, A. and Jørgensen, H., 1980. Fodermidlernes vaerdi til svin. Report no. 301, National Institute of Animal Science, Copenhagen.

Grandhi, R.R., Narendran, R., Bowman, G.H. and Slinger, S.L., 1979. Effects on performance of pigs fed steam-pelleted rapeseed meal diets. Can. J. Animal Sci. 59:323.

Gaardbo Thomsen, M., 1979. Stigends maengde rapsskrå eller rapsekspeller i slagtekyllingefoder. Report no. 226, National Institute of Animal Science, Copenhagen.

Hansen, V., Smedegård, K. and Jensen, Aa., 1978. Rapsskrå (Erglu) som delvis eller fuld erstatning for sojaskrå i slagtesvinenes foder. Report no. 244, National Institute of Animal Science, Copenhagen.

Hansen, V., Smedegård, K., Jensen, Aa., Andersen, Inger-Lise E., Barton, P., Olsen, O. and Sørensen, H., 1979a Rapsskrå (Erglu) som delvis eller fuld erstatning for sojaskrå i slagtesvinenes foder. Specielle undersøgelser. Report no. 263, National Institute of Animal Science, Copenhagen.

Hansen, V., Smedegård, K., Laursen, B., Jensen, Aa., Andersen, Inger-Lise E., and Bille, N., 1979b. Rapskager som en del af proteintilskudsfoderet til slagtesvin. Report no. 286, National Institute of Animal Science, Copenhagen.

Højland Fredriksen, J. and Andersen, P.E., 1979. Erglu-rapsekspeller til malkekøer. Report no. 280, National Institute of Animal Science, Copenhagen.

Kennelly, J.J., Aherne, F.X. and Lewis, A.J., 1978. The effects of levels of isolation or varietal differences in high fibre hull fraction of low glucosinolate rapeseed meals on rat or pig performance. Can. J. Animal Sci. 58:743.

Leeson, S., Slinger, S.J. and Summer, J.D., 1978. Utilisation of whole Tower rapeseed by laying hens and broiler chickens. Can. J. Animal Sci. 58:55.

Lewis, A.J., Aherne, F.X. and Hardin, R.T., 1978. Reproductive performance of sows fed low glucosinolate (Tower) rapeseed meal. Can. J. Animal Sci. 58:203.

March, B.E. and MacMillan, C., 1980. Choline concentration and availability in rapeseed meal. Poult. Sci. 59:611.

Papas, A., Ingalls, J.R. and Campbell, L.D., 1979. Studies on the effect of rapeseed meal on thyroid status of cattle, glucosinolate and iodine content of milk and other parameters. J. Nutr. 109:1129.

Singam, A.D.R. and Lawrence, T.L.J., 1979. Acceptability and nitrogen utilisation of Tower and Erglu rapeseed meals by the growing pig. J. Sci. Food Agric. 30:21.

Yapar, Z. and Clandinin, D.R., 1972. Effects of tannins in rapeseed meal on its nutritional value for chicks. Poultr. Sci. LI:222.

DISCUSSION

G. Röbbelen (FRG)

I wanted to ask you about this tannin question. Could this be a reason for the low crude fibre digestibility? Did you ever test yellow-seeded rapeseed?

B.O. Eggum (Denmark)

As I said, the low digestibility is partially due to crude fibre and tannin. As tannin is primarily in the crude fibre fraction, I would say that it does affect crude fibre digestibility. I have never tried the yellow type of rapeseed but I am quite convinced digestibility would be higher. Amongst all seeds, the digestibility is higher in varieties with yellow-type hulls.

U. Petersen (FRG)

I would like to ask a question with respect to tannin experiments. Is it the same situation with a supplementary tannin as compared with a native tannin? I think I can recall contents and there was no significant effect on protein quality. Could you comment on this?

B.O. Eggum

No, I am afraid you are wrong. There was a highly significant negative correlation to tannin levels in barley and protein digestibility. I can also refer to Swedish work on barley and tannin content which showed decreased digestibility, not only in protein but also in fat and carbohydrates, probably because the tannins also tied up the enzymes, the lipases, for fat digestibility, etc. So there was a general drop in dry matter digestibility due to tannin.

G. Röbbelen

German nutritionists tell us that in *Vicia faba*, in comparisons of brown and white seeded varieties, the tannins in the brown seed coat may inhibit the digestibility of high-value proteins in the rumen and keep the proteins intact for digestion in the small intestine.

B.O. Eggum

We know that the protein is protected in ruminant nutrition artificially. Tannin may be of some interest here because it is protected so that it can be digested in the small intestine and have a higher efficiency. But in sorghum there are 'bird resistant' varieties and in these varieties there is a very low digestibility of the protein because they are much higher in tannin.

G. Röbbelen

Then you can feed the rapeseed hulls to the cows to get this effect!

B.O. Eggum

If the tannin did not leak out of the hulls, we could protect the other protein sources coming in. That has to go on in the rumen because that is where you have the attack. I do not think it is that simple.

A. Rérat (France)

In Europe, and particularly in France, soyabeans are used a great deal in the diet of the cows, but very little rapeseed. It is possible to give a good protein concentrate from rapeseed to cows, and to ruminants in general, and reserve the soyabeans (without such things as tannins and glucosinolates) for monogastric animals.

W. Eeckhout (Belgium)

Why do tannins depress protein digestibility so tremendously in monogastric animals? Does it speed up the passage of the digesta in the small intestine?

B.O. Eggum

No, only to a minor extent if at all. I have some data where we measured amino acid composition of the faeces and all the proline was recovered in the faeces and 40% of glutamic acid and 60% of alanine, etc. So it is a reaction between tannin and the proteins. This is well known in chemistry because protein used to be calculated by tannin in the earlier methods. In the breweries they use it to remove the proteins.

A. Rérat

You do not think it is an increase of endogenous nitrogens? For example, proline is an amino acid of high content in the endogenous protein, particularly in mucus.

B.O. Eggum

I hesitated when I said it, because I know it is a problem. The main effect is that it reduces digestibility; it is a chemical reaction.

A. Rérat

I have two questions. You spoke about the problem of erucic acid in the animals. Do you really think this is a problem?

B.O. Eggum

I have not worked on this myself, but I have listened to several papers and heard details of data. I went to a conference in India last year. People in India had eaten the old winter types, or the high erucic acid rapeseeds, for thousands of years. The conclusion was that, so far, there had been no effect whatsoever of erucic acid in man. But you can provoke an effect in some animals. You can see some papers which show white muscle in the heart when animals are fed very high levels of erucic acid.

A. Rérat

The second problem is more theoretical; the problem of biological value. You said that rapeseed is interesting because the biological value is very high. But it is not necessary to have a high biological value; it is more important to have a good complementary value. What we need from a protein concentrate is not that all the amino acids have a very good biological value, but that they supplement the amino acids lacking in the cereals which are the basis of feeds for monogastric animals.

B.O. Eggum

That is why I compared soya and rapeseed meal. The main thing in soya is the high lysine level. The rapeseed seems to be high in lysine, threonine, sulphur and tryptophane. Today, when there is enough barley, then you are safe with soya. But we have a lot of by-products in Denmark: low in threonine, low in sulphur. Rapeseed would be a better match for them because there are so many essential amino acids in rapeseed. Primarily, soya is a lysine source and there are no inhibitor problems. Soya is very easy to use. However, I agree with you; you do not need everything in one source because you can mix them.

A. Rérat

The interest is in having good amounts of sulphur amino acids. But, once again, it depends on the type of cereal. With barley you do not need so much. You are right in saying that rapeseed is a very good protein, but we do not need all these amino acids - only 4 or 5 of them.

B.O. Eggum

That is right with barley. But soon we may be able to buy other cheaper sources which do not have as good a basic amino acid pattern as there is in barley. Then rapeseed will come into the picture.

In South America they use much more tapioca, which has very little protein. Therefore, they have to improve their staple food. It could be the same here.

A. Rérat

Are there anymore questions? Thank you very much, Dr. Eggum.

RAPESEED MEAL IN POULTRY RATIONS

H. Vogt
Institute for Poultry and Small Animals
Federal Research Centre for Agriculture (FAL)
Celle, Federal Republic of Germany

INTRODUCTION

Rapeseed oilmeal (RSM) is used as a protein supplement for poultry feeds although its use has been limited because of the presence of antinutritional or toxic factors.

There are five substances or groups of substances that are known to have, or may have, toxic effects - glucosinolates, tannins, saponine, sinapine and erucic acid - and there may be others.

Glucosinolates in rapeseed are hydrolysed by myrosinase enzyme (thioglucoside glucohydrolase) to a variety of aglucone products. Some of these products, especially the 1-5-vinyl-2-thiooxazolidenethione (VTO), have a goitrogenic effect and influence the development of the thyroids. In addition, some nitrile is formed when glucosinolates are hydrolysed and this could also account for the reduced weight gain with RSM feeding.

Quite large amounts of tannins (Fenwick et al. 1976: 1.57 to 3.13%) and saponines (March et al. 1974: 0.62 to 2.85%) have been reported in samples of RSM. High levels of both substances depress growth of chicks. Sinapine is involved in the egg taint problem.

The erucic acid content of rapeseed is a problem only when rapeseed oil is used in the diets.

RAPESEED MEAL IN BROILER DIETS

The nutritive value of RSM for broiler chickens has been intensively studied. The deleterious effects on weight gain, feed efficiency, thyroid development, and mortality, often associated with inclusion in broiler diets of RSM produced from the older varieties of rapeseed, are usually attributed to the high glucosinolate content of such meals. This led to the development of new rapeseed varieties, low in erucic acid and low in glucosinolate content. Experiments have been conducted to evaluate the feeding value of the new low-glucosinolate rapeseed varieties. Tests have been made with the *Brassica napus* varieties, Bronowski (Sharby et al. 1974; Olumo et al. 1975; Matsumoto et al. 1979; Campbell et al. 1979), Erglu (Vogt et al. 1974; Fuhrken, 1974; Ahlström et al. 1978; Ibrahim et al. 1980) and Tower (Slinger et al. 1978; Summers et al. 1978; Clandinin et al. 1978; Hulan et al. 1978; Matsumoto et al. 1978; Bhargava et al. 1979; Ibrahim et al. 1980), and with the *Brassica campestris* variety Candle (Slinger et al. 1978; Clandinin et al. 1978; Hulan et al. 1979; Bhargava et al. 1979). Rapeseed meal from the new low-glucosinolate varieties may be used at higher levels than those recommended for higher glucosinolate RSMs. From the results obtained it is suggested that a maximum level of 20% low-glucosinolate RSM can be included in broiler diets. Higher levels of low-glucosinolate RSMs partly depressed chicken growth nitriles and tannins may be involved in these growth depressions

Thyroid enlargement was much smaller with diets from the new varieties than from the old varieties but the effect from the new varieties was not negligible. Leg weakness assessments, however, were not greater with Tower and Erglu RSMs than with soyabean meal (Ibrahim et al. 1980). The findings of Hawrysh et al. (1980$_{a,b}$) suggest that the eating quality of chickens was not affected by the inclusion of 20% Tower RSM in the ration.

The true amino acid availability (TAAA) values ranged from 82 to 94% for Tower RSM, from 84 to 95% for Candle RSM and from 86 to 94% for both RSMs (Muztar et al. 1980). Protein isolates from seeds of the varieties Erglu and Lesira gave better chicken performance than the corresponding rapeseed meals; this is attributed to the favourable amino acid patterns of the isolates and the absence of glucosinolates (El Nockrashy et al. 1975, 197

Nwokolo et al. (1977) have shown that the phytic acid and fibre in RSM reduces the availability of some minerals. The same authors (Nwokolo et al. 1980) tested the biological availability of minerals in six samples of RSM from three cultivars of rapeseed (Span, Tower and Bronowski) and in one commercial RSM sample. The average availability was 68% for Ca, 75% for P, 62% for Mg, 54% for Mn, 74% for Cu and 44% for Zn.

METABOLISABLE ENERGY (ME)

A serious drawback to the use of rapeseed oilmeal (RSM) in rations for poultry has been its low metabolisable energy (ME) value. The ME content of RSM has been determined in many tests (Sibbald et al. 1963, 1977; Kubota et al. 1965; Sell, 1966; Jackson, 1969; Lodhi et al. 1969, 1970$_{a,b}$; Rao et al. 1970, 1972; Bayley et al. 1970, 1974, 1975; March et al. 1971, 1973, 1975; Yapar et al. 1972; Skotnicki et al. 1972; Seth et al. 1973; Leeson et al. 1977; Baudet et al. 1978; Muztar et al. 1978$_{a,b}$, 1979; Nwokolo et al. 1978; Smith et al. 1979).

Dependent on the composition of the RSM, the ME content varies over a wide range: March et al. (1971) determined ME values between 1 120 and 1 730 kcal/kg DM; March et al. (1973) obtained values between 1 185 and 1 600 with broiler chicks and between 1 788 and 2 385 kcal/kg with laying hens; Bayley et al. (1974) found ME values (N-corrected) between 1 340 and 2 000 kcal/kg and Nwokolo et al. (1978) ME values between 1 492 and 1 957 kcal/kg. Nwokolo et al. (1978) suggested two equations, based on the contents of sugar, starch and ether extract; both predicted the ME of RSMs with considerable accuracy. The value of 1 760 kcal (7.4 MJ) ME/kg seems a good average value for chickens (Blair et al. 1975; Clandinin et al. 1978). The ME of RSM diets increased with the age of the poultry when diets were given; several authors (Sell, 1966; Jackson, 1969; Lodhi et al. 1969; March et al. 1973) found higher values for layers. The value of 1 900 kcal (8 MJ) ME/kg seems more appropriate for adult poultry. The true metabolisable energy (TME) values of RSM for poultry was 2 420 (2 170 to 2 620) kcal/kg (Sibbald, 1977).

Although RSM in the feed influenced the fasting heat production, it had no influence on maintenance metabolisable energy requirements or on net availability of ME (Smith et al. 1979).

The determined ME value (average of 11 meals) was only 41% of the estimated catabolisable energy values, the corresponding proportion of catabolisable energy which is metabolisable was calculated to be 70% for soyabean meal (March et al. 1971). In the chick the nitrogen absorption from RSM is 5 to 8% lower than with soyabean oilmeal (Lodhi et al. 1970: 80 to 85%; Rao et al. 1972: 76 to 84%). Also, the availability of carbohydrates in RSM is much lower than in soyabean oilmeal (higher content of crude fibre? 8.4% pentosane (Clandinin et al. 1970)). The lowe ME of the RSM (730 kcal/kg lower than soyabean oilmeal) was fully accounted for by the effects on absorbable N (lower protein content and lower N absorbability) and available carbohydrates, and it was concluded that other factors had little or no effect (Rao et al. 1972); the goitrogen content was without effect on ME content (Lodhi et al. 1970; Rao et al. 1970).

But the high tannin content is a problem. The tannin content, estimated in 16 samples of RSM, averaged 3.56% (express as quercitannic acid) and that of sinapine, an hydroxylated cinnamic acid derivative, was 0.91%. When values were corrected for sinapine, the tannin content was 2.56% (Fenwick et al. 1976) This tannin level depressed ME content (Yapar et al. 1972).

The effect of antibiotic supplementation on the ME value of RSM was inconsistent (March et al. 1975). Detoxification of RSM with SO_2, H_2SO_4, SO_2 and urea, ammonia, and by heat, raised the ME value from 33 to 69% (Skotnicki et al. 1972).

Some workers have removed the hull from the seed before oil extraction and have been able to produce a high-protein product with higher ME content; an additional advantage lies in the fact that the hull-free meal would be essentially free of tannin

RAPESEED MEAL IN DIETS FOR LAYING HENS

The low energy content of rapeseed meal limits its possible employment in broiler rations, and it may be more interesting for use in layer rations. On the other hand, the longer life of laying hens may lead to a more distinct appearance of toxicity symptoms.

The results of earlier experiments with high-glucosinolate rapeseed meal (HG - RSM) indicated that increasing the rapeseed meal levels in the layer rations also increased mortality, decreased egg production, and affected egg size and egg Haugh units to a minor degree (Jackson, 1969, 1970; Vogt et al. $1969_{a,b}$, 1974, 1976, with Lesira RSM: Clandinin et al. 1970; Summers et al. 1971; Hill et al. 1974; Marangos et al. 1974, 1976; Calet, 1974; Overfeld et al. 1975; Calvarese et al. 1975; Leeson et al. 1976; Grandhi et al. 1977, with Span RSM: Smith et al. 1976; Thomas et al. 1978; Campbell, 1979, with Target RSM). Hence it may be recommended that not more than 5% HG - RSM should be used in the ration of laying hens.

The studies with the newly developed low-glucosinolate rapeseed meal (LG - RSM) have shown that at least twice as much LG - RSM may be included in rations for layers. LG - RSM depressed egg production less than HG - RSM (Marangos et al. 1974, 1976; Smith et al. 1976; Grandhi et al. 1977; Campbell, 1979), and Vogt et al. (1976) reported that feeding up to 15% Erglu LG - RSM had no significant influence on mortality or productive traits of layers. Similar results were reported by Slinger et al. (1978) for Candle LG - RSM, by Thomas et al. (1978) and Campbell (1979) for Tower Lg - RSM and by Hulan et al. (1980) for Tower and Candle LG - RSM. A reduction in egg size has been reported in some trials, but only when higher LG - RSM levels have been used.

Liver haemorrhage in laying hens

There are many reports indicating that the rate of mortality in laying birds is increased by the inclusion of rapeseed

meal in the diet (Jackson, 1969, 1970; Vogt et al. 1969$_{a,b}$;
Summers et al. 1971; Hall, 1972; Hill et al. 1974; Clandinin et
al. 1974; Marangos et al. 1974; Calet, 1974; Blair et al. 1975;
Olomu, 1975$_{a,b}$; March et al. 1975; Leslie et al. 1975; Leeson
et al. 1976; Yamashiro et al. 1975; Smith et al. 1976; Grandhi
et al. 1977; Pearson et al. 1978; Slinger et al. 1978; Campbell,
1979; Papas et al. 1979; Hulan et al. 1980). The effect on
mortality rate is not, however, consistent and there is evidence
that the genotype of the bird may affect its response to the
feeding of rapeseed meal (Jackson, 1969).

There are numerous reports of experiments in which mortality from liver haemorrhage in laying birds was high and closely related to the presence of rapeseed meal in the diet. Glucosinolates were implicated in the development of hepatic haemorrhage among hens receiving rapeseed meal, but the exact relationship was not clear (Campbell, 1979). Although haemorrhagic liver appeared to be evident among hens receiving low-glucosinol rapeseed meal as well as among those receiving high-glucosinolate rapeseed meal, in general the incidence was higher among hens receiving the high-glucosinolate rapeseed meal. Results of experiments by Papas et al. (1979) demonstrated that mortality rate could be reduced markedly by the addition of supplemental vitamin K to the feed or drinking water.

The results from Smith et al. (1976) show that when progoitrin is hydrolysed under conditions existing in the laying hen digestive tract, nitrile compounds are the predominant product. These authors speculate that the nitrile hydrolytic products might have an influence in the development of the connective tissue matrix of the liver, and hence predispose these birds to liver haemorrhage.

Egg taint

A further problem with layers has emerged in the last ten years, that of egg taint. Vogt et al. (1969) were the first to report a fairly high proportion of tainted eggs from birds given diets containing 10 to 20% rapeseed meal. In 1971 egg taint problems occurred in commercial flocks in the UK (Miller et al. 1972). There is a clear relationship between the presence of RSM in the diet and the fish or crab-like taint in the eggs (Hobson-Frohock et al. 1973, 1975; Vogt et al. 1974, 1976; Clandinin et al. 1974; Hawrysh et al. 1975; Bolton et al. 1976; Marangos et al. 1976; Leeson et al. 1978). In a test by Griffiths et al. (1979) the tainted eggs are described as 'fishy' (48%), 'crabby' (24%) or 'amine-like' (8%).

A very marked difference in incidence of tainted eggs was observed between types of birds: brown egg layers (mostly with Rhode Island Red ancestry) being particularly susceptible to taint, whereas in white-shelled eggs taint either has not been found, (Hobson-Frohock et al. 1973; Leslie et al. 1973; Clandinin et al. 1974; Blair et al. 1975; Fuhrken, 1975; Bolton et al. 1976; Vogt et al. 1976) or has occurred to only a slight degree (Vogt et al. 1974; Clandinin et al. 1974; Bolton et al. 1976; Marangos et al. 1976$_a$). Meals prepared from low-glucosinolate cultivars also produce taint (Vogt et al. 1974, 1976 (Erglu); Hobson-Frohock et al. 1977 (Tower)).

In a susceptible breed or strain only certain hens (6 - 27%) lay tainted eggs (Miller et al. 1972; Hobson-Frohock et al. 1973; Overfield et al. 1975), showing that tainting is under genetic control. Genetic studies by Bolton et al. (1976) have demonstrated that tainting is conditional on the presence in the hen, in the heterozygous or homozygous state, of an autosomal semi-dominant mutant gene that has variable expression, depending on environmental factors (including the rate of ingestion of rapeseed meal).

Studies on tainted eggs showed that there was a very close correlation with taint as detected by the senses and the trimethylamine (TMA) content of the egg. Tainted eggs contained more than 1.0 µg TMA/g whereas untainted eggs contained less than 0.1 µg/g; most of the TMA (~95%) was present in the yolk (Hobson-Frohock et al. 1973). The tests of Griffiths et al. (1979) showed that the taint occurs in eggs which contain more than 0.8 µg TMA/g.

A freeze-drying technique followed by gas chromotography of the distillate has been described by Hobson-Frohock (1979) for the analysis of TMA.

But, as thére is very little free TMA in RSM, some other substances must be involved. Hobson-Frohock (1975) demonstrated that the activity can be extracted from rapeseed meal with appropriate solvents and in 1977 Hobson-Frohock et al. reported that the source of the TMA in tainted eggs is sinapine. Sinapine is a cation and in rapeseed is associated with the glucosinolate anion, replacing some potassium ions. The mean sinapi content measured as thiocyanate in 16 samples of rapeseed meal, has been given as 0.91% (range 0.58 to 1.28) by Fenwick et al. (1976). Curtis et al. (1978) have examined the sinapine conten of over 130 seed samples and found none outside the recorded range of 0.6 - 1.5% and although the low-glucosinolate varietie Tower (0.7%) and Candle (0.6%), are much lower in sinapine than average, they also cause taint when fed.

The hydrolysis products of sinapine are choline and sinap acid, dimethoxy hydroxycinnamic acid. The results of the study of Mueller et al. (1978) led to the conclusion that bacteria ar present in the caeca of hens which are capable of converting the choline moiety of sinapine to TMA. The data of March et al (1979, 1980) provide evidence that the choline present in rapeseed meal is not efficiently utilised. Thus, choline may be released from a bound form during digestion but not readily absorbed from the intestine. The results of the third experiment of March et al. (1980) showed that unabsorbed choline in the intestine was associated with an increase in bacterial production of TMA. The caecal contents of chicks fed 28% dieta rapeseed meal contained significantly more choline and TMA (4.0 mg intestinal choline/g dry weight; 27.2 mg caecal TMA/g wet weight) than those of chicks fed a control soyabean meal diet (1.4 and 10.4 respectively) although both diets contained 2 310 mg choline/kg.

The taint of the eggs is caused by a gross impairment of

the hen's ability to convert high amounts of TMA to the odourless and tasteless N-oxide (TMA - O). This abnormality (tainted eggs) is associated with a decrease in TMA oxidase activity (Pearson et al. 1978) demonstrated by estimations of the activity of TMA oxidase in liver microsomes (Pearson et al. 1979$_a$) and measurements of the concentrations of ^{14}C-TMA.O in the blood plasma after the intravenous injection of a standard dose of ^{14}C-TMA (Pearson et al. 1979$_b$). This drastic reduction in the hen's capacity for oxidising TMA is brought about by substances present in the meal or derived from constituents. These include 5-vinyl-2-oxazolidinethione (Pearson et al. 1979$_c$), which is a potent goitrogen and probably depresses the synthesis of TMA oxidase through its anti-thyroid action, and an as yet unidentified constituent which strongly inhibits TMA oxidase *in vitro* (Pearson et al. 1979$_a$). Experiments with chickens (Butler, 1980) have shown that sinapine does not affect the activity of TMA oxidase *in vivo*.

The production of tainted eggs is caused by the interaction of three factors:

a) Choline, which originates from sinapine and is catabolised effectively by organisms in the gastro-intestinal tract to trimethylamine,

b) a genetic factor in the hen which adversely affects the synthesis of trimethylamine oxidase, and,

c) a substance in the rapeseed meal which drastically reduces the activity of hepatic trimethylamine oxidase.

Since no variety with a low sinapine content has been discovered, treatment of the meal to remove or decompose it, together with the selective breeding of stock for resistance to the effects of the meal on TMA oxidation, appears to be the most promising way of solving the egg taint problem.

To exploit the possibility of eliminating the tainting character from commercial flocks by selective breeding, a biochemical method of identifying the genotypes in males and females at an early age is required; for this purpose a TMA oxidation test which utilises ^{14}C-TMA has been developed (Pearson et al. 1979$_b$, 1979$_d$).

In the absence of a low-sinapine rapeseed for breeding purposes, methods of treating commercial rapeseed meal to decrease the sinapine to 'non-tainting' concentrations were sought. Treatment of rapeseed meal with calcium hydroxide suspension decreased the sinapine content by up to 90%. Smaller decreases were obtained by autolysis (64%), steaming (42%), and treatment with ammonia (36%). When this treated meal was fed to susceptible ('tainting') hens, the concentration of TMA in the eggs was decreased to much less (<0.4 µg/g) than that required to cause taint (Fenwick et al. 1979). But Vogt et al. (1976, 1979) still found tainted eggs in brown egg layers when fed with calcium-hydroxide treated rapeseed meal.

Rapeseed meal should be omitted from the diet of brown-egg laying hens until the taint problem has been solved.

OTHER POULTRY

In guineafowl, rapeseed meal produced more pronounced depression of growth than in chickens (Blum et al. 1974).

There are few reported studies on the effects of feeding RSM from the new rapeseed varieties to turkeys. In the experiments conducted by Moody et al. (1978), Salmon et al. (1979$_a$) and Hulan et al. (1980), up to 30% Tower or Candle LG - RSM was used with no negative effect on turkey broiler performance, and only one of the subsequent tests by Salmon et al. (1979$_b$) suggested a lower limit.

The inclusion of 5, 10 or 15% toasted RSM in feeds for broiler ducks depressed growth (Kozłowski, 1976$_a$). The same author fed Peking ducklings with potato silage prepared with RSM (Faruga et al. 1974; Kozłowski, 1975, 1976$_b$, 1976$_c$).

Rapeseed meal in the feed of geese depressed weight gain but did not affect egg production, fertility, or hatchability of the eggs (Bielinski et al. 1978).

FULL-FAT RAPESEED

Ordinarily full-fat rapeseed is not considered as a feedstuff for poultry. However, at certain times it could prove economically sound to include ground or unground full-fat rapeseed in rations for poultry.

Early work involving rapeseed with moderately high or high glucosinolate content showed that such full-fat rapeseed required heat-treatment before it was suitable for inclusion in broiler rations up to the 10% level, or in layer rations up to the 5% level (Leslie et al. 1972, 1973; Woodly et al. 1972; Ruszczyc et al. 1974; Splittgerber, 1975).

Adding full-fat Span rapeseed (LEAR), which had been steamed at 90°C, at 5, 10 or 15% to diets, had no significant effect on mortality, feed efficiency or total egg output, but there was a significant fall in production per hen-day with increase of rapeseed and death from haemorrhagic liver was found among hens given 10 or 15% rapeseed (Olomu et al. 1975).

Tower rapeseed fed at levels of 10 or 20% as whole seed, autoclaved whole seed, or autoclaved and ground seed in the laying ration, did not influence egg production but decreased egg weight. Tower rapeseed at 10% or 20% in the broiler ration had no influence on growth, but the 20% inclusion gave a slightly lower feed efficiency (Leeson et al. 1978).

Experiments with heated full-fat Tower rapeseed indicated that levels up to 15% in the broiler ration had no adverse effect on either bodyweight or feed efficiency; a significant depression in bodyweight was observed when 20% rapeseed was incorporated (Bhargava, 1978).

Feed intake and weight gain of broilers decreased in proportion to dietary rapeseed fed at 10, 20 or 30% levels, and more rapidly with Erra than with the Erglu variety. Feeds with 30% Erglu autoclaved for 20 min at 120°C had no adverse effect.

The nitrile N content of rapeseed products, which, unlike glucosinolates, decreased with heating, was regarded as a good index of suitability for use in feed (Brak et al. 1978).

The true metabolisable energy was determined in seeds of *Brassica* sp.; it averaged 4.52 kcal/g DM for nine varieties of *Brassica campestris*, 4.82 for four varieties of *B. hirta* and 5.08 for ten varieties of *B. napus* (Sibbald et al. 1977$_a$).

RAPESEED OIL

The first observations of growth depression after feeding rapeseed oil were made by Beznak et al. (1943$_{a,b}$) with rats. With chicks, ducks and turkeys, too, high erucic acid rapeseed oil (HEAR) caused growth depression, increased feed conversion and increased mortality (Salmon, 1969; Sheppard et al. 1971; Abdellatif et al. 1973; Vogtmann et al. 1974$_{c,d}$; Ratanasethkul et al. 1976; Clement et al. 1977; Clandinin et al. 1978; Renner et al. 1979). Many authors (Summers et al. 1966; Kondra et al. 1968; Sell et al. 1968; Biedermann, 1970; Vogtmann et al. 1973$_a$ 1974$_a$, 1975$_a$; Leslie et al. 1973; Lall et al. 1973$_b$; Bragg et a 1973; Karanajeewa, 1974; March et al. 1976) observed a reductio in egg production, egg weight and yolk weight, when a diet containing HEAR was fed to laying hens. Leclercq (1972) proved that the erucic acid content was responsible for the negative influence of HEAR on laying hens.

The development of new varieties brought in low erucic acid rapeseed oils (LEAR). With the use of Lear-oil little or no negative effects were observed. Oil from the varieties Oro and Zephyr (*Brassica napus*) did not affect performance, diets with Span (*Brassica campestris*) oil gave intermediate results (Vogtmann et al. 1973$_{a,b}$, 1974$_{a,c,d}$, 1975$_{a,b}$; Lall et al. 1973$_b$; Bragg et al. 1973; Abdellatif et al. 1973; March et al. 1976; Ratanasethkul et al. 1976; Clement et al. 1977; Clandinin et al. 1978; Renner et al. 1979). The ME values (kcal/g) determined by Lall et al. (1973$_a$, 1974) were:

	For chickens	For laying hens
Degummed rapeseed oil	7.89	8.85
Undegummed rapeseed oil	7.99	8.62
LEAR (Oro)	8.71	8.90

Refined rapeseed oils were lower in ME than the corresponding crude oils, (Lall et al. 1973$_b$, 1974). Sibbald (1977) determined the true metabolisable energy (TME) value of crude rapeseed oils as 9.26 kcal (38.8 kJ) /kg DM and of degummed rapeseed oils as 9.08 kcal (38 kJ)/kg DM.

The apparent digestibility of lipids in the HEAR-oil was 84/85% (Sell et al. 1963; Vogtmann et al. 1973$_a$); the digestibility of lipids in the HEAR-oil is higher (Vogtmann et al. 1973$_a$). Pressed rapeseed oil had only about 1 mg, and extracted oil 10 mg, isothiocyanate/100 g. Only extracted oil contained vinyl-thio-oxazolidone (VTO), at about 2 mg/100 g (Franzke et al. 1975).

RAPESEED GUMS

During the refining of RSO a fraction is removed by steam stripping and centrifugation known as gums. Basically rapeseed gums consist of glycolipids and phospholipids, with variable amounts of triglycerides, sterols, fatty acids, etc. Over 25 components have been recognised.

Gums from high or low erucic acid rapeseed, fed to chicks at 4 or 5% of the diet, resulted in growth rates and feed efficiency similar to those of oils (Salmon, 1970; March, 1977). Acidulated and non-acidulated soapstock from high and low erucic acid rapeseed, when compared with oils, gave favourable results for support of growth rate. The non-acidulated soapstock, probably as a result of their 6% mineral content, reduced feed efficiency slightly (March, 1977). Higher contents of RSO foots caused some growth depression (Cuidad et al. 1972; Lall et al. 1973$_a$).

Although low levels (up to 0.4%) of rapeseed gums in the diet appear to have no deleterious effects on the performance of laying hens (Leeson et al. 1977; Summers et al. 1978), the results from Hulan et al. (1978) indicate that at higher levels (2.0%) rapeseed gums significantly increase mortality, lower egg production and increase the amount of feed required for comparable egg production.

REFERENCES

Abdellatif, A.M.M. and Vles, R.O. 1973. Pathological effects of dietary rapeseed oils with high or low erucic acid content in ducklings. Poult. Sci: 52 (5): 1932-1936

Ahlström, B. 1978. Glucosinolate-poor rapeseed meal to broiler chicks. Proc. 5th Int. Rapeseed Conf. Malmö, 2: 292-294

Akiba, Y. and Matsumoto, T. 1973. Thyroid function of chicks after withdrawal of (-)-5-vinyl-2-oxazolidinethione, a goitrogen in rapeseed. Poultry Sci: 52 (2): 562-567.

Akiba, Y. and Matsumoto, T. 1976. Effect of goitrin on iodine absorption from the gastrointestinal tract of chicks. Jap. J. Zootechn. Sci: 47: 679-683

Akiba, Y. and Matsumoto, T. 1976. Antithyroid activity of goitrin in chicks. Poult Sci: 55 (2): 716-719

Akiba, Y. and Matsumoto, T. 1977. Effects of graded doses of goitrin, a goitrogen in rapeseed, on synthesis and release of thyroid hormone in chicks. Jap. J. Zootechn. Sci. 48 (12): 757-765

Akiba, Y. and Matsumoto, T. 1979. Relationship between chemically determined goitrin content in ration and physiologically active goitrin induced in chicks fed rapeseed meal or glucosinolates. Jap. J. Zootechn. Sci. 50 (1): 8-14

Akiba, Y. and Matsumoto, T. 1979. Physiologically active goitrin induced in chicks fed myrosinase-inactivated rapeseed meal. Jap. J. Zootechn. Sci. 50 (2): 73-78

Aleandri, M. and Olivetti, A. 1975. Impiego di miscele vegetali per le pollastre in batteria. Primi risultati sperimentali. Avicoltura 44 (3): 91-94

Baudet, J-J., Bourdon, D. and Guillaume, J. 1978. Valeur nutritionelle des tourteaux de colza sans glucosinolates influence du depelliculage essais sur porcs et volailles. Proc. 5th Int. Rapeseed Conf. Malmö, 2, 264-266

Bayley, H.S. and Hill, D.C. 1975. Nutritional evaluation of low and high fibre fractions of rapeseed meal using chickens and pigs. Can. J. Anim. Sci. 55 (2): 223-232

Bayley, H.S., Hill, D.C., Summers, J.D. and Slinger, S.J. 1970. The value of rapeseed meal as a poultry feed. Proc. 14th World Poult. Congr. Madrid, 221-222

Bayley, H.S., Slinger, S.J., Summers, J.D. and Ashton, G.C. 1974. Factors influencing the metabolizable energy value of rapeseed meal: level in diet, effects of steam-pelleting, age of chick, length of time on diet, variety of rapeseed and oil extraction procedure. Can. J. Anim. Sci. 54 (3): 465-480

Best. P. 1974. Rapeseed meal in laying hens. Poult. Int. 13 (4): 38-40

Beznák, A.v., Beznák, M.v. and Hajdu, I. 1943. Ernährungsphysiologische Wertmessung verschiedener Fette und Öle an weissen Ratten. Ernährung 8: 209-224

Beznák, A.v. and Beznak, M.v. 1943. Ernährungsphysiologische Werte des Rapsöls und einiger Rapsölprodukte. Ernährung 8: 236-244

Bhargava, K.K. and O'Neil, J.B. 1978. Evaluation of full-fat Tower rapeseed as a protein supplement for male broiler chicks. Poult. Sci. 57 (4) 1119

Biedermann, R. 1970. Der Einfluss von Art und Menge des Futterfettes auf den Verlauf des Fettstoffwechsels bei der Legehenne und auf die Zusammensetzung des Eierfettes. Thesis No. 4400, ETHZ, Switzerland

Bieliński, K., Bielińska, Jamroz D. and Kaszyński, J. 1978. Bobik, groch i poekstrakcyjna śruta rzepakowa jako źródło białka w pełnoporcjowych mieszankach granulowanych dla gęsi niosek. Roczniki Naukowe Zootechniki 5 (2): 223-233

Blair, R., Robblee, A.R., Dewar, W.A., Bolton, W. and Overfield, N.D. 1975. Influence of dietary rapeseed meals and selenium on egg production and egg tainting in laying hens. J. Sci. Fd. Agric. 26 (3): 311-318

Blair, R. and Scougall, R.K. 1975. Chemical composition, nutritive values of rapeseed meals. Feedstuffs 47 (6): 26-27

Blum, J.C., Guillaume, J. and Leclercq, B. 1974. Influence du tourteau de colza sur la croissance du pintadeau: comparaison avec le poulet. In: Conference on poultry and rabbit research, December, 1973. Paris, France: 179-182

Bock, H.D., Kracht, W. Völker, T. and Jähnke, M. Weitere Untersuchungen zur Verbesserung der Qualität von Raps und Rapsextraktionsschrot. Forschungszentrum fur Tierproduktion Dummerstorf. Rostock, GDR: 175-183

Bolton, W., Carter, T.C. and Morley, J.R. 1976. The hen's egg: genetics of taints in eggs from hens fed on rapeseed meal. Br. Poult. Sci. 17 (3) 313-320

Bowland, J.P. 1975. (Einsatz von Rapsextraktionsschrot in der Schweine und Geflugelernährung. 9th Nutrition Conf. for Feed Manufactuers, Nottingham, UK

Bragg, D.B. and Seier, L. 1974. Mineral content and biological activity of selenium in rapeseed meal. Poult. Sci. 53 (1): 22-26

Bragg, D.B., Sim, J.S. and Hodgson, G.C. 1973. Influence of dietary energy source on performance and fatty liver syndrome in White Leghorn laying hens. Poult. Sci. 52 (2): 736-740

Brahmakshetriya, R.D. and Sah, R.D. 1975. Replacement of groundnut cake with rapeseed meal in layers. Ind. Poult. Gazette 59 (4): 85-88

Brak, B. and Henkel, H. 1978. Untersuchungen über den Einsatz glucosinolatarmer Rapssaaten bzw. Extraktionschrote als Futtermittel für monogastrische Tiere. Fette, Seifen: Austrichmittel 80 (3): 104-105

Butler, E.J., Pearson, A.W. and Fenwick, G.R. 1980. The production of egg taint by rapeseed meal. Proc. 6th Europ. Poult. Conf., Hamburg, 3: 205-211

Calet, C. 1974. Some results on food efficiency of field beans and rapeseed oil meal for poultry. Wld's Poult. Sci. J. 30 (2): 142

Calet, C., Blum, J.C. and Guillaume, J. 1974. Le tourteau de colza dans l'alimentation des volailles. Roche Actualités (France), No. 1338

Calvarese, S., Restuccia, A. and Gramenzi, F. 1975. Effect of rapeseed meal on laying hens. Vet. Ital. 26 (9/12): 319-329

Campbell, L.D. 1979. Incidence of liver haemorrhage among white leghorn strains fed on diets containing different types of rapeseed meals. Br. Poult. Sci. 20: 239-246

Campbell, L., Cansfield, P., Ingalls, R. and Papas, A. 1978. Rapeseed meal quality as determined by glucosinolate analysis and balance trials with animals. Proc. 5th Int. Rapeseed Conf., Malmö, 2, 273-275

Campbell, L., Cansfield, P., Israels, E. and Papas, A. 1978. Antinutritional effects of rapeseed glucosinolates in poultry. Proc. 5th Int. Rapeseed Conf. Malmö, 2: 276-278

Campbell, L.D. and Smith, T.K. 1979. Response of growing chickens to high dietary contents of rapeseed meal. Br. Poult. Sci. 20: 231-237

Canada, Rapeseed Association of, 1978. Canadian rapeseed meal. Poultry and animal feeding. Publication No. 51

Chiasson, R.B., Sharp, P.J., Klandorf, H. Scanes, C.G. and Harvey, S. 1979. The effect of rapeseed meal and methimazole on levels of plasma hormones in growing broiler cockerels. Poult. Sci. 58, 1575-1583

Chiericato, G.M. and Filotto, U. 1977. Contributo sperimentale allo studio degli effeti di differenti livelli di 5-vinil-ossaxolidin-2-tione in diete per polli da carne. Riv. Zootecn. Vet. (1): 10-21

Ciudad, B.C. and Bravo, Z.E. 1972. Determinción de energía metabolizable de ácidos grasos provenientes de soap-stocks de pepas de uva y de raps, en broilers. Agric. Técn. 32 (1): 26-32

Clandinin, D.R. 1961. Rapeseed oil meal studies. 4. Effects of sinapin, the bitter substance in rapeseed oil meal, on the growth of chickens. Poult. Sci. 40: 484-487

Clandinin, D.R. Hawrysh, Z., Howell, J., Hanson, J.A., Christian, R.G. and Milne, G. 1974. Problems associated with the feeding of rations containing rapeseed meal to laying chickens. Proc. 4th Int. Rapeseed Congress, Giessen: 463-470

Clandinin, D.R. Ichikawa, S., Robblee, A.R. and Thomas, D. 1978. The use of low glucosinolate-type rapeseed meal in rations for layers and broilers. Proc. 5th Int. Rapeseed Conf. Malmö, 2: 284-286

Clandinin, T., Innis, S.M. and Renner, R. 1978. The effect of high and low erucic acid rapeseed oil on energy metabolism in chicks. Proc. 5th Int. Rapeseed Conf. Malmö, 2: 279-283

Clandinin, D.R. and Rao, P.V. 1970. Pentosans on prepress-solvent soved processed rapeseed meal. Poult. Sci. 49(6): 1741-1742

Clandinin, D.R. and Robblee, A.R. 1970. Rapeseed meal as a feedstuff for White Leghorn chickens. Proc. 14th Wld's Poult. Congress, Madrid: 699-703

Clandinin, D.R. and Robblee, A.R. 1978. Evaluation of rapeseed meal and protein for feed use. Proc. 5th Int. Rapeseed Conf. Malmö 2: 204-212

Clement, H. and Renner, R. 1977. Studies of the utilization of high and low erucic acid rapeseed oils by the chick. J. Nutrition 107(2): 251-260

Curtis, F., Fenwick, R., Heaney, R., Hobson-Frohock, A. and Land, D. 1978. Rapeseed meal and egg taint. Proc. 5th Int. Rapeseed Conf. Malmö, 2: 300-302

Darlington, K., Vogtmann, H., Robblee, A.R. and Clandinin, D.R. 1978. Influence of variety of rapeseed meal on egg shell membrane coloration in different breeds and strains of chickens. Arch. Geflugelk. 42, (6) 213-215

El Nockrashy, A.S. 1976. Protein isolates from new varieties of rapeseed. Fette, Seifen: Austrichmittel 78 (8): 311-317

El Nockrashy, A.S., Kiewitt, M. Mangold, H.K. and Mukherjee, K.D. 1975. Nutritive vale of rapeseed meals and rapeseed protein isolates. Nutr. Metabolism 19 (3-4): 145-152

Faruga, A., Kozłowski, M. and Kozłowska. 1974. Śruta rzepakowa i kiszonki z ziemniaków w zywieniu brojlerow kaczych. Roczniki Nauk Rolniczych, B 96 (1): 61-73

Fenwick, R.G., Hobson-Frohock, A., Land, D.G. and Curtis, R.F. 1979. Rapeseed meal and egg taint: treatment of rapeseed meal to reduce tainting potential. Br. Poult. Sci. 20: 323-329

Fenwick, R.G. and Hoggan, S.A. 1976. The tannin content of rapeseed meals. Br. Poult. Sci. 17(1): 59-62

Franzke, C., Göbel, R., Noack, G. and Seiffert, I. 1975. Uber den Gehalt an Isothiocyanaten und Vinylthiooxazolidon in Rapssaat und Rapsöl. Nahrung 19(7): 583-593

Fuhrken, E. 1974. Broiler Mastversuch mit Rapsschrot aus einer senfölarmen Sommerraps-Sorte. Dt.Geflügelw. Schweineprod. 26(45): 1121-1122

Fuhrken, E. 1975. Rapsschrot für braune Legehennen? Dt.Geflügelw. Schweineprod. 27(12): 277

Goh, Y.K. and Clandinin, D.R. 1977. Transfer of 125-J to eggs in hens fed on diets containing high-and low-glucosinolate rapeseed meals. Br. Poult. Sci. 18(6): 705-710

Goh, Y.K., Clandinin, D.R. and Robblee, A.R. 1979. Protein quality evaluation of commercial and laboratory heatdamaged rapeseed meals by the dye-binding technique and by biological assay with chicks. Can. J. Anim. Sci. 59(1): 195-201

Grandhi, R.R., Slinger, S.J. and Summers, J.D. 1977. Productive performance and liver lesions in two strains of laying hens receiving two rapeseed meals. Poult. Sci. 56(6): 1904-1908

Griffiths, N.M., Land, D.G. and Hobson-Frohock, A. 1979. Trimethylamine and egg taint. Br. Poult Sci. 20 555-558

Hawrysh, Z.J., Clandinin, D.R., Robblee, A.R., Hardin, R.T. and Darlington, K. 1975. Influence of rapeseed meal on the odor and flavor of eggs from different breeds of chickens. Can. Inst. Fd. Sci. Techn. J. 8 (1): 51-54

Hawrysh, Z.J., Steedman-Douglas, C.D., Robblee, A.R., Hardin, R.T. and Sam, R.M. 1980$_a$. Influence of low glucosinolate (cv Tower) rapeseed meal on the eating quality of broiler chickens. 1. Subjective evaluation by a trained test panel and objective measurements. Poult. Sci. 59: 550-557

Hawrysh, Z.J., Steedman-Douglas, C.D., Robblee, A.R., Hardin, R.T. and Sam, R.M. 1980$_b$. Influence of low glucosinolate (cv Tower) rapeseed meal on the eating quality of broiler chicken. 2. Subjective evaluation by a consumer panel. Poult. Sci. 59: 558-562

Hijikuro, S. and Takamasa, M. 1979. Effects of processing of rapeseed meal on the feeding value for chicks. 1. Comparison of the feeding value of some oil meals. Jap. Poult. Sci. 16: 172-176

Hijikuro, S. and Takamasa, M. 1979. Effects of processing of rapeseed meal on the feeding value for chicks. 2. Effects of heat treatments of rapeseed meal on the growth, feed efficiency and thyroid size of chicks. Jap. Poult. Sci 16: 177-183

Hijikuro, S. and Takamasa, M. 1979. Effects of processing of rapeseed meal on the feeding value for chicks. 3. Effects of autoclaving conditions on goitrin and amino acid content in rapeseed meal. Jap. Poult. Sci 16: 184-189

Hill, R. 1979. A review of the 'toxic' effects of rapeseed meals with observations on meal from improved varieties. Br. vet. J. 135(1): 3-16

Hill, R. and Marangos, A. 1974. The effects of a range of rapeseed meals and a mustardseed meal on the health and production on hybrid laying pullets. Proc. 15th Wld's Poult. Congr., New Orleans, 608-610

Hobson-Frohock, A. 1979. The quantitative determination of trimethylamine in egg. J. Fd. Technol. 14: 441-447

Hobson-Frohock, A., Fenwick, G.R., Heaney, R.K., Land, D.G. and Curtis, R.F. 1977. Rapeseed meal and egg taint - association with sinapine. Br. Poult. Sci. 18(5): 539-541

Hobson-Frohock, A., Fenwick, G.R., Land, D.G., Curtis, R.F. and Gulliver, A.L. 1975. Rapeseed meal and egg taint. Br. Poult. Sci 16(2): 219-222

Hobson-Frohock, A., Land, D.G., Griffiths, N.M. and Curtis, R.F. 1973. Egg taints - association with trimethylamine. Nature 243, 5405:304-305

Hoffmann, M., Schmiedinghoff, W. and Berger, H. 1974. Der Einfluss der Art des Futterfettes auf die Stickstoffretention bei wachsenden Geflügel. Arch. Tierernähr. 24(1): 57-66

Hulan, H.W. and Proudfoot, F.G. 1978. Nutritional efficacy of rapeseed meal as a source of dietary protein for broiler chicken. Proc. 5th Int. Rapeseed Conf. Malmö, 2: 287-291

Hulan, H.W. and Proudfoot, F.G. 1978. The effect of adding soybean or rapeseed gums to the diet on the performance of laying hens. Proc. 5th Int. Rapeseed Conf. Malmö, 2: 295-299

Hulan, H.W. and Proudfoot, F.G. 1980. The nutritional value of rapeseed meal for layer genotypes housed in pens. Poult. Sci. 59: 585-593

Ibrahim, I.K., Burdett, B.M. and Hill, R. 1980. The effects of rapeseed
meals, including those from new low-glucosinolate varieties on broiler
production. Proc. 6th Europ. Poult. Conf., Hamburg, 3: 388-393

Israels, E.D., Papas, A., Campbell, L.D. and Israels, L.G. 1979. Prevention
by menadione of the hepatoxic effects in chickens fed rapeseed meal.
Gastroenterology 16: 584-589

Jackson, N. 1969. Toxicity of rapeseed meal and its use as a protein supplement in the diet of two hybrid strains of caged laying hens. J.
Sci. Fd. Agric. 20(12): 734-740

Jackson, N. 1970. Algerian and French rapeseed meals as a protein source
for caged laying hens, with observations on their toxic effects. J.
Sci. Fd. Agric. 21(10): 511-516

Jeroch, H., Kracht, W. and Hennig, A. 1972/73: Untersüchungen zum Einsatz
von Rapssamenschrot als Energieträger in Broilermastfutter. Jahrb.
Tierernähr. Futt. 8: 389-393

Jones, J.D. and Sibbald, I.R. 1979. The true metabolizable energy values
for poultry of fractions of rapeseed (*Brassica napus* cv. Tower). Poult.
Sci. 58(2): 385-391

Kaminski, J. and Krełowska-Kulas, M. 1975. Wpływ śruty rzepakowej i
metylotiouracylu (MTU) na aktywnosc aminotransferaz (Asp AT, AIAT) w
wątrobie kurcząt. Acta Agr. Siv., Zootechn. 15(i): 53-59

Karunajeewa, H. 1974. Effect of plane of nutrition during the growing phase,
source of cereals, fat and the methionine content of laying diets on
the performance of crossbred pullets. Austrl. J. Exp. Agric. Anim.
Husb. 14(69): 461-469

Kondra, P.A., Choo, S.H. and Sell, J.L. 1968. Influence of strain of chickens
and dietary fat on egg production traits. Poult. Sci. 47: 1290-1296

Kozłowski, M. 1975. Ocena celowsći odgoryczania poekstrakcyjnej śruty
rzepakowej w procesie kisenia z paszami weglowodanowymi. Zesz. nauk.
Akad. Roln. Technicznej w Olsztynie, No. 142: 3-78

Kozłowski, M. 1976. Zastosowanie poekstrakcyjnej śruty rzepakowej toastowanej
w zywieniu brojlerów kaczych. Zesz. nauk. Akad Roln Technicznej w
Olsztynie, No. 162: 61-67

Kozłowski, M. 1976$_a$. Badania nad zastosowaniem kiszonek ziemniaczanych i
śrut rzepakowych w tuczu brojlerów kaczych. 1. Wynicki produkcyjne
oraz ocena paubojowa. Roczn. Nauk. roln. B 98(1):67-81)

Kozłowski, M. 1976$_b$. Badania nad zastosowaniem kiszonek ziemniaszanych i
śrut rzepakowych w tuczu brojlerów kaczych. 2. Aktywność wolotwórcza
śrut rzepakowych. Roczn. Nauk. roln. B 98(i): 83-89

Kozłowski, M., Chudy, J. and Cichon, R. 1976$_c$. Badania nad wartością odżywczą oleju rzepakowego. 4. Kwasy tłuszczowe żółci kaczek żywionych dietą z dodatkiem oleju rzepakowego i smalcu wieprzowego. Zesz. nauk. Akad. Roln.- Technicznej w Olsztynie, No. 181: 23-31

Kramer, M. 1973. Untersuchungen über die biologischen Wirkungen des Rapsöls. Nahrung 17(6): 643-681

Kuchta, M. and Malinowska, W. 1974. Śruta rzepakowa nieprepararowana i preparowana w dawce pokarmowej dla brojlerów. Zesz. nauk. Akad. Roln. w Krakowie. Zootechn., No. 87: 139-151

Kubola, D. and Morimoto, H. 1965. Nutritive value of feedstuffs for poultry and reliability of digestible nutrient of formula feeds calculated from the digestible crude protein and TDN contents of the ingredients. (II) Jap. Poult. Sci. 2: 63-68

Lall, S.P. and Slinger, S.J. 1973$_a$. The metabolizable energy content of rapeseed oils and rapeseed oil foots and the effect of blending with other foods. Poult. Sci. 52(1): 143-151

Lall, S.P. and Slinger, S.J. 1973$_b$. Nutritional evaluation of rapeseed oils and rapeseed soapstocks for laying hens. Poult. Sci. 52(5): 1729-1740

Lall, S.P. and Slinger, S.J. 1974. Effect of refining on the nutritional value of rapeseed oils for the chick and rat. J. Sci. Fd. Agr. 25: 423-432

Leclercq, B. 1972. Comparative utilization of erucic acid and oleic acid by the domestic fowl. Nutr. Rep. Intern. 6(5): 259-265

Leclercq, B. 1972. Influence de l'acide érucique alimentaire chez la poule pondeuse. Diminution du poids de l'oeuf. Ann. Biol. Anim. Bioch. Biophys. 12(3): 505-508

Leeson, S., Boorman, K.N., Lewis, D. and Shrimpton, D.H. 1977. Metabolisable energy studies with turkeys: nitrogen correction factor in metabolisable energy determinations. Br. Poult. Sci. 18(4): 373-379

Leeson, S., Slinger, S.J. and Summers, J.D. 1977. Performance of laying hens fed diets containing gums derived from Tower rapeseed. Can. J. Anim. Sci. 57(3): 479-483

Leeson, S., Slinger, S.J. and Summers, J.D. 1978. Utilization of whole Tower rapeseed by laying hens and broiler chickens. Can. J. Anim. Sci. 58 (1)55-61

Leeson, S. and Summers, J.D. 1976. Effect of rapeseed meal on the carcass grading of broilers. Poult. Sci. 55(6):2465-2467

Leeson, S. and Summers, J.D. 1978. Dietary gums and fishy odours in eggs. Poult. Sci. 57(1): 314-315

Leeson, S., Summers, J.D. and Grandhi, R. 1976. Carcass characteristics and performance of laying birds fed rapeseed meal. Can. J. Anim. Sci. 56(3): 551-558

Leslie, A.J., Hurnik, J.F. and Summers, J.D. 1973. Effects of color on consumption of broiler diets containing rapeseed meal and rapeseed. Can. J. Anim. Sci. 53(2): 365-369

Leslie, A.J., Pepper, W.F., Brown, R.G. and Summers, J.D. 1973. Influence of rapeseed products on egg quality and laying hen performance. Can. J. Anim. Sci. 53(4): 747-752

Leslie, A.J. and Summers, J.D. 1952. Feeding value of rapeseed for laying hens. Can. J. Anim. Sci. 52(3) 563-566

Leslie, A.J. and Summers, J.D. 1975. Amino acid balance of rapeseed meal. Poult. Sci. 54(2): 532-538

Leslie, A.J. and Summers, J.D. 1975. Influence of rapeseed products on reproductive performance of the hen and subsequent chick growth. Poult. Sci. 54(3): 916-917

Leslie, A.J., Summers, J.D., Grandhi, R. and Leeson, S. 1976. Arginine-lysine relationship in rapeseed meal. Poult. Sci. 55(2): 631-637

Leslie, A.J., Summers, J.D. and Jones, J.D. 1973. Nutritive vale of air-classified rapeseed fractions. Can. J. Anim. Sci. 53(1): 153-156

Leung, P. and March, B.E. 1976. The thyroidal response to chronic goitrogenic stimulation and the persistence of effects of early goitrogenic stimulation. Can. J. Phys. Pharmac. 54(4): 583-589

Lodhi, G.N., Clandinin, D.R. and Renner, R. 1970_a. Factors affecting the metabolizable energy value of rapeseed meal. 1. Goitrogens. Poult. Sci. 49(1): 289-294

Lodhi, G.N., Renner, R. and Clandinin, D.R. 1969. Studies on the metabolizable energy of rapeseed meal for growing chickens and laying hens. Poult. Sci. 48(3): 964-970

Lodhi, G.N., Renner, R. and Clandinin, D.R. 1970_b. Factors affecting the metabolizable energy value of rapeseed meal. 2. Nitrogen absorbability. Poult. Sci. 49(4): 991-999

Marangos, A.G. 1975. The effects of rapeseed meals on health and production of poultry and pigs. Thesis, London University, UK, 294 pp

Marangos, A. and Hill, R. 1974. The use of rapeseed meal as a protein supplement in poultry and pig diets. Vet. Rec. 96(17): 377-380

Marangos. A. and Hill, R. 1974. The hydrolysis and absorption of thioglucosides of rapeseed meal. Proc. Nutr. Soc. 33(3): 907

Marangos. A. and Hill, R. 1976. The use of rapeseed meals and a mustard seed meal as a protein source in diets for laying pullets. Br. Poult. Sci. 17(6): 643-653

Marangos, A., Hill, R., Laws, B.M. and Muschamp, D. 1974. The influence of three rapeseed meals and a mustard seed meal on egg and broiler production. Br. Poult. Sci. 15(4): 405-414

March, B.E. 1977. Response of chicks to the feeding of different rapeseed oils and rapeseed oil fractions. Can. J. Anim. Sci. 57(1): 137-140

March, B.E. and Biely, J. 1971. An evaluation of the supplementary protein and metabolizable energy value of rapeseed meals for chicks. Can. J. Anim. Sci. 51(3): 749-756

March, B.E., Biely, J. and Soong, R. 1975. The effects of rapeseed meal fed during the growing and/or laying periods on mortality and egg production in chickens. Poult. Sci. 54(6): 1875-1882

March, B.E., Bragg, D.B. and Soong, T. 1978. Low erucic acid, low glucosinolate rapeseed meal, with and without added gums in the layer diet. Poult. Sci. 57(6): 1599-1604

March, B.E. and MacMillen, C. 1979. Trimethylamine production in the caeca and small intestine as a cause of fishy taints in eggs. Poult. Sci. 58(1) 93-98

March, B.E. and MacMillen, C. 1980. Cholin concentration and availability in rapeseed meal. Poult. Sci. 59 611-615

March, B.E. and Sadiq, M. 1974. Rapeseed Association of Canada. Publ. 35 p. 70

March, B.E., Smith, F. and El-Lakany, S. 1973. Variation in estimates of the metabolizable energy value of rapeseed meal determined with chickens of different ages. Poult. Sci. 52(2): 614-618

March, B.E., Smith, T. and Sadiq, M. 1975. Factors affecting estimates of metabolizable energy values of rapeseed meal for poultry. Poult. Sci. 54(2): 538-546

March, B.E. and Soong, R. 1976. Mortality and production characteristics of laying chickens fed high- and low- erucic acid rapeseed oils. Poult. Sci. 55(4): 1557-1560

March, B.E. and Soong, R. 1978. Effects of added rapeseed gums in chick diets containing soybean meal or low-erucic acid, low glucosinolate, rapeseed meal. Can. J. Anim. Sci. 58(1): 111-113

Matsumoto, T. and Akiba, Y. 1979. Effects of rapeseed meal on growth and thyroid function of broiler chicks. Jap. Poult. Sci. 16(1): 1-9

Matsumoto, T., Akiba, Y., Hoshi, C. and Aoto, O. 1975. Antithyroid factors in rapeseed meal and counteracting the toxicity by autoclaving. Jap. Poult. Sci. 12(6): 265-270

Matsumoto, T., Akiba, Y. and Ikeda, W. 1979. Effects of feeding 'double low' rapeseed meal on performance of broiler chicks. Jap. Poult. Sci. 16(1): 49-81

Mazanowska, A. and Cholocinska, T. 1967. (The influence of rape-oil added to rations with different crude protein level on the performance of broilers). Postepy drobiarstwa 9(2): 7-9

Miller, W.S., Ewins, A. and Chubb, L.G. 1978. Egg taints. Vet. Rec. 91: 632-633

Mohyuddin, M. and Bowland, J.P. 1978. Nutritive value of some turkey diets and high protein meals assessed with the fungus *aspergillus flavus*. Qualitas Plantarum 28(3): 241-250

Moody, D.L., Slinger, S.J., Leeson, S. and Summers, J.D. 1978. Utilization of dietary Tower rapeseed products by growing turkeys. Can. J. Anim. Sci. 58(4): 585-592

Mueller, M.M., Coleman, R.N. and Clandinin, D.R. 1978. Trimethylamine production from sinapine by enteric bacteria from laying hens. Proc. 5th Int. Rapeseed Conf., Malmö, 2: 303-306

Muzlar, A.J., Likushi, H.J.A. and Slinger, S.J. 1978. Metabolizable energy content of Tower and Candle rapeseeds and rapeseed meals determined in two laboratories. Can. J. Anim. Sci 58(3): 485-492

Muzlar, A.J., Sadiq, M. and Slinger, S.J. 1978. Effect of length of excreta collection period and feed input level on the true metabolizable energy value of rapeseed meal. Nutr. Refs. Intern. 19(5): 689-694

Muzlar, A.J., Slinger, S.J., Likushi, H.J.A. and Dorell, H.G. 1980. True amino acid availability values for soybean meal and Tower and Candle rapeseed and rapeseed meals determined in two laboratories. Poult. Sci. 59: 605-610

Nwokolo, E.N. and Bragg, D.B. 1977. Influence of phytic acid and crude fiber on the availability of minerals from four protein supplements in growing chicks. Can. J. Anim. Sci. 57 475-477

Nwokolo, E.N. and Bragg, D.B. 1978. Factors affecting the metabolizable energy content of rapeseed meals. Poult. Sci. 57(4): 954-958

Nwokolo, E.N., Bragg, D.B. and Kitts, W.D. 1976. The availability of amino acids from palm kernel, soybean, cottonseed and rapeseed meal for the growing chick . Poult. Sci. 55(6): 2300-2304

Olomu, J.M., Robblee, A.R. and Clandinin, D.R. 1974. Effects of processing and amino acid supplementation on the nutritive value of rapeseed for broilers. Poult. Sci. 53(1): 175-184

Olomu, J.M., Robblee, A.R. and Clandinin, D.R. 1975. Effects of Span rapeseed on the performance, organ weights and composition of the carcass, heart and liver of broiler chicks. Poult. Sci. 54(3): 722-726

Olomu, J.M., Robblee, A.R., Clandinin, D.R. and Hardin, R.T. 1975$_a$. Effects of Span rapeseed meal on productive performance, egg quality, composition of liver and hearts and incidence of 'fatty livers' in laying hens. Can J. Anim. Sci. 55(1): 71-75

Olomu, J.M., Robblee, A.R., Clandinin, D.R. and Hardin, R.T. 1975$_b$. Evaluation of full-fat Span rapeseed as an energy and protein source for laying hens. Can. J. Anim. Sci. 55(2): 219-222

Olomu, J.M., Robblee, A.R., Clandinin, D.R. and Hardin, R.T. 1975$_c$. Utilization of full-fat rapeseed and rapeseed meals in rations for broiler chicks. Can. J. Anim. Sci. 55(3): 461-469

Overfield, N.D. and Elson, H.A. 1975. Dietary rapeseed meal and the incidence of tainted eggs. Br. Poult. Sci. 16(2): 213-217

Papas, A., Campbell, L.D. and Cansfield, P.E. 1979$_a$. The effect of glucosinolates on egg iodine and thyroid status of poultry. Can. J. Anim. Sci. 59(1): 119-131

Papas, A., Campbell, L.D. and Cansfield, P.E. 1979$_b$. A study of the association of glucosinolates to rapeseed meal induced hemorrhagic liver in poultry and the influence of supplemental vitamin K. Can. J. Anim. Sci. 59(1): 133-144

Pearson, A.W. and Butler, E.J. 1979$_c$. Rapeseed meal goitrogens and egg taint. Vet. Rec. 104(8): 168

Pearson, A.W., Butler, E.J., Curtis, R.F., Fenwick, G.R., Hobson-Frohock, A. and Land, D.G. 1979$_a$. Effect of rapeseed meal on trimethylamine oxidase activity in the domestic fowl in relation to egg taint. J. Sci. Fd. Agric. 30(3): 291-298

Pearson, A.W., Butler, E.J., Curtis, R.F., Fenwick, G.R., Hobson-Frohock, A. and Land, D.G. 1979$_b$. Effect of rapeseed meal on trimethylamine metabolism in the domestic fowl in relation to egg taint. J. Sci. Fd. Agric. 30: 799-804

Pearson, A.W., Butler, E.J., Curtis, R.F., Fenwick, G.R., Hobson-Frohock, A. and Land, D.G. 1979$_c$. Rapeseed meal and egg taint: demonstration of the metabolic defect in male and female chicks. Vet. Rec. 104(14): 318-319

Pearson, A.W., Butler, E.J., Curtis, R.F., Fenwick, G.R., Hobson-Frohock, A., Land, D.G. and Hall, S.A. 1978. Effects of rapeseed meal on laying hens *(Gallus domesticus)* in relation to fatty liver-haemorrhagic syndrome and egg taint. Res. Vet. Sci. 25(3): 307-313

Pearson, A.W., Butler, E.J. and Fenwick, G.R. 1979. Rapeseed meal and liver damage: effect on plasma enzyme activities in chicks. Vet. Rec. 105: 200-201

Peraza, C. and Rodriguez, G. 1974. Efecto de niveles altos de pasta de nabo en el pollo de engorda. Proc. 15th Wld's Poult. Sci. Congr. New Orleans, 610-612

Rao, P.V. and Clandinin, D.R. 1970. Effect of method of determination on the metabolizable energy value of rapeseed meal. Poult. Sci. 49(4): 1069-1074

Rao, P.V. and Clandinin, D.R. 1972$_a$. Chemical determination of available carbohydrates in rapeseed meal. Poult. Sci. 81: 1474-1475

Rao, P.V. and Clandinin, D.R. 1972$_b$. Role of protein content, nitrogen absorbability and availability of carbohydrates in rapeseed meal on its metabolizable energy value for chicks . Poult. Sci. 51(6): 2001-2006

Ratanasethkul, C., Riddell, C., Salmon, R.E. and O'Neil, J.B. 1976. Pathological changes in chickens, ducks and turkeys fed high levels of rapeseed oil. Can. J. Comp. Med. 40(4): 360-369

Renner, R., Innis, S.M. and Clandinin, M.T. 1979. Effects of high and low erucic acid rapeseed oils on energy metabolism and mitochondrial function of the chick. J. Nutrition 109(3): 378-387

Roos, A.J. and Clandinin, D.R. 1975. Transfer of 125-I to eggs in hens fed on diets containing rapeseed meal. Br. Poult. Sci. 16(4): 413-415

Ruszczyc, Z., Jamroz, O., Fritz, Z. and Piech, A. 1974. Opracowanie składu mieszanek paszowych dla kurczat rzeznych w oparciu o surowce krajowe. I .Zasto sowanie mieszanek zawierajacych nariona lub olej rzepakowy Postepy drobiarstwa 16(1): 5-14

Salmon, R.E. 1969. The relative value of rapeseed and soybean oils in chick starter diets. Poult. Sci. 48(3): 1045-1050

Salmon, R.E. 1970. Rapeseed gum in poultry diets. Can. J. Anim. Sci. 50: 211-212

Salmon, R.E. 1977. Effects of age on the absorption of fat by turkeys fed mixtures of beef fat and rapeseed oil. Can. J. Anim. Sci. 57(3): 427-431

Salmon, R.E. 1979$_a$. Rapeseed meal in turkey starter diets. Poult. Sci. 58(2): 410-415

Salmon, R.E., Klein, K.K. and Larmond, E. 1979$_b$. Low glucosinolate rapeseed meal in turkey broiler diets of varying nutrient density. Poult. Sci. 58: 1514-1523

Sell, J.L. 1966. Metabolizable energy of rapeseed meal for the laying hen. Poult. Sci. 45(4): 854-856

Sell, J.L., Choo, S.H. and Kondra, P.A. 1968. Fatty acid composition of egg yolk and adipose tissue influenced by dietary fat and strain of hen. Poult. Sci. 47: 1296-1302

Sell, J.L. and McKirdy, J.A. 1962. Comparative value of dietary rapeseed oil, sunflower seed oil, soybean oil and animal tallow for chickens. J. Nutrition 16: 113-118

Sell, J.L. and McKirdy, J.A. 1963. Lipids in chick urine: the influence of dietary rapeseed oil. Poult. Sci. 42: 380-383

Seth, P.C.C. and Clandinin, D.R. 1973$_a$. Metabolisable energy value and composition of rapeseed meal and of fraction derived therefrom by air classification. Br. Poult. Sci. 14(5): 499-505

Seth, P.C.C. and Clandinin, D.R. 1973$_b$. Effect of including rapeseed meal in the ration of broiler-type chickens on the incidence of perosis and the ineffectiveness of supplemental manganese. Poult. Sci. 52(3): 1158-1160

Seth, P.C.C., Clandinin, D.R. and Hardin, R.T. 1975. *In vitro* uptake of zinc by rapeseed meal and soybean meal. Poult. Sci. 54(2): 626-629

Sharby, T.F. and Bell, J.M. 1974. Effect of feeding diets containing Bronowski rapeseed meal and fababeans *(Vicia faba L)* on broiler chick performance. Poult. Sci. 53(5): 1977

Sheppard, A.J., Fritz, J.C., Hooper, W.H., Roberts, T., Hubbard, W.D., Prosser, A.R. and Boehne, J.W. 1971. Crambe and rapeseed oils as energy sources for rats and chicks and some ancillary data on organ weights and body cavity fat composition. Poult. Sci. 50(1): 79-84

Shu, J.C. 1975. (Studies on the feeding value of rapeseed meal for broilers). Taiwan Agr. Quarterly 11(1): 123-132

Sibbald, I.R. 1977. The true metabolizable energy values for poultry of rapeseed and of meal and oil derived therefrom. Poult. Sci. 56(5) 1652-1656

Sibbald, I.R. and Price, K. 1977. The true metabolizable energy values of the seeds of *Brassica campestris*, *Brassica hirta* and *Brassica napus*. Poult. Sci. 56(4): 1329-1331

Sibbald, I.R. and Slinger, S.J. 1963. Factor affecting the metabolizable energy content of poultry feeds. 12. Protein quality. Poult. Sci. 42: 707-709

Siddiqui, I.R. and Wood, P.J. 1977. Carbohydrates of rapeseed: a review. J. Sci. Fd. Agric. 28(6): 530-538

Sim, J.S., Bragg, D.B. and Hodgson, G.C. 1973. Effect of dietary animal tallow and vegetable oil on fatty acid composition of egg yolk, adipose tissue and liver of laying hens. Poult. Sci. 52(1): 51-57

Skotnicki, J. and Koreika, A. 1972. Wartość energi metabolicznej śruty rzepakowej i nasion rzepaku odgoryczonych róznymi metodami w zywieniu drobiu. Zesz. probl. post. nauk roln, 126: 131-199

Slinger, S.J., Leeson, S., Summers, J.D. and Sadiq, M. 1978. Influence of steam pelleting on the feeding value of Tower and Candle rapeseed products for broiler chickens. Anim. Fd. Sci. Technol. 3(3): 251-259

Slinger, S.J., Summers, J.D. and Leeson, S. 1978. Utilization of meal from a new rapeseed variety, *Brassica campestris* cv CAndle, in layer diets. Can. J. Anim. Sci. 58(4): 593-596

Smith, T.K. and Campbell, L.D. 1976. Rapeseed meal glucosinolates: metabolism and effect on performance in laying hens. Poult. Sci. 55(3) 861-867

Smith, W.K., MacLeod, M.G., Tullet, S.G. and Klandorf, H. 1979. The effect of rapeseed meal on the energy metabolism of laying hens. Br. Poult. Sci. 20: 453-462

Splittgerber, H. 1975. Rapssaat im Hähnschen-Mastalleinfutter. Ol. Geflugelw. Schweineprod. 27(14): 325-326

Srivastara, V.K. and Hill, D.C. 1976. Effect of mild heat treatment on the nutritive value of low glucosinolate - low erucic acid rapeseed meals. J. Sci. Fd. Agric. 27(10): 953-958

Srivastara, V.K., Philbrick, D.J. and Hill, D.C. 1975. Response of rats and chicks to rapeseed meal subjected to different enzymatic treatments. Can. J. Anim. Sci. 55(3): 331-335

Steedman, C.D., Hawrysh, Z.J., Hardin, R.T. and Robblee, A.R. 1978. Influence of rapeseed meal on the eating quality of chicken. Proc. 5th Int. Rapeseed Conf. Malmö, 2: 307-311

Steedman, C.D., Hawrysh, Z.J., Hardin, R.T. and Robblee, A.R. 1979a. Influence of rapeseed meal on the eating quality of chicken. 1. Subjective evaluation by a trained tasted panel and objective measurements Poult. Sci. 58(1): 148-155

Steedman, C.D., Hawrysh, Z.J., Hardin, R.T. and Robblee, A.R. 1979b. Influence of rapeseed meal on the eating quality of chicken. 2. Subjective evaluation by a consumer taste panel. Poult. Sci. 58 (2): 337-340

Summers, J.D. 1974. (Aminosäure - Verfügbarkeit aus Rapsekstraktionsschrot für junge Leghorn-Hähnschen). 10th ann. Nutrition Conf. for Feed Manufacturers, Guelph, Ontario, (Can).

Summers. J.D. and Leeson, S. 1977a. Effect of thyroxine and thiouracil additions to diets containing rapeseed meal on chick growth and carcass composition. Poult. Sci. 56(1): 25-35

Summers, J.D. and Leeson, S. 1977b. Performance and carcass grading characteristics of broiler chickens fed rapeseed gums. Can. J. Anim. Sci. 57(3): 485-488

Summers, J.D. and Leeson, S. 1978. Feeding value and amino acid balance of low-glucosinolate *Brassica napus* (Cv Tower) rapeseed meal. Poult. Sci. 57(1): 235-241

Summers, J.D., Leeson, S. and Slinger, S.J. 1978. Performance of egg-strain birds during their commercial life cycle when continuously fed diets containing Tower rapeseed gums. Can. J. Anim. Sci. 58(2): 183-189

Summers, J.D., Pepper, W.F. and Wood, A.S. 1971. The value of rapeseed meal in broiler breeder diets. Poult. Sci. 50(5): 1387-1391

Summers. J.D., Slinger, S.J. and Anderson, W.J. 1966. The effect of feeding various fats and fat by-products on the fatty acid and cholesterol composition of eggs. Br. Poult. Sci. 7: 127-134

Tao, R., Belzile, R.J. and Brisson, G.J. 1971. Amino acid digestibility of rapeseed meal fed to chickens: effects of fat and lysine supplementation Can. J. Anim. Sci. 51(3): 705-709

Teuteberg, W. 1974. Einsatz von glucosinolatarmen Rapsschrot in der Fütterung. Proc. 4. Int. Rapskongress, Giessen, 447-452

Thomas, D., Robblee, A.R. and Clandinin, D.R. 1978. Effects of low and high glucosinolate rapeseed meals on productive performance, egg quality, composition of liver and incidence of haemorrhagic liver syndrome in laying birds. Br. Poult. Sci. 19(4): 449-454

Vogt. H. 1973. Versuche über den Einsatz von senfölarmen und behandelten Raps extraktionsschrot im Geflügelfütter. Landw. Forschg. 30(I. Sonderheft): 210-216

Cog. H. 1974. Versuche uber den Einsatz von senfölarmen und behandelten Rapsextraktionsschrot im Geflügelfütter. Prc. 4 Int. Rapskongress, Giessen, 453-462

Vogt. H. 1977. The use of rapeseed meal produced from a rape sort low in erucic acid and glucosinolate in poultry feeding. In: Protein quality from leguminous crops. Kirchberg, Luxembourg, Commission of the European Communities, 282-283

Vogt. H. and Rauch, H.W. 1979. Reduziert die Behandlung des Rapsextraktionschrotes mit Calciumhydroxid die Beeinflüssung des Eigeruchs und geschmacks. (Unpublished)

Vogt. H., Schubert, H.J., Stute, K. and Rauch, W. 1969$_a$. Futterwert und Einsatz von Rapsschrot in der Geflügelfütterung. 3. Mitteilung: Rapsschrot in der Legehennenfütterung. Arch. Geflügelk. 33: 119-124

Vogt. H., Schubert, H.J. and Stute, K. 1969$_b$. Futterwert und Einsatz von Rapsschrot in der Geflügelfütterung. 4. Mitteilung: Zweiter Legeheenenversuch mit Rapsschrot. Arch. Geflügelk. 33: 392-395

Vogt. H. and Stute, K. 1974. Führt eine Senkung des Vinyloxazolidinethion-Gehaltes zu einer Verbesserung des Futterwertes von Rapsextraktionsschrot im Geflügelfütter. Arch. Geflügelk. 38(4): 127-138

Vogt. H. and Torges, H.G. 1976. Rapsextraktionsschrot aus einer erucasaeure und glucosinolatarmen Sommerrapssorte im Legehennenfutter. Arch. Geflügelk. 40(6): 225-231

Vogtmann, H. and Clandinin, D.R. 1974$_c$. Vitamin E and high or low erucic acid rapeseed oils in broiler diets. Can. J. Anim. Sci. 54(4): 669-677

Vogtmann, H. and Clandin, D.R. 1974$_d$. Low erucic acid rapeseed oils in rations for broiler chickens: Oro and hydrogenated Oro oil. Poult. Sci. 53(6) 2108-2115

Vogtmann, H. and Clandinin, D.R. 1975. The effects of crude and refined low erucic acid rapeseed oils in diets for laying hens. Br. Poult. Sci. 16(1): 55-61

Vogtmann, H., Clandinin, D.R. and Hardin, R.T. 1973$_a$. Utilization of rapeseed oils of high and low erucic acid contents. 1. Digestibility and energy utilization. Nutr. Metabol. 15(4-5): 252-266

Vogtmann, H., Clandinin, D.R. and Hardin, R.T. 1974$_a$. The influence of high and low erucic acid rapeseed oils on the productive performance of laying hens and on the lipid fraction of egg yolk. Can. J. Anim. Sci. 54(3): 403-410

Vogtmann, H., Clandinin, D.R. and Hardin, R.T. 1974b. Utilization of rapeseed oils of high and low erucic acid content. 2. Influence on tissue Nutr. Metabol. 17(3): 136-147

Vogtmann, H., Clandinin, D.R. and Hardin, R.T. 1975. The effects of crude and refined low erucic acid rapeseed oils in diets for broiler chicks. Br. Poult. Sci. 16(1): 63-68

Vogtmann, H., Clandinin, D.R. and Robblee, A.R. 1973b. Low and high erucic acid rapeseed oils in rations for laying hens. Poult. Sci. 52(3): 955-962

Vogtmann, H., Clandinin, D.R. and Thompson, J.R. 1978. Time course changes in total lipid content and fatty acid distribution in various tissues of young chickens after feeding rapeseed oil or soybean oil. Intern. J. Vit. Nutr. Res. 48(1): 90-100

Vogtmann, H., Thompson, J.R., Clandinin, D.R., Fenton, T.W. and Turner, B.V. 1974. Metabolism of ^{14}C- Erucic acid. Proc. 4. Int. Rapskongress, Giessen, 709-717

Wartenberg, L., Kinal, S., Trebusiewicz, B. and Kroliczek, A. 1975. Wplyw odgoryczonyck nasion rzepaku w paszy na sklad kwasow tluszczowyck Huszczu zapasowego kurczal. Roczniki Naukowe Zootechniki, Monografie i Rozprawy, no. 2: 111-120

Wartenberg, L., Trebusiewicz, B. and Kroliczek, A. 1975. Poziom witaminy a w watrobie kurczat raeznych w zaleznosci od zawartosci nasion rzepaku w paszy. Roczniki Naukowe Zootecyniki, Monografie i Rozprawy, no. 2, 121-129

Woodly, A., Summers, J.D. and Bilanski, W.K. 1972. Effects of heat treatment on the nutritive value of whole rapeseed for poultry. Can. J. Anim. Sci. 52(1): 189-194

Yamashiro, S. and Bast, I. 1978. Ultrastructure of livers of broiler chickens fed diets containing rapeseed meal. Res. Vet. Sci. 25(1): 21-24

Yamashiro, S., Bhatnagar, M.K., Scott, J.R. and Slinger, S.J. 1975. Fatty haemorrhagic liver syndrome in laying hens on diets supplemented with rapeseed products. Res. Vet. Sci. 19(3): 312-321

Yamashiro, S., Umemura, T. Bhatnagar, M.K., David, L., Sadiq, M. and Slinger, S.J. 1977. Haemorrhagic liver syndrome of broiler chickens fed diets containing rapeseed products. Res. Vet. Sci. 23(2): 179-184

Yapar, Z. and Clandinin, D.R. 1972. Effect of tannins in rapeseed meal on the nutritional value for chicks. Poult. Sci. 51(1): 222-228

Yule, Z. and McBride, R.L. 1976. Lupin and rapeseed meals in poultry diets: effect on broiler performance and sensory evaluation of carcasses. Br. Poult. Sci. 17(2): 231-239

Yule, W.J. and McBride, R.L. 1978. Rapeseed meals in broiler diets: effect on performance and sensory evaluation of carcasses. Br. Poult. Sci. 19(4): 543-548

DISCUSSION

C. Calet (France)

I agree with Dr. Vogt on the importance of anti-nutritional factors in rapeseed, but I am not sure that all the results are the same on this subject. The main effects of the reduction in glucosinolates are to increase the energy value of the rapeseed oil meal and to reduce mortality levels in poultry.

As for other aspects, I am not sure that all the authors are in agreement. For instance, some people say that they have achieved the same results with OO-rapeseed meal as with soyabean meal; others do not get the same results. Perhaps the answer is that we have had not just one sample of rapeseed oil meal, but many. One of the most important problems which has not been discussed here is the technological effect - the processing effect - on the feed value of the rapeseed.

B.O. Eggum (Denmark)

With all the inhibitors that Dr. Vogt has been discussing, do we know where they are located in the kernel? How much is in the hull fraction? I know that tannin is primarily in the hull fraction. If we can solve the energy problem, we can solve the tannin problem.

H. Vogt (FRG)

I have some data on the content of the various factors, but no specified values for the hull and the kernel separately.

G. Röbbelen (FRG)

With respect to the glucosinolates in the cotyledon and embryo, I would like to give you some data from an experiment done by Professor Henkel in co-operation with us. The low glucosinolate rapeseed meal is low because of one or two genetic factors. The reduction is in respect to the glucosinolates derived from methionine. It is almost certain that all the other glucosinolates are unaffected. It is well known that the indole glucosinolates are not affected and, in all probability, the others will not be reduced or affected by the low glucosinolate characteristic. The question is: can we get a rough, general measure of the toxic principle in total? Henkel was of the opinion that if the glucosinolates were broken down through nitriles, you might find something. We know, from what Dr. Sørensen told us, that the break-down procedure differs with different molecules; some break down quicker, some go through the nitriles and are transformed even further. It is a question of which variety of glucosinolates will show up in the seed.

This is what we did: we collected seeds of *Brassica* groups which had one dominant glucosinolate. For example, *Brassica napus* is very high in progoitrin, whereas *Brassica campestris* is relatively high in gluconapin but low in progoitrin, though the sum total is high in both cases. *Brassica oleracea gongylodes*, on the other hand, is very low in total glucosinolates. When we measured the nitrile which was found in mg/kg of fat free DM, it corresponded roughly with the sum of the glucosinolates. Broilers were fed with a control diet containing soyabean meal, and with isocaloric, isonitrogenous diets containing about 20% of different rapeseed meals. There was a big reduction in corrected final weight of broilers when they were fed with rapeseed meals containing high amounts of epigoitrin (in *Crambe abysinnica*), and progoitrin. The second worst effect, giving only 86% of the average broiler weight from the control diet, was with seed meals from *Brassica napus*, high in progoitrin. The figures for the thyroid glands correspond well to the weights. It is interesting to note that the erucin, which gave a high nitrile content, did not decrease final weight and had a rather low effect on increasing the thyroid gland compared with the progoitrin and the epiprogoitrin figures. This indicates the direction that Dr. Sørensen was showing us yesterday.

A. Rérat   (France)

Thank you very much for these indications. It is interesting for us to know what type of anti-nutritive factor is involved. Now, Mr. Vogt, I would like to ask you about the haemorrhage of the liver. Can you say what is the origin?

H. Vogt

The substances which cause the haemorrhage are not really known, but the work of Smith and Campbell in Manitoba demonstrates that the nitrile compounds have an effect.

A. Rérat

Thank you very much, Dr. Vogt.

RAPESEED MEAL AS A PROTEIN SUPPLEMENT FOR DAIRY COWS -
RESULTS FROM FEEDING EXPERIMENTS IN SWEDEN

L. Lindell
Department of Animal Husbandry, University of Agricultural
Sciences, Uppsala, Sweden

INTRODUCTION

Rape is an important agricultural crop in Sweden. Winter varieties of oil rape (*Brassica napus*) dominate but a certain amount of winter turnip rape (*Brassica campestris*) is also grown. Summer varieties of both species are grown in certain areas.

Most of the seeds are processed in Sweden by solvent extraction. The commercial product of rapeseed meal (RSM) is a mixture of meals of different varieties. Percentages of meals of different origin in the marketed product are not available but meal of *B. napus* constitutes the major part.

The effect of long-term feeding of different amounts of Swedish commercial rapeseed meal has been studied in two experiments with individually fed cows (Lindell and Knutsson, 1976; Lindell, 1976).

Rapeseed as a source of fat in straw-based diets has been studied in three experiments with individually fed cows. In these experiments commercial rapeseed meal and low-glucosinolate rapeseed was fed (Frank, 1979a; Frank, 1979b; Frank, 1980).

EXPERIMENTS WITH DIFFERENT LEVELS OF RSM

Each of the two experiments with different levels of RSM covered two lactations with each cow. The first ten weeks after calving in the first experimental lactation were used as a preperiod and the experimental period started 11 weeks after calving and continued to 40 weeks after calving in the second experiment.

In the first experiment, with 30 first-lactation cows of the Swedish Friesian breed (SLB), three different concentrate mixtures containing 0% (group 1), 4.2% (group 2) or 8.05% (group 3) of RSM were fed (Table 1).

TABLE 1

COMPOSITION OF THE CONCENTRATE MIXTURES (%). EXPT. 1 (LINDELL AND KNUTSSON, 1976).

	Group 1	Group 2	Group 3
Oats	30.00	29.50	28.50
Barley	30.00	29.50	28.50
Cottonseed cake	8.00	8.40	9.20
Peanut cake	5.00	5.25	5.75
Rapeseed meal	0.00	4.20	8.05
Soyabean meal	7.00	3.15	0.00
Dried beet pulp	20.00	20.00	20.00

In the second experiment, with 22 SLB cows, two different concentrate mixtures containing 0% (group 1) and 10% (group 2) of RSM were fed (Table 2).

TABLE 2

COMPOSITION OF THE CONCENTRATE MIXTURES (%). EXPT. 2 (LINDELL, 1976)

Ingredients	Preperiod	Experimental period Group 1	Experimental period Group 2
Oats	29.50	31.00	30.00
Barley	29.50	31.00	30.00
Cottonseed cake	8.40	5.40	6.00
Peanut cake	5.25	3.60	4.00
Rapeseed meal	4.20	0.00	10.00
Soyabean meal	3.15	9.00	0.00
Dried beet pulp	20.00	20.00	20.00

The maintenance ration consisted of 5 kg of hay and 2.5 kg dry matter of beet tops silage in experiment 1 and of 3.5 kg of hay, 2.0 kg of straw and 2.0 kg dry matter of grass silage in experiment 2.

The concentrate was fed according to the milk yield during the previous week. Corrections for differences in body weights were also made. The calculation of requirements for metabolisable energy and digestible crude protein was based on the current Swedish standard. All cows were individually fed.

RESULTS

Feed Consumption

There was no difference in consumption of offered feeds between the groups and feed refusals were small. No palatability problems connected with feeding of RSM were observed during any part of the lactations. The change from the concentrate fed during the preperiods to those fed during the experimental periods could be made without any adaptation problems (Table 3 and 4).

The consumption of concentrates during the second lactation represented an average daily intake of 0.32 kg and 0.61 kg RSM in groups 2 and 3 respectively in experiment 1; and 0.78 kg RSM in group 2 in experiment 2. The highest consumption of RSM in experiment 2 was 1.39 kg per day measured as an average over seven days.

The content of isothiocyanates was 4.1 mg (SD 0.4) and of oxazolidinethiones 9.7 mg (SD 1.5) per g dry matter of RSM fed in experiment 1. Traces of myrosinase were found in 3 of 27 samples.

The content of isothiocyanates in RSM fed in experiment 2 was 4.3 mg (SD 0.5) and of oxazolidinethiones 9.5 mg (SD 1.5) per g dry matter. Traces of myrosinase were found in 3 of 29 samples in this experiment.

TABLE 3

FEED CONSUMPTION, EXPERIMENT 1.

	Lactation 1, weeks 11-40			Lactation 2, weeks 2-40		
	Group			Group		
	1	2	3	1	2	3
Hay, kg	5.00	5.00	4.98	4.99	4.95	4.93
Silage, kg	11.84	12.15	11.95	13.38	13.08	12.84
Concentrate, kg	6.38	6.48	6.40	7.18	7.64	7.60
ME, MJ	133	135	133	141	144	143
DCP, g	1491	1519	1503	1640	1685	1695

TABLE 4

FEED CONSUMPTION, EXPERIMENT 2.

	Lactation 1, weeks 11-40		Lactation 2, weeks 2-40	
	Group		Group	
	1	2	1	2
Hay, kg	3.50	3.50	3.50	3.50
Straw, kg	1.56	1.64	1.97	1.94
Silage, kg	7.13	7.14	7.31	7.52
Concentrate, kg	7.17	6.97	7.80	7.80
ME, MJ	142	139	145	144
DCP, g	1383	1341	1301	1277

Milk yield

The differences in milk yield, yield of FCM and content of fat and protein in the milk were small and in no case significant in the two experiments (Tables 5 and 6). There was no difference in reaction to the experimental treatment between low yielding cows (about 3 800 kg FCM) and high yielding cows (about 5 900 kg FCM).

Health of the cows

A control of the general health of the cows, including the determination of a number of blood components as well as the total iodine and thiocyanate levels in the milk, was carried out by the Department of Clinical Biochemistry of the Faculty of Veterinary Medicine, Swedish Univ. Agr. Sci. The general condition of the cows was normal for the herd. The values obtained in haematological and chemical analyses of blood were within the normal range. Cows fed RSM had on the average higher thiocyanate and lower total iodine concentration in their milk than cows fed no RSM. It was also found that all groups showed an increasing total iodine concentration in milk during the lactation.

In experiment 2 a mineral supplement containing potassium iodide was fed, which led to a general increase in milk iodine concentration as compared with experiment 1 where no extra iodine was fed.

Average calving intervals and services per conception were normal and no significant differences between the groups were observed. However, cows with the highest level of RSM did show a slightly but not significantly greater number of services per conception during the second lactation in both experiments.

TABLE 5
MILK YIELD, EXPERIMENT 1.

	Lactation 1									Lactation 2		
	Preperiod Weeks 2-10				Experimental period Weeks 11-40					Weeks 2-40		
	Group				Group					Group		
	1	2	3		1	2	3		1	2	3	
Milk, kg	21.96	21.62	20.93		14.50	14.81	14.51		16.90	18.66	17.97	
FCM, kg	21.11	20.58	20.41		14.36	14.53	14.33		16.48	17.88	17.68	
Fat, %	3.75	3.67	3.86		3.92	3.85	3.97		3.82	3.71	3.90	
Protein, %	3.28	3.28	3.45		3.46	3.54	3.43		3.36	3.35	3.46	

TABLE 6
MILK YIELD, EXPERIMENT 2

	Lactation 1					Lactation 2	
	Preperiod Weeks 2-10			Experimental period Weeks 11-40		Weeks 2-40	
	Group			Group		Group	
	1	2		1	2	1	2
Milk, kg	25.39	25.56		15.97	15.41	17.71	17.69
FCM, kg	24.44	24.61		15.78	15.11	17.23	17.14
Fat, %	3.75	3.76		3.92	3.89	3.80	3.80
Protein, %	3.26	3.30		3.48	3.61	3.48	3.55

Experiments with rapeseed as a source of fat in straw-based diets

The experiments where rapeseed was studied as a source of fat in the diets (experiments 3,4,5) covered one lactation with each cow. In all three experiments the first 8 weeks after calving were used as a preperiod and the experimental period covered weeks 9 - 40 (experiments 4 and 5) or weeks 9 - 32 (experiment 3). Experiment 3 was a change-over experiment where the treatment of the cows changed after 20 weeks. Individually fed cows of the Swedish Friesian Breed were used in all experiments.

In experiment 3, two concentrate mixtures containing 5.7% commercial rapeseed meal with 4.3% glucosinolate in fat free dry matter (group 1) and 8.4% low-glucosinolate (1.5%) rapeseed (group 2) respectively, were compared. In group 1, 16 g digestible crude fat was supplied per kg fat-corrected milk (FCM), corresponding to the Swedish standard, and in group 2, 27 g digestible fat per kg FCM, the amount calculated on the total diets. The maintenance ration consisted of 1 kg hay, 4 kg straw and 2.5 kg dry matter of beet tops silage (Table 7).

TABLE 7

COMPOSITION OF THE CONCENTRATE MIXTURES (%). EXPERIMENT 3 (FRANK, 1979a).

	Group 1	2
Oats	30.0	25.4
Barley	30.0	28.5
Dried beet pulp	18.7	19.0
Cottonseed cake	10.6	11.1
Soyabean meal	5.0	7.6
Rapeseed meal	5.7	-
Rapeseed	-	8.4

In experiment 4, three concentrate mixtures, containing 5.2% commercial rapeseed meal with 4.4% glucosinolate (group 1), 9.0% low-glucosinolate rapeseed with 0.35% glucosinolate (group 2) and 19.0% expeller cake from commercial rapeseed with 5.2% glucosinolate were studied. The amount of digestible fat per kg FCM was 16 g in group 1 and 30 g in groups 2 and 3. The maintenance ration consisted of 2 kg hay, 3 kg straw and 2.8 kg dry matter of beet tops silage (Table 8).

TABLE 8

COMPOSITION OF THE CONCENTRATE MIXTURES, %. EXPERIMENT 4 (FRANK, 1979b)

	Group 1	Group 2	Group 3
Oats	28.0	26.0	28.5
Barley	34.0	28.0	26.5
Dried beet pulp	20.0	20.0	20.0
Cottonseed cake	7.3	9.5	4.0
Soyabean meal	5.5	7.5	2.0
Rapeseed meal	5.2	-	-
Rapeseed	-	9.0	-
Rapeseed expeller cake	-	-	19.0

The roughage in experiment 5 consisted of 4 kg hay (groups 1 and 2) or 4 kg straw (group 3). All groups were also fed 3.5 kg dry matter of grass silage. The concentrate mixture fed to group 1 contained 5.0% commercial rapeseed meal and in groups 2 and 3, 8.0% low-glucosinolate rapeseed (Table 9).

The rapeseed fed in all three experiments was unheated.

The concentrate was fed according to the milk yield during the previous week. Corrections for differences in body weights were also made. The calculation of requirements for metabolisable energy and digestible crude protein was based on the current Swedish standard.

TABLE 9

COMPOSITION OF THE CONCENTRATE MIXTURES (%). EXPERIMENT 5 (FRANK, 1980)

	Group 1	Group 2	Group 3
Oats	28.0	24.0	22.0
Barley	38.0	36.0	30.0
Dried beet pulp	20.0	20.0	20.0
Cottonseed cake	4.0	5.0	8.0
Soyabean meal	5.0	7.0	12.0
Rapeseed meal	5.0	-	-
Rapeseed	-	8.0	8.0

RESULTS

Feed consumption

No negative effect on feed consumption was observed in any of the experiments (Tables 10, 11 and 12). The average consumption of rapeseed in group 2, experiment 3, was 0.75 kg per day and 0.78 kg in group 3, experiment 5. In experiment 4, the average consumption of expeller cake was 1.6 kg which corresponded to about 14% rapeseed meal in the concentrate mixture, or around 1.1 kg per day.

TABLE 10

FEED CONSUMPTION, KG/DAY DURING WEEKS 9-40. EXPERIMENT 3.

	Group 1 weeks 9-20	Group 1 weeks 21-32	Group 2 weeks 9-20	Group 2 weeks 21-32
Straw, kg	3.3	3.8	3.3	3.8
Hay, kg	1.0	1.0	1.0	1.0
Silage, kg	13.6	13.3	13.4	13.3
Concentrate, kg	10.1	7.8	9.9	8.6
ME, MJ	164	146	168	150
DCP, g	1772	1505	1798	1560
Digestible crude fat, g	341	493	581	299

TABLE 11

FEED CONSUMPTION, KG/DAY DURING WEEKS 9-40. EXPERIMENT 4.

	Group 1	Group 2	Group 3
Straw, kg	3.8	3.8	3.8
Hay, kg	1.0	1.0	1.0
Silage, kg	15.6	15.4	15.5
Concentrate, kg	8.3	8.1	8.4
ME, MJ	147	151	153
DCP, g	1415	1461	1435
Digestible crude fat, g	251	503	491

TABLE 12

FEED CONSUMPTION, KG/DAY DURING WEEKS 9-40. EXPERIMENT 5.

	Group 1	Group 2	Group 3
Straw, kg	-	-	3.7
Hay, kg	4.0	4.0	-
Silage, kg	13.3	13.2	13.2
Concentrate, kg	9.1	8.6	9.7
ME, MJ	173	173	176
DCP, g	1631	1621	1678
Digestible crude fat, g	320	544	585

Milk yield

The higher fat levels in the concentrate mixtures containing rapeseed in experiments 3 and 4 gave a higher milk production compared to the lower fat levels in the concentrate mixtures containing rapeseed meal (Table 13).

In experiment 3, the difference in milk yield and FCM yield between group 2 and group 1 was significant ($p < 0.01$), but in experiment 4 (Table 14) the differences between the groups were not significant ($0.3 > p > 0.2$).

In experiment 5 the milk production and the production of FCM was very similar in all three groups (Table 15). The content of fat and protein in the milk was not affected in any of the experiments.

The feeding of rapeseed resulted in a higher iodine number in milk fat, i.e. the softness of the butter was improved by adding fatty products to the ration.

Health of the cows

The control of the general health of the cows included even in experiment 3, 4 and 5, the determination of a number of blood components.

In experiment 3 and 4 the level of oxazolidinethione in the milk was determined.

The general condition of the cows was normal for the herd. The values obtained in haematological and chemical analyses of blood were within the normal range. There was, however, a tendency to higher level of cholesterol in the blood from cows fed fatty rape products, but the level was still within the normal range.

In experiment 4 the level of protein-bound iodine in blood was significantly ($p < 0.05$) higher in the group fed low-glucosinolate rapeseed, indicating an increased thyroid function.

Average calving intervals and services per conception were normal for the herd also in these three experiments and no differences between the groups were observed.

TABLE 13

MILK YIELD, EXPERIMENTAL PERIOD. EXPERIMENT 3.

	Group 1		Group 2		Sign.*
	weeks 9-20	weeks 21-32	weeks 9-20	weeks 21-32	
Milk, kg	21.95	17.62	23.05	18.08	$p < 0.01$
FCM, kg	20.92	17.18	22.02	17.78	$p < 0.01$
Fat, %	3.69	3.83	3.70	3.64	NS
Protein, %	3.23	3.30	3.20	3.34	
Iodine number	33.2	36.8	37.0	34.2	

* calculated as differences in decrease between weeks 9-20 and weeks 21-32

TABLE 14

MILK YIELD, DURING WEEKS 9-40. EXPERIMENT 4.

	Group 1	Group 2	Group 3
Milk, kg	16.80	17.29	17.88
FCM, kg	16.12	16.96	17.21
Fat, %	3.73	3.87	3.75
Protein, %	3.57^a	3.58^a	3.40^b
Iodine number	31.0^a	35.4^c	36.6^c

a - b $p < 0.05$
a - c $p < 0.001$

TABLE 15

MILK YIELD, DURING WEEKS 9-40. EXPERIMENT 5.

	Group 1	Group 2	Group 3
Milk, kg	20.81	21.13	21.62
FCM, kg	20.72	20.67	20.97
Fat, %	3.97	3.85	3.80
Protein, %	3.56	3.48	3.42
Iodine number	27.8	31.7	33.2

DISCUSSION

The present investigations do not indicate any differences in palatability between a concentrate mixture containing 10% RSM and a concentrate where RSM was replaced by soybean meal. Neither milk production nor milk composition seemed to be negatively affected by this level of RSM. The total milk yield of individual cows during the second lactation in experiment 2 (weeks 2-40) ranged from about 3 800 kg to 5 900 kg FCM. High yielding cows did not react differently from low yielding ones, nor could any difference in persistency of milk production be discovered between the two groups when comparing the slope of the lactation curves.

When using straw-based diets for dairy cows, a positive effect on milk yield and total feed utilisation was observed when fatty rape products with low content of glucosinolate were used as a source of fat and the fat level in the total diet was increased to 26 - 29 g/kg FCM.

When substituting hay for barley straw the average milk yield normally decreases by about 7% according to previous Swedish experiments. However, in this investigation an increase of the fat content in the total diet from 3.2% to 5.1% in dry matter, or from 15 g to 26 - 28 g digestible crude fat per kg FCM, eliminated the differences in milk yield between hay and straw based diets. The higher fat level in the hay based diet influenced the milk production very little (experiment 5).

The results from the experiments indicate that concentrates containing up to 10% Swedish commercial RSM can be fed to dairy cows over long periods without adverse effects on milk production and milk composition. The experiments also indicate that unheated rapeseed with a low content of glucosinolates can be used as a source of fat for dairy cows fed low-quality roughage. The material is too small, however, to permit any conclusions about possible effects of RSM on fertility.

REFERENCES

Frank, B., 1979a. Fatty rape products for dairy cows. 1. Rapeseed in a straw-based diet. Swedish Univ. of Agric. Sci. Dep. of Animal Husbandry, Report 70.

Frank, B., 1979b. Fatty rape products for dairy cows. 2. Rapeseed and rapeseed expeller cake in straw-based diets. Swedish Univ. of Agric. Sci. Dep. of Animal Husbandry, Report 71.

Frank, B., 1980. Användning av rapsfrö och andra fettrika produkter till mjölkkor. 3. Rapsfrö som tillstas i olika grovfoderstater. Unpublished.

Lindell, L. and Knutsson, P.G., 1976. Rapeseed meal in rations for dairy cows. 1. Swedish J. Agric. Res., 6: 55-63.

Lindell, L., 1976. Rapeseed meal in rations for dairy cows. 2. Swedish J. Agric. Res., 6: 65-71.

DISCUSSION

B.O. Eggum   *(Denmark)*

Did you look at the zinc status in the animals? Did you analyse the zinc content of the diets? Zinc deficiency is a problem in laboratory animals.

L. Lindell   *(Sweden)*

No, we did not.

U. Petersen   *(FRG)*

I was surprised at the effects you found with a straw based diet. Energy content of straw is very low for high yielding cows. The good results with full fat rapeseed could be only an energy effect.

L. Lindell

This could be so, but it might also be due to depressed methane production and increased propionic acid production in the rumen when this fat is added. I am not sure about this, but I have read papers on the subject suggesting this explanation. The amount of fat which we add to this regime is not very much.

C. Calet   *(France)*

You say that under the conditions of your experiment you were unable to reach a conclusion. However, as regards the effect of rapeseed oil meal on palatability, do you think that your results are conclusive?

L. Lindell

You must remember that feed consumption does not depend only on the concentrate mixture, but also on the quality of the roughage. The amount of roughage compared with the amount of concentrate fed will also influence consumption.

B.O. Eggum

I am not sure if I understood and I would like to get it straight. Your highest level of rapeseed meal (RSM) consumption was approximately 1 kg/cow per day. Can you conclude from this that there is no influence on palatability? I must refer to the Danish experiments: we fed RSM at levels of 25%, 50% and 75%. At the 25% level, which was 1.2 kg/cow per day, there were no problems. When we went to 50% there were severe palatability problems. We went much higher with RSM in the Danish experiments compared to yours.

L. Lindell

Yes, I said there were no problems up to the level used. I do not know if we could use a higher level. It would depend on possible problems with fertility.

T.M. Thomas *(Ireland)*

For those of us who are not experts in nutrition, could you clarify a point. In you experiments you talk about a 10% level. The Danish people were talking about a 25% level - possibly 50%. Where is the real value? How high can we go with rapeseed in ruminant diets?

L. Lindell

There could be problems in translating different feeding experiments if they are expressed as a percentage of a concentrate mixture. It then depends upon how much concentrate is being fed. In our Swedish experiment we fed a roughage diet at a maintenance level for the cow, with enough energy to produce 1 - 2 kg of milk. Then we added the concentrate mixture. If more roughage is fed, therefore adding less concentrates for milk production, perhaps there could be a higher percentage of rapeseed meal in the concentrate.

A. Rérat *(France)*

We have some very good papers on rapeseed in animal nutrition. We can conclude that rapeseed has an extraordinary potential in animal nutrition. It can be used well by polygastric animals, except under certain conditions when palatability is a problem. However, we must improve rapeseed for monogastric animals. We can grow and develop rapeseed in Europe and it could represent a very high potential for monogastric animals if we could solve the anti-nutritional problems. Therefore, there are two kinds of work to do: technological work and breeding work. The conclusion is classical - we have a lot to do!

SESSION V

CLOSING SESSION

Chairman : M. Dambroth

CLOSING SESSION

M. Dambroth *(FRG)*

Before closing our colloqium, a short summary might be appropriate. After consultation with the co-ordinator, Mr. Calet, and with Mr. Gillot, I would like to present certain aspects for discussion. Afterwards, the administrative point of view will be given by Mr. Gillot. We will then have the opportunity to discuss future activities. Of course it will not be possible to come to a definite conclusion today. This will be reached after further discussions in the Working Group. I would like to ask Mr. Calet, to conduct the discussion after my comments and those of Mr. Gillot.

First of all I would like to sum up the possibilities of utilisation and the use of oilseed crops. These can be illustrated as follows:

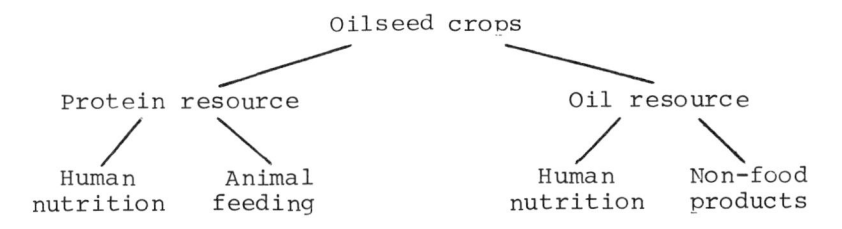

```
                    Oilseed crops
                   /             \
         Protein resource        Oil resource
          /        \              /         \
      Human      Animal        Human      Non-food
    nutrition   feeding      nutrition    products
```

We can use the oilseed crops as a protein resource and as an oil resource. The protein resource can be used for human nutrition and for animal feeding. The oil can also be used for human nutrition and for non-food products. As these are for different purposes, we feel there will be a need for special cultivars to meet the different purposes. For this, we need research activities in different fields. We need research activities in plant breeding, agricultural practice, utilisation and economic aspects. There may be other factors, but from the findings of this group I think that plant breeding and utilisation aspects have the highest priority. For the plant breeders, the main points for support by the EEC - not listed in order of priorities - are high yield, quality, tissue culture and crop physiology. In the field of utilisation, human nutrition, animal feeding and industrial purposes seem to be the important aspects for further support.

The activities of this Working Group are considered deserving of support by the Commission, as suitable oilseed species can be found for both northern and southern European production. In this way each of the EEC member countries can select those species that are most beneficial for their own territories without running a risk of competition in the distribution of research funds. This is very important for the Commission. It will be possible, for example, to have work on soyabeans for Italy and the south of France, and further studies on rapeseed in the northern plane, etc. With funds from the Commission, it will be possible for all the countries to take part in this research.

For the long-term strategy of this Working Group it may be necessary to focus main attention either on the production of food and feed, or on the production of replaceable resources for industrial purposes. It is my personal opinion that the employment of oil from plants is of greater future importance for industrial purposes than for food and feedstuff. However, this requires more consideration and discussion.

For the cultivation of oil plants it is important to improve the respective productive potential. To this end, contributions are required from both breeding research and practice. In the future it will no longer be possible for unproductive crops to be subsidised by the Commission or national governments. Therefore, breeding for yield potential and stability will be the most important tasks. Needless to say, the aspect of quality should receive particular attention. However, in my opinion, the different aspects of quality must not be investigated in great detail if a satisfactory yield has not been achieved. With this proviso, I hold the view that both aspects require attention in the promotion of any research by the European Commission. For this reason, many details need to be discussed in order to develop a clear strategy for research activities that the member countries are expected to contribute.

Such reflection also means that the Commission needs to recognise the long-term aspects of these attempts.

Please believe me, Mr. Gillot, that in this regard you have the backing of all the scientists, and we hope that all the negotiations with the Commission can reach a satisfactory conclusion. Everything will be done to support you by myself, Dr. Sommer and Dr. Glöy - the Federal Minister of Agriculture - and to help you in this direction.

That is all I have to say, ladies and gentlemen, and I will ask Mr. Gillot to express the administrative point of view.

Thank you very much.

J. Gillot *(CEC/Brussels)*

Ladies and gentlemen, it is a great pleasure for me to be able to thank the audience for the very efficient meeting that we have had. It is always agreeable for the Commission when a good investment is made - and I think that this meeting was a really good investment, not only because you have been able to exchange views, data, results, methods, etc., but also because I believe you have made progress in clarifying some of the aspects in the research work which needs to be done.

As a member of the Commission, I have always tried to help you and to help science as much as I could, because I am an old scientist myself. But the Commission cannot decide anything; above us there is a council of Ministers, composed of National Ministers of Agriculture. It is necessary that you keep in mind the important role you can play in communicating your research activities to your politicians. It helps us at the level of the Council when we feel that national preparation has been well done. We do our best, but you have close contact with the officials who come to Brussels regularly and who take decisions on proposals made by the Commission to the Council. If we can combine our efforts in persuading our politicians to do something I am sure we will succeed.

I have been listening carefully for three days to the presentation of papers and to the discussions. I must confess that I am not an expert in this field, but I feel that I have at least gained a knowledge of the true research problems in your field of interest.

I feel that several political factors could influence the behaviour of oilseed crop production in the European Community. I will not go into too much detail but three factors immediately spring to mind. The first is the gap - a gap which exists at the level of protein production, at the European level. This gap induces a dependence of the Community on soyabean production in the United States. This factor influences Commission decisions as regards the efforts to be made to improve the development of oilseed crop production in the European Community. But the only instrument that the Commission has, for the moment, is the financial instrument. They can try to develop some activity by supporting prices at the production level. This instrument can be complimented with research. I think there is a slight change in the Commission as regards the relative importance of research in decision-making. Some progress has been made there and, in the future, we may expect more consideration from the people who make political decisions. This is the first factor which could influence the behaviour of oilseed production.

The second factor is the energy crisis with which we are faced, and which could have an impact in the near future on agricultural production. My feeling is that the Commission might consider more carefully the problem of energy consumption in agriculture than has previously been the case. Having heard what you have said at this meeting, I believe that you have oilseed crops which can be grown everywhere in Europe. This is

an important factor which contributes to energy saving. If our
natural soil potentiality can be used better, we will improve
the energy balance in agriculture. I would look very carefully
at the adaptations of these crops, taking into consideration
the natural potentiality of the soils and also climatical
factors, availability of water, better water management - all
these factors can contribute greatly to a reduction of energy
consumption. Bearing in mind that we are working with fairly
new crops, I would also look carefully at the introduction of
these crops into the various cropping systems. We must consider
the reaction of the crop which is being introduced and how well
it can be fitted in with other crops.

The third political impact which could play a very
important role in oilseed production is, in my opinion, the
enlargement of the Community by three new partners, all
located in the south of the European Community. This is an
important factor. For instance, a factor that is lacking in
Braunschweig is the sun. As we know, the sun is one of the
major components for seed production. I learned that when I
was a young student at university. The sun is becoming more
and more important for the future, due to the fact that solar
energy may become one of the few remaining sources of energy.
In the south we have an ecological region where many things
could be done in the field of agriculture. We would like to
create a part of the Common Market which would not be
competitive with the rest of Europe. Perhaps we could find
interesting oilseed crops for that area, which could be grown
successfully using the sun as basic energy and trying to get
water where it is available.

These are my three major remarks. However, before I
hand over to Mr. Calet, I would like to thank, on behalf of the
Commission, Mr. Dambroth and Mr. Sommer for having organised
a very efficient meeting. I hope that we will have another
opportunity to meet again in the future.

Thank you very much.

FINAL DISCUSSION

C. Calet *(France)*

Thank you, Dr. Dambroth and Mr. Gillot. There is much to discuss and I hope there is enough time to exchange views. I agree with what Mr. Gillot has said, but perhaps too much has been said about energy. I hope this problem will not mask the other problems of the protein programme. I am a nutritionist and I have learned that it is necessary always to have a good balance between protein and energy to feed both animals and humans. The economical problem is the same. Personally, I believe it is impossible to consider energy without taking into account the problem of protein.

L. Toniolo *(Italy)*

I would like to start the discussion by saying that I agree with Dr. Dambroth when he says that we should try to take different species into consideration in Europe.

The second point is that when we introduce new species they should be leguminous: I am not talking just about soyabeans, there is *Vicia faba* and so on.

Energy is very important, but we could save much of the energy used to maintain soil fertility in the different countries. Related to this is the fact that, 15 days ago, the price of fertiliser went up again. We should bear in mind that we need economical methods.

As regards research and the aid of politicians: I can only say that I would be very happy to do research but I do not feel I can find a close relationship with the politicians. The ideas of the technicians are not always those of the politicians!

G. Röbbelen *(FRG)*

I would like to add something to Professor Toniolo's final remarks. I think it is most important that scientists can explain their data in a way that is understandable to the layman. Having someone like Mr. Gillot and others listening and talking to our group is very important. During the coffee break we had a discussion about how the official publications of the Commission should be written in a manner easier to understand for those who have to take the decisions. For the scientists, this means that they should always consider what they are saying and writing: they should be scientifically accurate but understandable. There is no doubt that the scientist becomes engrossed in his current work, but the politician may not be able to see things clearly. I am sure scientists are very grateful for all the Commissioners are doing to bring the two sides together.

D.J. Morgan *(UK)*

I thought the remarks made by Dr. Dambroth and Mr. Gillot were very useful. I would like to supply a little ammunition for the arguments to be put to the politicians. The success of breeding in cereals is something on which we can draw here. In an experiment at the Plant Breeding Institute, Cambridge, they compared new varieties with the old ones, growing side by side with exactly the same fertiliser treatment and the same irrigation treatments. What they discovered was that over the years they had increased the harvest index: more of the resources were going into the economic product. A message comes out of this: if you want to use energy more sensibly, you must put more of what the plant produces into what you want. There are a number of examples of this; we have seen it in rapeseed, in sunflower and we saw it yesterday in the demonstrations of Dr. Dambroth with his sugar beet. It is a question of getting the economic product to be a more efficient sink. The input may well be the same, but what we must do is to get more of the resources into the economic product. The success story of the cereals is there and it applies equally well to many of the crops about which we have heard today, and which are in the early stages of improvement.

C. Calet

We have worked on production of yield to increase the amount of crude material, but perhaps it is necessary now to find another orientation for our work. I agreed with the second proposition of Mr. Gillot and the remarks made by Professor Toniolo when they said that we must find crops which are advantageous for the farmers. Perhaps it is not only a question of yield, but rather a problem of returns. We have many solutions to solve the problems of protein, but the solutions are different for the north and the south. Perhaps we should consider the social problems of the production of protein, and see that each farmer gets a good return.

M. Dambroth (FRG)

I agree with this, Mr. Calet, and I think this will increase if other states, like Spain and Greece, join the EEC. In one country we need protein production more, and in another we need oil production for industrial purposes. That is what I meant; in this group we cannot follow only one road - it must be divided into different sections. The money comes from the corner with the power. There is no difference between the government and Brussels. The interests of the member states are quite different; we always have these discussions in these groups, and in other groups, because they have different interests. However, in this group, and perhaps that of the irrigation group, we can find a level of interest for all the member states. Therefore, it is necessary to have a wide scale of research activities.

L. Toniolo

I think it is important in research for us to exchange young people from one institute to another - young researchers. We must think about the future and hopefully these young researchers will do better than we have done in the past: the future belongs to them. It would be good if they could meet each other now and begin to work together.

G.J. de Jong *(The Netherlands)*

Perhaps my remark is not relevant but, considering the situation in our country, the accessibility of new varieties is greatly determined by private breeders, sometimes resulting in prospective new varieties not becoming available. I am not familiar with all the rules which govern this field, but some prospective varieties, especially foreign varieties, do not arrive in our country or they arrive too late. On our farm we have managed to escape some of these restrictive practices and the potential yield capacity of our cereal varieties compared to those normally grown by the farmers in our country, shows a difference of 5% - 6%.

Is it possible to take measures to speed up the availability of new varieties, and especially foreign varieties, to farmers in all the countries? I think a 5 - 6% average of agricultural value is important and an enormous sum of money.

G. Röbbelen

I do not have an answer but I would like to make two comments. I am convinced that plant breeding owes quite a bit of its success to its structure; in many countries there is a close interaction between state institutions and private activities. Therefore, my question is, in what way can the Community, plus our group, include private activity? I know it is difficult because of the different situations in different countries. In some countries most of the activities with the main crops are private; in others, most are controlled by the state.

I do not really understand what Mr. de Jong's problem is. Since 1971, when the first Paris convention started to think about the availability of varieties all over Europe, the Community has regulated the introduction of new varieties. The varieties are tested in each country, but if a country is hesitant to take foreign varieties into its own official tests, then two years after licensing of these varieties they get into the European lists and can be used freely.

However, I would hesitate to ask someone to advertise varieties which have not been through these tests. Although I am a plant breeder, I am convinced that quite a bit of the varietal challenge is due to the adaptation of the variety to agronomic practices. A sudden changeover is not a good idea; agronomists must have some knowledge about handling the variety. Therefore, I do not see the situation in the same light as Mr. de Jong and I do not think the state should become more involved. I think the administration is well organised at this level.

C. Calet

I think it is time to close this session. I have been very impressed by the efficiency of this meeting. I think this is due to Dr. Dambroth and Dr. Sommer and their staff. I would like to thank them very much for their hospitality and for their organisation.

Now that we have reached the end of our meeting, my feeling is that we have received much information to help to orientate research in each of our countries. This is important for those of us who direct the actions of the EEC group on protein. In this group we try to supply financial aid to support work on the subjects which are most likely to reduce the dependence of Europe on imported protein resources.

Until now, this work has been to increase yield, and stability of yield; in other words, to increase the productivity of seed and feed. I thank Dr. Dambroth for his proposal that we should take account of other problems in our objectives, e.g. quality of the product for the human nutrition or animal nutrition, plus the characteristics of the product for industrial purposes.

I think we need such proposals and _other_ proposals, and perhaps the people in this group could send them to us. Perhaps we should take into account the technological aspect. I think it is impossible to make progress in selection for genetic purposes if we do not take into account the technological interactions.

After this, and perhaps other, meetings, it will be necessary for us to consider the conclusions and advise members of other EEC groups on the work of each country.

Thank you very much, Dr. Dambroth, for your hospitality.

M. Dambroth

Ladies and gentlemen, thank you all for coming. We hope you enjoyed your stay in Braunschweig and that we will see you all again in a future programme on protein and oilseed crops.

At this point, let me thank Dr. Sommer for his part in the organisation of this colloqium; my part in this was very small.

LIST OF PARTICIPANTS

BELGIUM

Dr. W. Eeckhout Rijksstation voor Veevoeding
Scheldeweg 68
B-9231 Melle-Gontrode

Dr. L. van Hee Rijksstation voor Plantenveredeling
Burg. van Gansbergerlaan 109
B-9220 Merelbeke

Dr. L. van Holm Katholieke Universiteit Leuven
Fakulteit der Landbouwwetenschapen
Lab. voor Bodemvruchtbaarheid
de Croylaan 42
B-3030 Leuven

DENMARK

Dr. E. Augustinussen Statens Forsøgsstation
Ledreborg Allée 100
DK-4000 Roskilde

Dr. B.O. Eggum Statens Husdrybrugsforsøg
Rolighedsvej 25
DK-1958 Copenhagen V

Dr. H. Sørensen Royal Veterinary and Agricultural University
Chemistry Department
40 Thorvaldensvej
DK-1871 Copenhagen

FEDERAL REPUBLIC OF GERMANY

Dr. J. Böhm Institut für Pflanzenbau und Pflanzenzüchtung
Ludwigstrasse 23
D-6300 Giessen

Dr. A. Bramm Institut für Pflanzenbau und Pflanzenzüchtung
Bundesforschungsanstalt für Landwirtschaft
 Braunschweig-Völkenrode (FAL)
Bundesallee 50
D-3300 Braunschweig

Prof. Dr. M. Dambroth Institut für Pflanzenbau und Pflanzenzüchtung
Bundesforschungsanstalt für Landwirtschaft
 Braunschweig-Völkenrode (FAL)
Bundesallee 50
D-3300 Braunschweig

Dr. A. Gland Institut für Pflanzenbau und Pflanzenzüchtung
von-Siebold-Strasse 8
D-3400 Göttingen

Dr. I. Kübler   Institut für Pflanzenbau und Pflanzenzüchtung
                       Ludwigstrasse 23
                       D-3600 Giessen

Dr. G. Mix   Institut für Pflanzenbau und Pflanzenzüchtung
                       Bundesforschungsanstalt für Landwirtschaft
                         Braunschweig-Völkenrode (FAL)
                       Bundesallee 50
                       D-3300 Braunschweig

Prof. Dr. H.J. Oslage Institut für Pflanzenbau und Pflanzenzüchtung
                       Bundesforschungsanstalt für Landwirtschaft
                         Braunschweig-Völkenrode (FAL)
                       Bundesallee 50
                       D-3300 Braunschweig

Dr. C. Paul   Institut für Pflanzenbau und Pflanzenzüchtung
                       Bundesforschungsanstalt für Landwirtschaft
                         Braunschweig-Völkenrode (FAL)
                       Bundesallee 50
                       D-3300 Braunschweig

Dr. U. Petersen   Institut für Pflanzenbau und Pflanzenzüchtung
                       Bundesforschungsanstalt für Landwirtschaft
                         Braunschweig-Völkenrode (FAL)
                       Bundesallee 50
                       D-3300 Braunschweig

Prof. Dr. Dr. G. Röbbelen   Institut für Pflanzenbau und Pflanzenzüchtung
                       von-Siebold-Strasse 8
                       D-3400 Göttingen

Dr. C. Sator   Institut für Pflanzenbau und Pflanzenzüchtung
                       Bundesforschungsanstalt für Landwirtschaft
                         Braunschweig-Völkenrode (FAL)
                       Bundesallee 50
                       D-3300 Braunschweig

Dr. E. Schulz   Institut für Pflanzenbau und Pflanzenzüchtung
                       Bundesforschungsanstalt für Landwirtschaft
                         Braunschweig-Völkenrode (FAL)
                       Bundesallee 50
                       D-3300 Braunschweig

Dr. F. Schenk   Institut für Pflanzenbau und Pflanzenzüchtung
                       von-Siebold-Strasse 8
                       D-3400 Göttingen

Dr. C. Sommer   Institut für Pflanzenbau und Pflanzenzüchtung
                       Bundesforschungsanstalt für Landwirtschaft
                         Braunschweig-Völkenrode (FAL)
                       Bundesallee 50
                       D-3300 Braunschweig

Dr. H. Vogt	Institut für Pflanzenbau und Pflanzenzüchtung
	Bundesforschungsanstalt für Landwirtschaft
	  Braunschweig-Völkenrode (FAL)
	Bundesallee 50
	D-3300 Braunschweig

FRANCE

Dr. C. Calet	Inspecteur General INRA
	149 rue de Grenelle
	75341 Paris Cedex 07

Dr. M.A. Merrien	INRA Toulouse
	B.P. 12
	Auzeville
	3120 Castanet-Tolosan

Dr. G. Pelletier	Laboratoire d'Amélioration des Plantes
	Université de Paris-Sud
	Bâtiment 360
	91405 Orsay Cedex 05

Dr. A. Rérat	INRA
	Laboratoire de Physiologie de la Nutrition
	CNRZ
	78350 Jouy-en-Josas

IRELAND

Mr. J.P. Dunne	Department of Agriculture and Fisheries
	Backweston
	Leixlip
	Co. Kildare

Mr. T.M. Thomas	Crop Husbandry Department
	Oak Park Research Centre
	Carlow

ITALY

Dr. I. Greco	Istituto di Miglioramento Genetico
	Università degli Studi
	Via Amendola 165/A
	I-70126 Bari

Prof. Monotti	Istituto di Agronomia Generale
	  e Coltivazione Erbacee
	Facoltà di Agraria
	Borgo XX Giugno
	Perugia

Dr. G. Mosca	Istituto di Agronomia
	Università degli Studi
	Via Gradenigo
	I-35100 Padua

Prof. L. Toniolo   Istituto di Agronomia
                        Università degli Studi
                        Via Gradenigo
                        I-35100 Padua

THE NETHERLANDS

Dr. G.J. de Jong   RYP Smedinghuis
                        8200 AK Lelystad

Ir. W.J.M. Meijer   Proefstation voor de Akkerbouw en de
                           Groenteteelt in de Vollegrond
                        Postbus 430
                        NL-8200 AK Lelystad

SPAIN

Dr. J.D. Gimenez   CRIDA 10 INIA
                        Finca de la Alameda
                        Apartado de Correos del Obispo 240
                        Cordoba

Dr. J.M.F. Martinez   CRIDA 10 INIA
                        Finca de la Alameda
                        Apartado de Correos del Obispo 240
                        Cordoba

SWEDEN

Dr. L. Lindell   Department of Animal Husbandry
                        Division of Ruminant Research
                        Swedish University of Agricultural Sciences
                        Kungsangensgard
                        S-75590 Uppsala

UNITED KINGDOM

Dr. E.S. Bunting   Plant Breeding Institute
                        Maris Lane
                        Trumpington
                        Cambridge CB2 2LQ

Mr. P. Goetz   United Oilseeds
                        Parnella House
                        Market Place
                        Devizes
                        Wiltshire

Mr. S.E.W. Hallam   Janssen Services
                        33a High Street
                        Chislehurst
                        Kent BR7 5AE

Mr. D.G. Morgan   St. John's College
                        Cambridge CB2 1TP

Mrs. M.J. Robins   Janssen Services
                          33a High Street
                          Chislehurst
                          Kent BR7 5AE

Dr. E. Thomas   Department of Biochemistry
                          Rothamsted Expermental Station
                          Harpenden
                          Herts Al5 2JQ

Mr. K.J. Thompson   Plant Breeding Institute
                          Maris Lane
                          Trumpington
                          Cambridge CB2 2LQ

Manuscript prepared by:
JANSSEN SERVICES, 33a High Street, Chislehurst, Kent BR7 5AE, UK.